Food Science and Technology

nova science publishers

www.novapublishers.com

Food Science and Technology

Food Byproducts: Valorization Through Nutraceutical Production
Sandra Teresita Martín-del-Campo, PhD,
Anaberta Cardador-Martínez, PhD
and Jessica del Pilar Ramírez-Anaya, PhD (Editors)
2023. ISBN: 979-8-88697-612-0 (Hardcover)
2023. ISBN: 979-8-88697-876-6 (eBook)

Bioprospecting of Natural Compounds in Food, Pharmaceutical and Biomedical Science
Siddhartha Pati, PhD, Tanmay Sarkar, PhD
and Dibyait Lahiri, PhD (Editors)
2022. ISBN: 979-8-88697-361-7 (Hardcover)
2022. ISBN: 979-8-88697-431-7 (eBook)

Proceedings of BIOSPECTRUM: The International Conference on Biotechnology and Biological Sciences: Biotechnological Intervention Towards Enhancing Food Value
Sanket Joshi, PhD, Susmita Mukherjee
and Moupriya Nag, PhD (Editors)
2022. ISBN: 978-1-68507-985-7 (Hardcover)
2022. ISBN: 979-8-88697-069-2 (eBook)

Nutraceuticals: Food Applications and Health Benefits
Anita Kumari, PhD
and Gulab Singh, PhD (Editors)
2022. ISBN: 978-1-68507-488-3 (Hardcover)
2022. ISBN: 978-1-68507-512-5 (eBook)

A Closer Look at Polyphenolics
Peter Bertollini (Editor)
2022. ISBN: 978-1-68507-480-7 (eBook)
2021. ISBN: 978-1-68507-434-0 (Hardcover)

Food Processing: Advances in Research and Applications
Myriam Huijs (Editor)
2022. ISBN: 978-1-68507-570-5 (Hardcover)
2022. ISBN: 978-1-68507-581-1 (eBook)

More information about this series can be found at https://novapublishers.com/product-category/series/food-science-and-technology-series/

Sandra Teresita Martín-del-Campo, PhD
Anaberta Cardador-Martínez, PhD
and Jessica del Pilar Ramírez-Anaya, PhD
Editors

Food Byproducts

Valorization through Nutraceutical Production

nova
science publishers

www.novapublishers.com

DOI: https://doi.org/10.52305/LIXJ7605

Library of Congress Cataloging-in-Publication Data

Names: Teresita Martín-del-Campo, Sandra, editor. | Cardador-Martínez, Anaberta, editor. | Del Pilar Ramírez-Anaya, Jessica, editor.
Title: Food byproducts : valorization through nutraceutical production / editors, Sandra Teresita Martín-del-Campo, PhD, Independent consultant, Food Chemistry, Mexico, Anaberta Cardador-Martínez, PhD, Professor, Tecnologico de Monterrey, Campus Querétaro, México, Jessica del Pilar Ramírez-Anaya, PhD, Researcher, Food Sciences and Technology, University of Guadalajara, Mexico.
Description: New York : Nova Science Publishers, [2023] | Series: Food science and technology | Includes bibliographical references and index. |
Identifiers: LCCN 2023021197 (print) | LCCN 2023021198 (ebook) | ISBN 9798886976120 (hardcover) | ISBN 9798886978766 (adobe pdf)
Subjects: LCSH: Food industry and trade--By-products. | Food industry and trade--Waste minimization.
Classification: LCC TP373.8 .F63 2023 (print) | LCC TP373.8 (ebook) | DDC 664/.08--dc23/eng/20230516
LC record available at https://lccn.loc.gov/2023021197
LC ebook record available at https://lccn.loc.gov/2023021198

Published by Nova Science Publishers, Inc. † New York

Contents

Preface

An important volume of byproducts is produced during the harvesting, distribution, and industrialization of fruits, vegetables, cereals, pseudocereals or animal origin raw materials. Byproducts such as food with low-quality standards, seeds, peels, hulls, pulps, serum, among others, have different physicochemical properties and compositions depending on the properties of their origin.

During harvesting and distribution, some products are discarded since they do not accomplish with countries' regulations for human consumption, but others are discarded because they do not fulfill the marked requirements of size, shape, color, etc. or because the market price has fallen to nonprofitable levels. There are no exact numbers about the amount of food lost or wasted. The FAO estimate that one-third of food world production is lost or wasted.

As a result of the industrial transformation of superfoods, the peels, seeds, hulls, some pulp, and serum are discarded since they are not part of the final product due to their characteristics or must not be present due to regulations, depending on the commodity and the different countries regulations. Additionally, wastewater represents an important and underestimated byproduct.

The produced byproducts are often discarded, losing important compounds and additionally producing pollution issues. Often the byproducts are discarded in terrains, rivers, or incinerated. Even though these byproducts are organic and biodegradable, depending on the nature of the original food and the volume discarded, different phenomena could be observed where the byproducts are placed, such as eutrophication, soap formation, acidification, and salinization, as well as greenhouse gas production.

On the other hand, since these byproducts could present similar or higher concentrations of important nutrients and bioactive compounds than the original food, they could be recovered to produce new items that could be used for restitution, enrichment, or development of new products developed to present specific nutraceutical properties. Some recovered compounds could also be used in the cosmetic and pharmaceutical industries. Valorization of food byproducts could help to decrease food waste and pollution.

Using byproducts to produce other higher-value products could be important to the UN 2030 agenda for Sustainable Development. Valorization of byproducts and wastewater has the potential, among others, to contribute to reducing per capita food waste (SDG Target 12.3), improve nutrition by providing nutrients and secure additives to be used in foods (SDG 2), as well as reducing the environmental impact by reducing the tons of organic materials discarded (SDG 6, SDG 13, and SDG 15).

This book shows information about the various byproducts obtained from the harvesting and industrialization of the different superfoods previously mentioned. It is addressed the different bioactive and nutritive compounds that could be recovered from each byproduct.

Different recovery and stabilization processes adapted for each compound are also presented. Finally, some applications for the obtained products are proposed.

The book focuses on superfoods byproduct valorization and is divided into three sections. In the first section, the valorization of selected fruits is reviewed. In the second section, the emphasis is on foods such as cereals, pseudocereals, and legumes. Finally, in the third section, the revision focus on animal origin products.

Section 1. Fruits

Chapter 1

Avocado Byproducts as a Source of Functional and Nutraceutical Compounds

Mariano Otero López-Anguiano[1]
Adriana Cavazos-Garduño[2]
Julio César Serrano-Niño[2]
Ma. Claudia Castañeda-Saucedo[3,*]
Jessica Elizabeth Pineda-Lozano[4]
Sandra Teresita Martín-del-Campo[5]
Daniel Guajardo-Flores[6]
and Jessica del Pilar Ramírez-Anaya[7,†]

[1]Master in Public Health, CUSur-University of Guadalajara. Ciudad Guzman, Mexico
[2]Department of Pharmacobiology, CUCEI-University of Guadalajara, Guadalajara, Mexico
[3]Department of Natural Sciences, CUSUR-University of Guadalajara, Ciudad Guzmán, Mexico
[4]Institute for Research on Eating Behavior and Nutrition (IICAN), CUSUR-University of Guadalajara, Ciudad Guzmán, Mexico
[5]Tecnologico De Monterrey, Querétaro, Mexico
[6]FEMSA Biotechnology Center, Tecnológico de Monterrey, Monterrey, Mexico
[7]Department of Computer Science and Technological Innovation, CUSUR-University of Guadalajara, Ciudad Guzmán, Mexico

Abstract

Due to the adverse environmental impact, agroindustrial residues have been a focus of interest. Fruits and vegetable wastes are usually discarded or underused despite the vast reports regarding their use as a source of high-value compounds with beneficial properties for health. The revalorization of these materials depends on the technological approaches suitable to improve sustainability and reduce environmental issues in an economically feasible manner. The production of avocados has constantly been increasing due to the progressive higher demand for this fruits whose processing generates large amounts of residues. Due to its content of valuable bioactive compounds, byproducts such as seeds,

[*] Corresponding Author's Email: claudia.saucedo@cusur.udg.mx.
[†] Corresponding Author's Email: jessicar@cusur.udg.mx.

In: Food Byproducts
Editors: S. Teresita Martín-del-Campo, A. Cardador-Martínez and J. del Pilar Ramírez-Anaya
ISBN: 979-8-88697-612-0

peels, wastewater, and pomace are investigated for producing functional and nutraceutical ingredients and food products. By this means, using avocado waste materials to develop new ways to improve health is relevant to minimizing waste disposal and, thus, the environmental impact as a strategy to promote the circular economy and sustainability of the avocado industry.

Keywords: seed, peel, pomace, wastewater, extraction methods, delivery technologies

Introduction

Avocado is a dicotyledonous tropical plant from the Lauraceae family, which includes 50 genera and 2,500 to 3,000 species. Some of them are commercially relevant as laurel and cinnamon; however, the avocado is the member with the most significant commercial impact worldwide (Schaffer, Wolstenholme, and Whiley 2013). The avocado origin is fixed in Mesoamerica but currently grows under cultivation conditions around the world; Barrientos-Priego and López-López (2000) considered that avocado fruit appeared 12,000 years ago. There is archeological evidence of cotyledons found in the Coaxcatlan cave in Tehuacan, Mexico, dated from 8,000 B.C. and 7,000 B.C. Avocado figured among the main foods of Mesoamerican populations; in the years 200 B.C to 600 A.C. there were populations called Ahuacatlan dedicated to the cultivation of avocado (Téliz-Ortiz and Marroquín-Pimentel 2007). The emergence of different avocado varieties is linked to commerce in pre-Columbian America. With the arrival of Spaniards to America, avocado extended to Cuba and Jamaica, whit the first record in 1657, while in Europe, the first tree was found in 1601. However, the crop arrived late in the United States of America until 1871, when was done the first seed recollection; from this moment, the avocado industry started its development. This development led to the creation of new featured varieties more adapted to various environmental conditions (Schaffer, Wolstenholme, and Whiley 2013).

Avocado fruits are described as a berry with a single large seed; they can be round, elongated, or pear-shaped, and the size depends on the variety, metabolism, and cultivation practices, reaching weights from 50g to 2 kg. On average, 18 to 31% of the total weight are, respectively, peel and seed; the rest of the fruit is edible pulp or mesocarp. The peel or exocarp may be green, yellow, brown, or purple, have a waxy cuticle, and have a single-layered epidermis. The mesocarp consists of parenchymal tissue with a buttery consistency; a high quantity of oil is stored in its vacuoles and specialized storage cells called idioblasts. The seed parenchymal cells store starch, proteins, or oil; 50% of this oil and the reserve nutrients are present in the embryo (Wang, Bostic, and Gu 2010, Chanderbali et al. 2013, García-Fajardo, Ramos-Godínez, and Mora-Galindo 1999).

According to the Food and Agricultural Organization (FAO 2022), avocado is grown under cultivation conditions in 68 countries around the five continents. The world production in 2020 reached more than 8 million tons harvested in a total of 825,532 Ha. That year, Mexico, Colombia, Dominican Republic, and Peru produced 56.3% of the world total, with the former contributing 29.3%.

Avocado fruit is mainly consumed fresh; therefore, industrialization is based on surplus production of fresh fruit (García-Fajardo, Ramos-Godínez, and Mora-Galindo 1999). In countries such as Mexico, the volume of avocado for industrial uses is limited to 16% of the fruit production (Secretaría de Economía 2012). During the last years, industrial applications have been developed for the extraction of avocado oil, elaboration of processed food such as purees, guacamole, frozen, dried products, cosmetic, personal care, and nutraceutical products (Ramírez-Anaya et al. 2021). During food production and transformation, a certain proportion of the food is lost or wasted; according to Gustavsson et al. (2013), food wastes are those food losses of quality and quantity by the supply chain process taking place during production, postharvest, and processing stages (Rguibi 2021). In this context, the primary cultivation for the production of fresh avocado fruits generates agricultural waste such as biomass in the form of leaves and branches, while the industrial processing of avocado leads to a large volume of seeds and peels in a proportion of 21 to 30% of the processed raw material (Kosińska et al. 2012); other residues are those from avocado oil industry namely wastewater and pomace (Permal, Chang, Seale, et al. 2020).

The efficient use of agricultural and industrial byproducts is essential due to environmental concerns that contribute to ecological, social, and economic sustainability. Therefore, the reduction of food waste and its valorization have become a topic of increasing research interest over the past years and have become an issue of political priority all over the world (Kringel et al. 2020, Aysu and Durak 2015). In the case of avocado, it is recognized that the waste materials from fruit production and industrialization have a high commercial potential as raw materials for producing commodities for food, pharmaceutical, and cosmetic use (Figure 1).

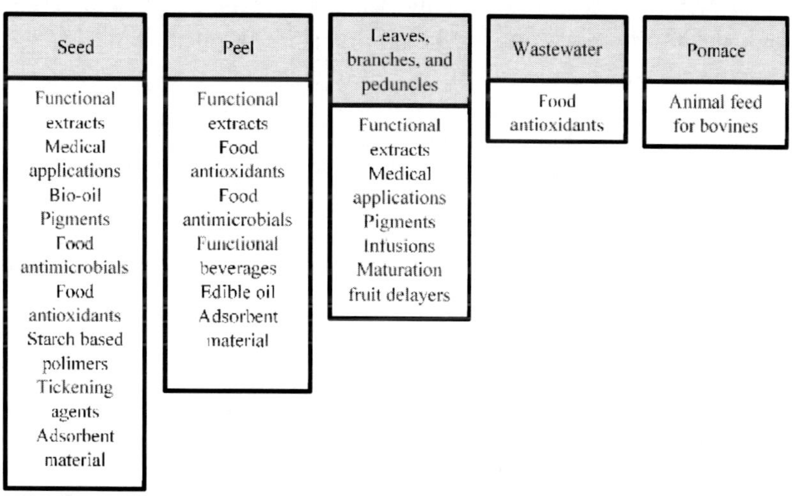

Figure 1. Actual and potential uses of avocado byproducts.

This potential originates from their chemical composition and biological properties. The main components of avocado byproducts are carbohydrates that can be used as functional ingredients and also to make biodegradable polymers, as drug delivery systems, or for the production of food packaging materials (Dávila et al. 2017); high-quality fats or extracts with high biofunctional power are also obtained (Araújo et al. 2018). But it is by far that the production of extracts has been the main focus of interest since it is based on the recovery of bioactive molecules, which sometimes are in higher content in residues than in the edible fraction of the fruits (Colombo and Papetti 2019). In the case of avocado, those molecules consist of pigments, arabinogalactans, a wide diversity of polyphenolic compounds, sterols, terpenoids, and acetogenins, among others; which show properties such as larvicidal, insecticide, antimicrobial, antifungal, inhibition of different viruses; and health benefits as an antioxidant, proapoptotic, neuroprotector, immunomodulatory, anti-inflammatory, antithrombotic; and auxiliary in the reduction of diabetes, hypertension, and dyslipidemic conditions.

Also, contribute to the reduction of lipid and protein deterioration in foods. However, although those compounds can be successfully exploited, their use in food production is limited by challenges arising from their technological properties during extraction and isolation (Wang, Bostic, and Gu 2010, Figueroa et al. 2018a, Ge et al. 2017, Leite et al. 2009, Oliveira et al. 2019, Salinas-Salazar et al. 2017, Figueroa et al. 2018b).

Composition of Avocado Byproducts

Seed

The seed represents 14-24% of the total weight of the fruit. This structure has a composition of 56% moisture, 2-7% protein, 2-14% lipids, 32-36% carbohydrates, 1.8-2.43% soluble sugars, and 3-6% fiber. Among the minerals are sodium, magnesium, potassium, calcium, manganese, iron, copper, and zinc (Ge et al. 2017). The fiber comprises soluble hemicelluloses, pectins, cellulose, and insoluble hemicelluloses. Content of essential amino acids in the seed protein is adequate compared to the levels in egg albumin (Table 1), outstanding leucine, threonine, tryptophan, and histidine, whose contents exceed or equal those of egg albumin (Ferreira da Vinha, Moreira, and Barreira 2013, Flores et al. 2019, Bressani, Rodas, and Ruiz 2009, Barbosa-Martín et al. 2016, Ceballos and Montoya 2013).

Table 1. Amino acid content in avocado seed

Amino acid	Avocado seed (g/100 g protein)[a]	Egg albumin (g/100 g protein)[b]
Valine	5.0-6.79	7.78
Isoleucine	3.57-4.78	5.69
Threonine	3.34-5.09	5.29
Tryptophan	0.45-1.96	1.75
Phenylalanine	5.05-5.44	6.78
Leucine	6.93-9.15	1.02
Lysine	5.81-8.73	9.53
Methionine	1.73-2.58	4.54
Histidine	1.99-2.60	2.40
Cysteine	1.14-1.53	3.80
Tyrosine	2.66-3.12	4.35
Aspartic acid	9.52-11.91	12.42
Serine	5.11-6.27	8
Glutamic acid	11.76-13.96	16.44
Proline	4.70-5.69	4.61
Glycine	4.75-6.12	3.94
Alanine	5.27-7.66	6.67
Arginine	6.20-7.56	6.45

[a]Bressani, Rodas, and Ruiz (2009), [b]USDA (2021).

The lipids found in the avocado seed are mainly neutral lipids (77.1-80.3%), followed by glycolipid (12.0-13.2%) and phospholipid (7.4–10.9%) fractions. The main neutral lipids are triacylglycerols, followed by mono- and diacylglycerols. Also are found free sterols, fatty acids, and sterolesters. The glycolipid fraction includes monogalactosyl-diacylglycerol (56.3-57.7%), acylsterylglucoside (17.5-18.6%), steryl glucoside (10.1-10.8%) and cerebroside (9.8-10.7%). While phospholipidic fraction include phosphatidylethanolamine (31.3-31.9%), phosphatidylcholine (28.9-31.7%) phosphatidic acids (14.5-17.6%) and phosphatidylglycerol (10.5-13.4%), among others (Takenaga et al. 2008). Of the main fatty acids in the seed lipids in decreasing order of abundance are linoleic ($C_{18:2}$), palmitic ($C_{16:0}$), oleic ($C_{18:1}$), and linolenic ($C_{18:3}$) acids; followed by other less abundant such as pentadecanoic ($C_{15:0}$), palmitoleic ($C_{16:1}$), margaric ($C_{17:0}$), lignoceric ($C_{24:0}$), behenic ($C_{22:0}$), and stearic ($C_{18:0}$) among others. Saturated fatty acids (SFA) constitute around 32.45%, monounsaturated (MUFA) 20.71%, and polyunsaturated (PUFA) 46.72%. Its concentrations vary widely depending on the avocado fruits' maturation, variety, and geographical distribution (Bora et al. 2001, De Sousa Galvão, Narain, and Nigam 2014).

Other apolar compounds in the seed are polyhydroxylated fatty alcohols being identified five; the most abundant are 1-acetoxy-2,4-dihydroxy-hepta-dec-16-ene and 1-acetoxy-2,4-dihydroxy-heptadec-16-yne accounting for 32 and 53.8%, respectively; the other three sum the remaining 14.2%. It has been postulated that avocado seed polyhydroxylated fatty alcohols can play an essential function as a photoprotective agent in UV-induced skin damage. Introducing these compounds to keratinocytes exerts a protective effect against UVB exposure, increasing cell viability, decreasing the secretion of IL-6 and PGE_2, and enhancing DNA repair (Rosenblat et al. 2011). Carotenoids are fat-soluble bioactive compounds with antioxidant and pro-vitamin A activity. Avocado seeds contain lower carotene concentrations than those reported in pulp, reaching values between 1.98 to 2.02 mg/100 g (Ge et al. 2017), including lutein and β-carotene (Bressani, Rodas, and Ruiz 2009). Tocopherols are organic compounds with vitamin E activity. The α, β, γ, and δ isomers may be present in the seed in concentrations that depend on the avocado variety; the α isomer contents range between 30.54 ± 0.37 to 75.57 ± 0.59 mg/100 g, and the β + γ tocopherol have been reported only in the Margarida variety in contents of 9.21 ± 0.43 mg/100g; while the δ form ranges from 8.06 ± 0.37 to 71.55 ± 0.61 mg/100 g (Amado et al. 2019). The sterol β-sitosterol and the triterpene 1,2,4-trihydroxy-nandecane are other valuable compounds reported in seed hexane extracts (Leite et al. 2009).

In the seed and seed coat were identified 83 polar compounds. Among them are organic acids such as citric, quinic, malic, succinic, and isocitric; the perseitol sugar and the terpenoids pentesmide and hydroxyabscisic acid glucoside from the abscisic acid family. Nevertheless, the polyphenolic compounds are the most varied polar molecules, with a total content of 328 to 1303 mg gallic acid equivalents/g dry basis (mg GAE/g d.w.) in this way, the seed can be considered a value-added product (Tables 2, 3, 4).

Seed phenolics are grouped into many classes, such as phenolic acids (hydroxybenzoic and hydroxycinnamic acids), phenolic alcohol derivatives, flavonoids, catechins, and condensed tannins. Additionally, in the seed is found hydroxyphenylacetic acid. In extracts, the epicatechin is the more abundant phenol; although flavonoid concentration also is high around 2.32 mg quercetin equivalents/g (mg QUE/g), it is comparatively lower than that measured in peel and leaves (Figueroa et al. 2018a, Araújo et al. 2020, Rosero et al. 2019, López-Cobo et al. 2016, Figueroa et al. 2018b, Tremocoldi et al. 2018, Oboh, Adelusi, and Akinyemi 2013).

Table 2. Phenolic acids and alcohol derivatives in avocado seed and peel

Phenolic compound	Compound name	Avocado structure	Authors
Phenolic acids			
Hydroxybenzoic acids	4-Hydroxybenzoic acid	Seed, peel	a, b
	Protocatechuic acid	Seed peel	a, b
	Gentisic acid	Seed, peel	a, b
	Benzoic acid	Seed, peel	a, b
	Syringic acid	Seed, peel	a, c
	Vanillic acid	Seed, peel	a, b
Hydroxyphenylacetic acids	3,4-Dihydroxy-phenylacetic acid	Peel	b
Hydroxycinnamic acids	3-O-Caffeoylquinic acid	Seed, peel	a, b, d
	4-O-Caffeoylquinic acid	Seed, peel	a, b, d
	5-O-Caffeoylquinic acid	Seed, peel	a, b, c, e
	3-p-Coumaroylquinic acid	Seed	a
	5-p-Coumaroylquinic acid	Seed	a, d
	4-p-Coumaroylquinic acid	Seed	a
	p-Coumaric acid	Seed, peel	a, b, c
	Caffeic acid	Seed, peel	a, b, c
	Sinapic acid	Seed, peel	c
Methoxycinnamic acids	Ferulic acid	Seed, peel	a, c
	Hydroxyferulic acid	Seed	a
	3-Feruloylquinic acid	Seed	a
	4-Feruloylquinic acid	Seed	a
	Dihydrocaffeic acid	Seed	a
	Ethyl protocatechuate	Seed	a
Phenolic alcohol derivatives	Hydroxytyrosol-glucoside	Seed, peel	a, b, d
	Tyrosol glucoside	Seed, peel	a, b
	Tyrosol-glucosiyl-pentoside	Seed, peel	a, b

[a]Figueroa et al. (2018a), [b]Figueroa et al. (2018b), [c]Rosero et al. (2019), [d]Araujo et al. (2021), [e]Tremocoldi et al. (2018).

Table 3. Flavonoids and catechins in the avocado seed and peel

Phenolic compound	Compound name	Avocado structure	Authors
Flavonoids			
	Quercetin	Seed, peel	a, b, c
	Quercetin-3-β-glucoside	Seed, peel	a, b
	Quercetin-diglucoside (2)	Seed, peel	a, b, c
	Quercetin-O-arabinosyl-glucoside	Peel	a, c
	Quercetin glucoronide	Peel	b
	Multinoside A	Peel	b
	Rutin	Peel	b, c,
	(±)-Naringenin	Seed, peel	a, b
	Kaempferol-O-glucosyl-rhamnoside	Peel	b, c
	Luteolin-7-O-(2”-O-pentosyl) hexoside	Peel	b
	Sukaranetin	Seed	a
Catechins			
	(+)-Catechin	Seed, peel	a, b, c, d, e
	(-)-Epicatechin	Seed, peel	a, b, c, d, e
	(Epi)gallocatechin	Seed	a
	(Epi)catechin glucopyranoside	Seed	a
	(Epi)Catechin gallate	Seed	a, d

[a]Figueroa et al. (2018a), [b]Figueroa et al. (2018b), [c]Rosero et al. (2019), [d]Araujo et al. (2021), [e]Tremocoldi et al. (2018).

Table 4. Condensed tannins and other polar compounds in the avocado seed and peel

Phenolic compound	Compound name	Avocado structure	Authors
Condensed tannins			
Procyanidin dimers	Procyanidins type A dimers (5)	Seed, peel	a, b, c
	Procyanidins type B dimers (11 seed, 7 peel)	Seed, peel	a, b, c, d
Procyanidin trimers	Procyanidin type A trimers (7 seed, 2 peel)	Seed, peel	a, b, c, d
	Procyanidin type B trimers (2)	Seed, peel	a, b, c, d
Procyanidin tetramers	Procyanidin type A tetramers (5)	Seed	a
	Procyanidin type B tetramer	Seed, peel	a
Procyanidin pentamers	Procyanidin type A pentamer	Seed	a
	Procyanidin type B pentamer (2)	Seed	a
Other polar compounds			
	Pyrogallol	Seed	a
	Pyrocatechol	Seed	a
	Vanillin	Seed, peel	a, b, c
	Dihydrocaffeic acid	Peel	b
	Chinchonain I	Peel	b
	Nudiposide	Peel	b
	Flavalignan isomers	Seed	b
	Phlorizin	Seed, peel	c
	Apigenin	Seed, peel	c

[a]Figueroa et al. (2018a), [b]Figueroa et al. (2018b), [c]Rosero et al. (2019), [d]Araujo et al. (2021).

Table 5. Acetogenin content in the avocado fruit (Rodríguez-López, Hernández-Brenes, and de la Garza 2015)

Structure	Acetogenin content* (mg/g fresh basis (f.w.))
Pulp	0.49-9.58
Seed	1.09-8.33
Peel	0.22-12.5

In the avocado seed, specifically in the idioblast oil cells, have been identified acetogenins, molecules derived from fatty acids with a long unsaturated aliphatic chain of 19 to 23 carbons and a highly oxygenated end. Of the purified and studied acetogenins, only 8 structures have been identified: AcO-avocadenyne (seed), AcO-avocadene, an unknown putative acetogenin, persediene, persenone C, persenone A, persin, and persenone B; where persenone A and persenone B are the compounds in the highest concentration (40 and 23%, respectively). The distribution and concentrations of acetogenins widely change according to the avocado fraction (seed, peel, and pulp) and variety (Table 5); however, in Hass avocados, the seed can be 1.6 times more than the pulp (Salinas-Salazar et al. 2017, Rodríguez-López, Hernández-Brenes, and de la Garza 2015).

Peel

The peel is part of the average weight of 8-26% of the fruit, its color and appearance can be light green, dark green, purple or black, and it can present a smooth, rough or lustrous texture; its composition is formed by 37-65% moisture, 2-3.9% protein, 0.9-7% lipids and 13-20% of fiber, 11.3% of carbohydrates, and 0.5% of ash. It is stated that the peel surpasses the fiber content compared with the seed (Bressani, Rodas, and Ruiz 2009, De Sousa Galvão, Narain, and Nigam 2014, Rotta et al. 2016). The presence of minerals such as calcium, magnesium, potassium, sodium, iron, copper, manganese, and zinc has been reported. Potassium and calcium are the most abundant minerals, and peel can be considered a good source of copper and magnesium (Bressani, Rodas, and Ruiz 2009, Morais et al. 2017),

The main fatty acids in the lipids of the peel in descending order of abundance are oleic ($C_{18:1}$), palmitic ($C_{16:0}$), linoleic ($C_{18:2}$), palmitoleic ($C_{16:1}$), and linolenic ($C_{18:3}$) acids; this order coincides among varieties such as Hass, Quintal, Fortuna, and Margarida. Other minor fatty acids found in the peel are myristic (C_{14}:0), cis-10 heptadecenoic ($C_{17:1}$), stearic ($C_{18:0}$), arachidic ($C_{20:0}$), gadoleic ($C_{20:1}$), behenic ($C_{22:0}$), DHA ($C_{22:6n-3}$), and lignoceric ($C_{24:0}$). As occurs with the seed, the contents vary widely depending on the maturation, variety, and geographical distribution (Table 6). The peel has a similar proportion of SFA, MUFA, and PUFA to that of pulp, which makes them a good source of omega-3 fatty acids (Amado et al. 2019).

Functional molecules with apolar nature as tocopherols reach total contents ranging from 72.94 to 230.7 mg/100 g as the sum of α-, β-, γ-, and δ- forms (Table 7), with higher concentrations than those measured in the seed. It was reported that the predominant tocopherol isomer changes depending on the studied avocado variety (Amado et al. 2019).

Other functional compounds in the peel found in lower amounts than seed and pulp are acetogenins (Table 5); the fractionation to obtain these molecules may show the biological properties reported for acetogenins, although to a lesser extent. In this way, it is possible to propose using peels as a source of bioactive compounds with biological properties beneficial to human health (Rodríguez-Carpena et al. 2011).

Referring polar compounds have been identified 61 in the peel. As in the seed, also are present organic acids (citric, quinic, malic, succinic, and isocitric), perseitol, the terpenoids pentesmide, and hydroxyabscisic acid glucoside Araújo 20 (Figueroa et al. 2018a, Araújo et al. 2020, Rosero et al. 2019, López-Cobo et al. 2016, Figueroa et al. 2018b). Nevertheless, polyphenols are the more varied polar compounds in total contents of 363.8 to 1058.00 mg GAE/g d.w.

Table 6. Content values of the main fatty acids found in the peel lipids (Amado et al. 2019)

Fatty acid	Content (mg/100 g)
Oleic	2283-665.8
Palmitic	1344-485.3
Linoleic	842.2-266.2
Palmitoleic	450.9-78.21
Linolenic	132.5-57.03
Saturated (SFA)	1407-536.5
Monounsaturated (MUFA)	2485-767.7
Polyunsaturated (PUFA)	993.5-335.7

Table 7. Content of tocopherols in the avocado peel (Amado et al. 2019)

Tocopherol isomer	Content (mg/100 g)
α-Tocopherol	75.41-77.28
β+γ-Tocopherol	54.07-89.74
δ-Tocopherol	76.80-82.01

The concentration of phenolic compounds in the peel is higher than in the seed when it comes from raw extracts; however, when the extraction of fractions rich in phenolic compounds is carried out, the seed shows a higher concentration. The polyphenolic compounds reported for the peel (Table 2, 3, 4) are molecules such as hydroxybenzoic acids, hydroxycinnamic acids, flavan-3-ols, catechin and epicatechin, procyanidin dimers, quercetin and some of its glycosylated derivatives, quercetin diglucoside, quercetin 3-O-arabinosyl-glucoside, quercetin 3-O-glucoside, quercetin-3-O-rutinoside, quercetin-3-O-arabinoside, and quercetin 3-O-rhamnoside. Most of the phenols present in peed also are present in the seed, whit the exception of the 3,4-dihydroxy-phenylacetic acid. Procyanidin B_2 and epicatechin have been recognized as the main compounds with significant contributions to antioxidant activity (Tremocoldi et al. 2018, Kosińska et al. 2012, Rosero et al. 2019, Trujillo-Mayol et al. 2021, Bressani, Rodas, and Ruiz 2009, Figueroa et al. 2021, Figueroa et al. 2018b).

Leaves, Branches, and Peduncles

The avocado leaves have lignin, mucilage, fats, starch, and calcium oxalate. Calcium, magnesium, and potassium are in high quantities, followed by phosphorous, sodium, and microelements such as manganese, iron, copper, and zinc (Yamassaki et al. 2017).

Table 8. Phenolic compounds in avocado leaves

Phenolic compound	Compound name	Authors
Phenolic acids		
	Chlorogenic acid	
Flavonoids		
	Quercetin	a
	Quercetin O-arabinosyl-glucoside	b, c
	Quercetin-3-O-7-rhamnoside	c
	Quercetin-3-O-galactoside	c
	Quercetin-3-O-glucoside	c
	Quercetin-3-O-arabinoside	c
	Quercetin-3-O-rhamnoside	c
	Rutin	a, c
	Kaempferol-3-O-glucoside	c
	Kaempferol-3-O-arabinoside	c
	Kaempferol-3-O-rhamnoside	c
	Tri-glycosylated kaempferol	c
	Luteolin-8-C-glucoside	a, c
Condensed tannins		
Procyanidin trimers	Procyanidin type A trimers (7 seed, 2 peel)	c
	Procyanidin type B trimers (2)	c
Other polar compounds		
	Isorhamnetin	a
	Apigenin	a

[a]Owolabi, Coker, and Jaja (2010), [b]López-Cobo et al. (2016), [c]Yamassaki et al. (2017).

The presence of phenols, alkaloids, cardiac glycosides, saponins, tannins, pigments, and terpenoids brings the leaves the potential to develop nutraceutical products (Lawal et al. 2021). Avocado leaves have higher contents of total phenols (43.82 mg GAE/g) and flavonoids (5.52

mg QUE/g) and a higher radical scavenging ability than the peel, seed, and flesh (Oboh, Adelusi, and Akinyemi 2013). The presence of 19 polyphenolic compounds has been referred outstanding chlorogenic acid, flavonoids, and two procyanidin type B and C trimers (Table 8) (Yamassaki et al. 2017, Owolabi, Coker, and Jaja 2010).

The leaves also contain antioxidant pigments in amounts of 221.81, 318.49, and 36.55 μ/g d.w. for carotenoids, a chlorophyll, and b chlorophyll, respectively. Additionally, water-soluble molecules in the leaves with polysaccharide and glycoconjugate nature have interesting biological effects. The arabinogalactan-protein-rich fraction (FRAGP) of aqueous extracts has been chemically characterized to contain 4.6%, 22.5%, and 76.7% of total protein, arabinogalactan-protein, and carbohydrates, respectively. Arabinose and galactose are the main monosaccharide constituents included in a structure of a type II arabinogalactan (Yamassaki 2013, Yamassaki et al. 2018).

Oil Production Residues

Extracting extra virgin avocado oil by cold pressing leads to a significantly high volume of residues such as pomace, seeds, peels, and wastewater. Those wastes have potential commercial applications in developing conventional, functional, and nutraceutical commodities for human, animal, or agricultural uses (Figure 1). The harvesting time of fruits and the technology used for oil extraction affects the oil yield and the proportion of waste. As seen in Table 9, the proportion of outputs produced changes depending on using a two- or three-phase decanter. In the first case, the oil production is higher, with lower pomace recovery and higher wastewater production. The proximal composition of the wastewater on a wet basis (% w.w.) from a three-phase process consists of 88.3% moisture and 6.3% lipids. 2.6 of dietary fiber, 2.1% ash, 1.2 protein, and 0.1% available carbohydrates; while that of the pomace presents 82.8% of moisture, 1.6% of lipids, 12.4% fiber,1.2% ash, 2.2% protein, and 0.1% of carbohydrates (Permal, Chang, Seale, et al. 2020, Wong, Eyres, and Ravetti 2013).

Table 9. Outputs during avocado cold-pressed oil production using two- or three-phase decanter technology

Technology	Two-phase decanter[a]				Three-phase decanter[b]	
Season	Early		Late		-----	
	Mass (kg)	proportion (%)	Mass (kg)	proportion (%)	Mass (kg)	proportion (%)
Inputs						
Fruit	796		673		1000	
Outputs						
Stone and skin	195	17.1	113	9.4	274	27.5
Pomace	120	10.5	152	12.7	150	15
Oil	73	6.4	129	10.8	75	7.5
Wastewater	750	65.9	805	67.1	448	44.9
Residual water	----	----	----	----	50	5
Total Outputs	1138	99.9	1199	100	997	99.9

[a]Wong, Eyres, and Ravetti (2013), [b]Permal, Chang, Chen, et al. (2020).

Uses of Avocado Byproducts

Seed

Some of the reported uses for avocado seed are found in treating hypertension, diabetes, hypercholesterolemia, and inflammatory conditions, in addition to its use in traditional medicine as antiparasitic, insecticidal, fungicidal, and antimicrobial (Dabas et al. 2013). Seed flour has been shown to significantly lower total cholesterol and LDL-C levels, reducing the prediction of the atherogenic index in a hypercholesterolemic mice model. Protocatechuic acid was the main phenolic compound found in seed flour, followed by kaempferide and vanillic acid (Pahua-Ramos et al. 2012). The long-chain lipids from avocado seed, namely long-chain fatty acids, aliphatic acetogenins, and polyhydroxylated fatty alcohols, are cytotoxic toward Caco-2 cells inducing the caspase-dependent apoptosis and simultaneously inducing arrest of the cell cycle and the modulation of cytokines related to the inflammatory response (Lara-Márquez et al. 2018). Additionally, methanol and hexane seed extracts and their constituents have been proposed as candidates for use as an alternative dengue control agent from their larvicidal effects. It was determined to have higher mortality of the hexane than methanol extract against *Aedes aegypti* since its respective lethal concentration (LC_{50}) were 8.87 and 16.7 mg/mL. The higher mortality of larvae is explained by the presence of β-sitosterol and 1,2,4-trihydroxynonadecane in the extract composition (Leite et al. 2009).

The antioxidant activity of the avocado seed compounds is the main functional property studied. The phenolic compounds (+)-catechin, (-)-epicatechin, kaempferide, rutin, and the chlorogenic, protocatechuic, vanillic, and syringic acids have been targeted as the molecules most contribute to the antioxidant activity. Those antioxidants have shown better protective activity than synthetic ones such as butylated hydroxyanisole (BHA) or butylated hydroxytoluene (BHT) (Moure et al. 2001, Soong and Barlow 2004, Segovia Gómez et al. 2014, Pahua-Ramos et al. 2012).

The evaluation of the avocado seed as a natural antioxidant for meat fat conservation through the prevention of free radicals attacks has been carried out in meat models under chilled storage conditions (4°C). The addition of seed extract (5%) or seed powder (0.1 to 0.5%) almost inhibited meat lipid oxidation (90%) and increased the protein protection by interfering with the formation of carbonyls, also lowered color deterioration by reducing the loss of redness through the retarded metmyoglobin formation. In oil-in-water emulsions of sunflower oil, using concentrations of 0.11% of the seed extract inhibited lipid oxidation by 30% and by 60% when combined with protein (egg albumin), showing a synergistic effect. The main phenolic compounds in the avocado seed that can scavenge free radicals are flavonols, catechins, hydroxycinnamic acids, quercetin glycosides, dimers, and trimers of proanthocyanins; being the condensed tannins the most antioxidant fractions. According to the investigations, the authors consider seed extract a source of antioxidants that decrease the deterioration of valuable compounds, mainly lipids and proteins, and improve the sensory and nutritional quality of meat foods (Segovia Gómez et al. 2014, Rodríguez-Carpena et al. 2011).

The antimicrobial activity of avocado seed extracts has also been studied. The first evaluation of the activity was done in strains of *Staphylococcus aureus, Bacillus subtilis, Aspergillus glaucus, Penicillium notatum*, and *Achromobacter perolens,* then was evaluated the activity of isolated fractions of avocado seed against strains *Micrococcus pyogenes* and

Sarcina lutea (Jensen 1951). The antimicrobial effect against yeasts and fungi of relevance for human and veterinary medicine, such as *Candida parapsilosis, Candida tropicalis, Candida albicans, Candida krusei, Malassezia pachydermatis,* and *Cryptococcus neoformans*, was evaluated too, the maximum concentration used to achieve the minimum inhibitory concentration was 0.65 mg of extract/mL and *Cryptococcus neoformans* was the most sensitive species requiring lower concentration of the extract (Leite et al. 2009). The inhibition of microorganisms commonly found in meat products was evaluated for *Bacillus cereus, Staphylococcus aureus, Listeria monocytogenes, Escherichia coli, Pseudomonas* spp., the mold *Aspergillus niger,* and the yeast *Yarrowia lipolytica.* The results have shown that the seed extracts are more effective against bacteria, being more sensitive those Gram-positive than Gram-negative. The highest inhibitory effect was observed against *B. cereus* and *L. monocytogenes*, whereas for Gram-negative bacteria, the strain *E. coli* was the most sensitive; the mold was found to be the most resistant microorganism and the yeast only showed moderate sensitivity (Rabadán et al. 2017, Rodríguez-Carpena et al. 2011).

Acetogenins are compounds that have shown functional biological activities such as antimicrobial, insecticide, antioxidant, antiplatelet, antithrombotic, and pro-apoptotic (Rodríguez-López, Hernández-Brenes, and de la Garza 2015, Brenes et al. 2019, Rodriguez Saona and Trumble 2000). The antimicrobial effect of isolated molecules of acetogenins has been performed in strains such as *Listeria monocytogenes, Clostridium sporogenes, Bacillus subtillis; Bacillus cereus; Staphylococcus aureus; Salmonella typhi, Shigella dysenteriae, Candida albicans* among others; observing that antibacterial activity increased in parallel with the number of unsaturations of the chain (Rodríguez-Sánchez et al. 2013, Pacheco et al. 2017, Salinas-Salazar et al. 2017, Neeman, Lifshitz, and Kashman 1970, Knapp and Melly 1986). There is a patented avocado seed oil enriched in acetogenins that contain 70% of acetogenins in its composition. AcO-avocadene, persenone, and persin are the main acetogenins at 31%, 28%, and 15% (w.w.), respectively. The product has been classified as a natural antimicrobial agent for ready-to-eat food. Its bactericidal activity was proved against *Listeria monocitogenes* since it is an extract that possesses a lytic effect on cells, showing after 4.3 h, a minimum inhibitory and bactericidal concentrations of 15.6 and 31.2 mg/L, respectively (Salinas-Salazar et al. 2017). This product also is effective against Gram-positive bacteria, with higher efficacy towards spore-formers (*Clostridium sporogenes, C. perfringens, Bacillus subtilis, and Alicyclobacillus acidocaldarius*) and is recommended to substitute or complement common additives used in foods based on milk, meat, or poultry (Villarreal-Lara et al. 2019, IICA 2022).

Obtaining pigments from avocado seeds has been an option for the food industry, which is looking for natural compounds with additional functional activities to their use. The perseoranginine is the pigment present in the avocado seed; it was identified due to the colorful reaction when exposing the avocado seed to oxygen during a cut. This compound results from the reaction catalyzed by the enzyme polyphenol oxidase on the phenolic compounds, leading to the formation of quinones that are red-orange colored compounds (Krokida and Philippopoulos 2005). Its chemical structure belongs to a glycosylated benzotropone with a molecular formula of $C_{29}H_{30}O_{14}$. As the color products form, polyphenol concentration diminishes in a direct relationship; this coloration increases as the pH rises from 2 to 11 but doesn't under a nitrogen atmosphere. Regarding the safety due to the consumption of perseoranginine, it has been tested in mice for 28 days without showing effects of depression, writhing, diarrhea, and hepatotoxicity. The use of this compound as a dye has been suggested for products stored at low temperatures under an inert atmosphere to preserve the stability of the color (Dabas et al. 2011, Hatzakis et al. 2019, Ozolua et al. 2009).

The treatment of the seed to obtain fiber by chemical methods and the elucidation of its compositional and functional characterization have allowed it to be proposed as a source of dietary fiber. The proximate composition of fiber residue has a protein content of 5%, which even becomes higher than the protein present in the pulp, 2% fat, and crude fiber with variations of 3-24%; the compositional variability may be due to differences in agro-climatic growing conditions of avocado fruit (Barbosa-Martín et al. 2016, Ceballos and Montoya 2013). Avocado seed fiber has considerable water trapping and oil adsorption capacities with uptake values of four or six times the weight of the fiber; therefore, its use is proposed to be used in the formulation of functional foods and the development of commercial applications such as maintenance of freshness, softness, and viscosity in baked products. In addition, based on fiber composition, its use is suggested as a supplement for health problems related to increased intestinal transit, bile acid retention, and cholesterol-lowering (Barbosa-Martín et al. 2016).

The native avocado seed starch granules have an oval shape and an A-type crystallinity pattern, and their solubility increases with temperature. The structure of the native avocado seed starch has been modified to develop technological ingredients for the improvement of the functional properties of viscous foods. Modification through acetylation proceeds by the esterification of the starch hydroxyl groups, replacing the hydroxyl moieties with acetyl groups leading to different degrees of substitution. Starches with a 0.07 degree of substitution yield granules with a round bell shape and C-type pattern, with higher swelling power and solubility at increasing temperatures, displaying reduced breakdown and synergy during freezing and a higher oil absorption than native starch (Silva et al. 2017). Avocado seed starch modification through cross-linking method using sodium tripolyphosphate leads to starch with phosphate group that strengthens the starch chain to prevent viscosity breakdown. Cross-linking using 6% sodium tripolyphosphate during 1 hour of reaction renders the best-modified starch according to moisture and ash content, paste clarity, gel strength, swelling power, solubility, yield, and degree of whiteness. When used in a cream soup, the resulting product had better sensorial acceptance and higher viscosity stability after 5 hours of storage than the product added with native starch or the commercial cream soup (Cornelia and Christianti 2018).

Oil seed extracted with hexane has a high *in vitro* antioxidant activity associated with compounds in its hydrophilic fraction. They show higher anti-inflammatory activity than ibuprofen when administered topically in a paw edema model in Wistar albino rats (Zavala-Guerrero et al. 2020). The biochemical properties of crude avocado seed oil have been studied since they are essential for application into cosmetic commodities such as soap fabrication. The values of refractive index (1.469), specific gravity (0.93 g/cm^3), and peroxide values (1.37 mEq O_2/kg) are very close to those reported for pulp oil; however, saponification value (231.6 mg KOH/g), iodine value (69.4 g I_2/100 g) and acid value (2.06 mg Eq oleic acid/kg) are higher than pulp oil but lower than other unconventional bio-oils such as mango seed, rambutan seed, *M. oleifera,* and pumpkin seed (Wu et al. 2015, Ge et al. 2018, Anwar and Bhanger 2003, Lourith et al. 2016, Rezig et al. 2012).

It is broadly recognized that the waste materials originating from avocado fruit industrialization (seed, peels, and other materials) are an excellent source of bioactive compounds with biological properties such as polyphenolic compounds and their associated antioxidant activity. The transformation of avocado wastes into storable raw materials with high antioxidant properties that may have different food uses has been approached mainly from drying strategies. Optimization of drying has focused on minimizing the loss of healthy compounds that provide added value to these byproducts taking into account the structural properties and desorption of moisture of the food matrix. In this context, air drying is the most feasible technique for commercial purposes. The drying procedure at a laboratory scale begins with reductions of the raw material to a coarse piece (≤4 x 4 cm), followed by drying (around 40-75°C), grounding into a powder of defined particle size (0.5-0.84 mm), packaging into bags to avoid light, oxygen and moisture exposure, and finally storage (-20°C) (Saavedra et al. 2017, Leite et al. 2009, Pahua-Ramos et al. 2012). During conventional drying at different temperatures (45 to 75°C), air velocities (0.8 to 1.6 m/s) and loading densities (1-3 kg/m^2), it took to the seed 60-75 min to reach the critical moisture and 500-700 min until constant weight. Once dried, the proportions of total phenolic contents (TPC) retained reach 54.89% (Saavedra et al. 2017). The chemical composition of the seed accounts for a higher hygroscopicity than the peels (Saavedra et al. 2017).

Peel

To date, the peel applications are smaller than the seed. Peels from different avocado varieties have been used to obtain oils through hexane extraction with broad perspectives for human consumption. Contrasting peel and pulp oil through their physical and physicochemical characteristics, the first oil displays higher specific gravity and values of acid and saponification but a lower refractive index and peroxide value, which means better oxidative quality (Table 10). Additionally, peel oil has ratios of PUFA/SFA from 0.61 to 1.13, surpassing the minimum of 0.4 necessary to reduce cardiovascular diseases (Bora et al. 2001, Amado et al. 2019).

A functional beverage prepared from sachets of dehydrated avocado peel was developed and analyzed, resulting in a high content of major phenolic compounds standing out of flavonoids, which displayed an antioxidant activity higher than apple tea. The good sensorial acceptability of the infusion and the absence of microbiological contamination in the sachets reinforced the product's commercial potential (Rotta et al. 2016). Another functional beverage made by brewing powdered avocado peel and red ginger in hot water (96°C) proved to possess significant TPC levels, antioxidant activity, and a potential antidiabetic activity, this last from its inhibitory activity of α-amylase enzymes (Putra, Sukesi, and Sulistyadewi 2020).

Table 10. Physicochemical characteristics of peel oils (Bora et al. 2001)

Physicochemical parameter	Range
Specific gravity	0.9116-0.9265
Refractive index	1.4062-1.4467
Acid value (%)	0.98-1.61
Saponification value (g KOH/kg)	200.5-200.9
Iodine value (g I2/100 g)	70.0-71.6
Peroxide value (mEq O2/kg)	1.48-1.7

Indeed, the practical uses of the peel are concentrated more on exploiting the antioxidant activity through the development of complex extracts or the recuperation of purified compounds for the food industry as innocuous additives. Peel extracts have been used for food preservation in porcine meat. Those extracts show limited antimicrobial activity against

Bacillus cereus, Staphylococcus aureus, Listeria monocytogenes, Escherichia coli, Pseudomonas spp., *Yarrowia lipolytica,* and *Aspergillus niger*; however, effectively reduced lipid and protein deterioration, and lowered inconvenient color changes in the meat (Rodríguez-Carpena et al. 2011).

Additionally, they have been assayed as antifungal additives through their incorporation in gelatin and carboxymethylcellulose coatings at a 200 mg/L concentration. The extracts added to the biopolymeric coatings decreased by 45% the proliferation of *Rhizopus stolonifer* and *Aspergillus niger* strains and extended the shelf life for 15 days (Vargas-Torrico et al. 2022). Mixed with nisin (an antimicrobial peptide) maximized the antioxidant and antimicrobial response against *Listeria* when combined in proportions of 39% of nisin and 61% peel. It was seen that the peel has the most potent antioxidant activity than nisin, seed, and their combinations (Calderón-Oliver et al. 2016). Adding a novel peel extract in beef and soy-based burgers in a proportion of 0.5% effectively increased the content of monomeric and oligomeric (epi)catechin and quercetin forms. Also, it reduced lipid and protein oxidation and the presence of urease and carbonic anhydrase activity associated with *Helicobacter pylori* infection (Trujillo-Mayol et al. 2021). Those results were consistent after in vitro digestion, although it reduced the recovery of other phenolic compounds such as catechin, procyanidin B1, procyanidin B2, rutin, and chlorogenic, caffeic, coumaric, and ferulic acids (Trujillo-Mayol et al. 2021).

As in the case of the seed, the drying procedure to condition the peel as a raw material for different uses also includes operations of drying throw convection (45-75 °C) and lyophilization, ground, packaging, and storage (-20 to -80°C) (Trujillo-Mayol et al. 2021, De Sousa Galvão, Narain, and Nigam 2014, Rotta et al. 2016). It has been seen that the peels dry faster than the seeds at different temperatures, air velocities, and loading densities. It takes to the peel 40-50 min to reach the critical moisture and 250-350 min until constant weight. After drying, the peel retains 62.82% of the initial phenolic content (Saavedra et al. 2017), obtaining the highest TPC during drying at 85 °C until a final moisture content of 7%; the most stable compounds are the phenolic acids and procyanidins the more thermolabile ones (Figueroa et al. 2018b).

Leaves, Branches, and Peduncles

As in the case of the peel, the use of leaves is limited; nevertheless, avocado leaves have had an ethnomedical interest since ancient times. Phenolics from the leaves of *Persea americana* have antioxidant activity, which may help prevent the progression of various oxidative stress-related diseases (Owolabi, Coker, and Jaja 2010). Leaf phenolic extracts obtained through hydromethanolic extraction have a protective effect on the pancreas against Fe^{2}-induced lipid peroxidation in vivo (Oboh, Adelusi, and Akinyemi 2013). It has been proved that the infusions strongly inhibited the herpes simplex virus type 1 (HSV-1), Aujeszky's virus (ADV), and adenovirus type 3 (AD3) in cell cultures, being more active in the mixture of infused substances than the identified compounds alone (De Almeida et al. 1998).

From its numerous health properties, the exploitation of the leaves has focused on the development of functional extracts based on the high antioxidant activity correlated with the polyphenolic, carotenoid, and chlorophyll content. Those studies pretend to increase the preservation of the compounds through conditioning of raw material (drying process), thermal incubation, and storage of the extract in solution (Yamassaki et al. 2017). The encapsulation of the resulting extracts has also been studied to improve the delivery of target compounds into food and pharmaceutical products; as in the study of Plazola-Jacinto et al. (2019), where the pigments from the *Hass* and *Drymifolia* varieties were extracted using vegetable oils (sunflower and corn oil). In this work, the resulting encapsulates differed in the pigment content depending on the avocado variety used, being the highest in the case of the *Drymifolia* variety; however, the antioxidant activity was higher when assayed the extract of the Hass variety with the corn oil.

Although the biological properties of phenolic fractions of hydroalcoholic leave extracts have attracted the most attention, the arabinogalactan-protein-rich-fraction has been studied since the FRAGP exhibits *in vitro* immunomodulatory properties as an inhibitor of the effect into the classical pathway of the complement system; being the side chains formed mainly by Ara*f* and Gal*p* probably related to the anti-complementary activity of FRAGP (Yamassaki et al. 2018, Yamassaki 2013).

When drying is done to get conditioned raw material, microwaving for 3 min or freeze-drying for 24 h show an insignificant change in TPC; in both cases, the polyphenol oxidase (PPO) and polyphenol peroxidase (POD) enzymes inactivation occurs because of the extremely high or low temperatures used, respectively (Loh and Lim 2018). The air-drying methods are the modalities that most reduce the phenolic content, mainly in chlorogenic acid and rutin levels. At indoor ambient drying conditions (24-26°C, 60% humidity) takes as long as 6 days to attain the final moisture content; while hot air oven drying (40-100°C) lasts from 43.5h (40°C) to 1 h (100°C) to get the constant weight. When hot air is used, temperatures from 40 to 60°C better preserve the TPC and antioxidant activity. Nowadays, the longer the drying times used, the higher the phenolic compounds losses in the air drying modality, mainly due to the enzymatic action (Yamassaki et al. 2017, Loh and Lim 2018). Short blanching pretreatments (1 min) at high temperatures (100°C) have been proved to ensure the inactivation of PPO and POD and the retention of phenolic compounds in avocado leaves, whit minimum leaching (5.6%) into the boiling water (Loh and Lim 2018).

The least studied avocado residues originate from harvest practices and packaging activities, as is the case of branches and peduncles. The nonpolar extracts from those residues and leaves were investigated to delay the maturation of avocado fruits, resulting in no significant effect on firmness or fruit respiratory rate but with an increase in the green color stability (Tochihuitl-Martiñón et al. 2018). To the best of our knowledge, no studies are using the mentioned structures to develop food, pharmaceutical, or cosmetic applications.

Wastewater and Pomace from Oil Production

Biofunctional compounds such as phenolics, carotenes, and tocopherols have been found in both pomace and wastewater, leading to the development of functional ingredients for the preservation of human food. In the case of a spray-dried powder obtained from wastewater, it was added to pork fat and sausages, effectively preventing lipid peroxidation. The effect was comparable to commercial additives such as BHT, BHA, and sodium erythorbate. The powder contained different amounts of α-tocopherol (95.5 to 181.6 mg/kg) and β-carotene (0.7 to 15.1 mg/kg) depending on the carrier material used (Permal, Chang, Chen, et al. 2020). Until now, the use of pomace has been referred to as animal feed (Wong, Eyres, and Ravetti 2013).

Recovery and Protection Technologies of Bioactive Compounds

Extraction Alternatives

The main focus in exploiting avocado residues is undoubtedly the recovery of bioactive compounds from different waste materials to produce phytochemicals used as food, pharmaceutic, or cosmetic ingredients. Avocado byproducts possess interesting biological properties with great potential as a source of ingredients for food, cosmetics, and other industrial fields (Colombo and Papetti 2019).

The methods for obtaining extracts with functional activities have commonly been solvent extraction. For antioxidant and antimicrobial compounds, solvents such as acetone, ethanol, methanol, isopropanol, hexane, diethylether, and water have been used in concentrated or diluted form. In some works, the authors previously grind the raw material to be processed (seed or peel) to reduce the particle size using mill equipment and, in some cases, standardize the particle size (around 40 mesh). The extraction material is processed in a fresh, lyophilized, or dehydrated form in drying ovens at temperatures ranging from 40-50°C and subsequently contacted with the solvent using the reflux method (Soxhlet) or homogenized under conditions of 10-16000 rpm for a few minutes (about 3) or even hours; some works have subsequently used an ultrasound bath to guarantee more penetrability of the solvent in the plant tissue (Pahua-Ramos et al. 2012, Sivanathan and Adikaram 1989, Rodríguez-Sánchez et al. 2013, Segovia Gómez et al. 2014, Rodríguez-López, Hernández-Brenes, and de la Garza 2015, Hatzakis et al. 2019, Rodríguez Carpena et al. 2011). Experimental designs have been carried out to evaluate the extraction efficiency of bioactive compounds by combining extraction factors such as solvent concentration (26-4%), temperature (20-90°C), and contact time (5-55 min). In those experiments, it was seen that solvent concentrations in the range of 10 to 60% allowed a good compound extraction, while the temperature significantly increased the solubility and diffusion of the molecules without affecting their stability; finally, the contact times did not have a significant effect for the extraction when combined with temperatures that improve the diffusion of the compounds, in a way that contact times of 5 minutes may be sufficient to achieve adequate extraction. However, the combination of variables must be considered essential to achieve extraction optimally; in this way, maceration for hours or even days could be avoided, reducing the extraction time (Segovia Gómez et al. 2014).

Once the homogenization with the solvent has been carried out, the mixture is filtered through filter paper, centrifuged, and the supernatant is lyophilized, or the solvent is evaporated to dryness at 40-80°C under vacuum using a rotary evaporator or under nitrogen (Pahua-Ramos et al. 2012, Rodríguez-López, Hernández-Brenes, and de la Garza 2015, Hatzakis et al. 2019, Raymond Chia and Dykes 2010, Salinas-Salazar et al. 2017, Rodríguez-Carpena et al. 2011).

The recovery of high-value compounds from agricultural byproducts has become a relevant topic, in which many studies have focused on their extraction strategies. Conventional extraction, such as maceration, liquid-liquid, and Soxhlet, has been mainly used in high-value compounds' industrial processes. For instance, Ferreira, Falé, and Santos (2022) studied the ethanolic extract of avocado peels for their antioxidant and antibacterial properties against the *Staphylococcus* family. The phenolic-rich extracts were further incorporated in emulsions to develop cosmetic formulations as a strategy to generate antimicrobial and antioxidant agents. However, several limitations are still to be solved for its industrial processing, such as large amounts of organic solvents, considerable time, and thermal degradation of phenolic compounds. Novel technologies have emerged to reduce these limitations and enhance the extraction and stability of phenolic compounds. Green technologies such as ultrasound-assisted extraction (UAE), microwave-assisted extraction (MAE), and pressurized liquid extraction have been studied to increase thermal efficacy and selectivity and reduce extraction times and solvent consumption. Rodríguez-Martínez et al. (2021) applied the UAE to optimize the extraction of phenolic compounds from avocado peels. The ethanolic extract obtained presented flavanols, flavonols, flavones, flavanones, and mainly phenolic acids. It proved to be a sustainable extraction technique as it considerably reduces the amount of organic solvent and time compared to conventional extraction.

Similarly, Monzón et al. (2021) used a response surface methodology and artificial neural network to optimize the UAE conditions from avocado seed and peel. This study represents an efficient, economical, and ecological strategy for extracting phenolic compounds from avocado residues. Furthermore, the MAE has also been studied as a green alternative to extracting phenolic compounds from avocado peels (Araujo et al. 2021). The fiber-bonded phenolics obtained (phenolic acids mainly) exhibited higher antioxidant activity compared to conventional extraction. Figueroa et al. (2021) optimized the extraction of bioactive polyphenols from avocado peel using MAE in another study; this avocado peel MAE extract exhibited higher matrix metalloproteinases inhibitory and antioxidant capacities, making it suitable for applications in the food industry. Multiple studies have reported the optimization process using a different solvent with MAE technology to obtain fiber-bonded phenolic acids from avocado seeds (Araujo et al. 2021, Weremfo, Adulley, and Adarkwah-Yiadom 2020). MAE technology exhibited advantages over conventional extraction methods regarding extraction efficiency and antioxidant activity. Finally, Mazyan et al. (2021) studied the optimum extraction of phenolics from avocado fruit flesh (*Persea americana*) using subcritical water extraction. The authors concluded that extracting with subcritical water enhances the phenolic compounds extraction yields.

Protection and Delivery Technologies

The residues of the avocado fruit derived from the pharmaceutical and food industry consist mainly of peel and seeds and, to a minor extent, leaves, branches, wastewater, and pomace. Those residues have in their composition phytochemical compounds with numerous biological activities, among which are anti-inflammatory, antimicrobial, antidiabetic, antihypertensive, anticancer, and antioxidant; however, these compounds are unstable and sensitive to different environmental conditions, so their concentration and activities may be reduced during processing, storage and shelf life (Brai et al. 2020, Cheikhyoussef and Cheikhyoussef 2022).

Once extracted, an alternative to avoid damage to the compounds of interest is encapsulation; this strategy provides an improvement in physicochemical stability, controlled release, bioavailability, and extending shelf life for long periods due to the formation of a cover that brings protection against oxygen, light, moisture, and other undesirable interactions (Nguyen et al. 2021). The encapsulation of these compounds facilitates their application into different systems that could be used to formulate products, mainly in the food and pharmaceutical sectors. The available encapsulation methods and systems are numerous; some include forming and stabilizing emulsions and using biopolymers for inclusion in gel or forming solid structures. Those systems have also been developed at different size scales where the compounds are included in various micro and nano-carriers. In either case, the type of system developed depends on the characteristics of the bioactive compound, the final formulation, and the area of application (food or pharmaceutical) (McClements 2018, Sarabandi, Gharehbeglou, and Jafari 2019).

Methodologies have been developed to synthesize and characterize particles from aqueous extracts of avocado residues on a nanometric scale. Silver nanoparticles of seed extracts produced in a facile one-pot method showed radical scavenging activity, potential antimicrobial activity against *Candida albicans*, and larvicidal activity against *Aedes aegypti* (Nallal, Prabha, and Razia 2022).

In the case of avocado derivatives, spray drying is the more used encapsulating method. In this case, the compounds of interest are protected and simultaneously transformed into powders, allowing more straightforward incorporation into different formulations. The first step is to prepare a stable emulsion with the compound of interest in the core; the core material is mixed with a polymer which acts as wall material, and finally, the mixture is exposed to a hot gas for seconds allowing solvent removal and microcapsule formation (Plazola-Jacinto et al. 2019).

The principal perspective of microencapsulation of avocado extracts is extending their efficacy as antioxidants for their application in the formulation of functional food products and supplements. Microencapsulation of oils (corn and safflower) used to extract pigments from avocado leaves was carried out through spray drying after emulsification. The resulting emulsion presented larger globule sizes when corn oil was used, whit a negative effect on the resulting microencapsulates hygroscopicity and moisture content and a change in the coloration (white to pale yellow). In all cases, the encapsulation efficiency was close to 50% (Plazola-Jacinto et al. 2019).

The encapsulation of the wastewater of cold pressed avocado oil production also was investigated. It was microencapsulated in powder form for its use as a preservative in lipid oxidation of fat and pork sausages. During the spray drying encapsulation, different carriers were evaluated, being whey protein concentrate added at 5%, the one that allowed a greater α-tocopherol and β-carotene protection, allowing a 49% yield (Permal, Chang, Chen, et al. 2020, Permal, Chang, Seale, et al. 2020). Calderón-Oliver et al. (2016) incorporated avocado skin extracts and nisin in microcapsules by complex coacervation; then evaluated their antioxidant and antimicrobial effect on the shelf life of ground beef. The variables analyzed for the formation of microcapsules were matrix wall material (collagen-alginate and collagen-pectin), drying methods (freeze and spray drying), and core dispersion systems (water in oil emulsion or suspension). The microcapsules with the higher encapsulation yield were formed by spray drying using collagen-pectin as the matrix wall material (Calderón-Oliver et al. 2016, Calderón-Oliver et al. 2017, Calderón-Oliver, Escalona-Buendía, and Ponce-Alquicira 2020).

Other technologies have been studied in addition to spray drying, such as homogenization followed by a freeze-drying process to obtain microcapsules from avocado seed extracts elaborated with different solvents. In this case, cassava and corn starch were compared as wall materials, with cassava starch showing the highest antioxidant activity and corn starch the highest digestibility; the antioxidant activity (DPPH and hydroxyl free radical scavenging ability) decreased up to 20% after 30 days of storage; however, the content of polyphenols increased. The best treatment was formulated with corn starch and Arabic gum (Awolu et al. 2022).

On the other hand, Regar, Nuningtyas, and Natsir (2022) formed microcapsules with a mixed extract of avocado seeds and leilem leaves (1:1) using a modified microwave equipped with a forced-air blower. They evaluated 5 different materials as encapsulants (Arabic gum, whey, chitosan, maltodextrin, and zeolite) at 25% of the formulation. Once the mixture was formed, it was dried in a microwave operating at 60°C. The authors analyzed the properties of the microcapsules, finding Arabic gum and chitosan as good encapsulants based on the yield (>45%), and the microcapsules with maltodextrin presented the highest solubility value (85%).

The protection of bioactive compounds by encapsulation also can be carried out by emulsification. In an emulsion, a surfactant surrounds the lipophilic compounds to form stable globules within a hydrophilic phase. Some advantages are the protection of the bioactive compounds and the increment of their solubility and availability. Recently was reported that the presence of avocado peel and seed extracts increases the emulsifying activity of low methoxyl pectin, leading to an enhancement in the stability of oil in water emulsions, with a lowering of the secondary oxidation products formation; with the improvement of a higher colloidal and oxidative stability compared with Tween 80 (Velderrain-Rodríguez et al. 2021). The emulsification method has already been used for nonpolar extracts of branches, peduncles, and leaves of *Persea americana* using aqueous solutions of the extracts at 100 and 200 ppm and Tween 20 as stabilizers. These solutions were used to test the effect on fruit ripening (Tochihuitl-Martiñón et al. 2018).

Conclusion

The avocado industry continues growing in response to consumer demand around the world; however, the global trend toward sustainable production and conscious consumption demands the development of products with health benefits and the elimination of environmental impacts, so the generation of lower volumes of waste and the use of green technologies in industrial processes continue to be the guideline for technological development in this agro-industrial sector.

Until now, a vast knowledge has been generated about the properties and uses of avocado waste, mainly from the seed; however, the pulp, the residues from the oil industry (wastewater and pomace); and the large volumes of biomass from avocado cultivation, remains as opportunity areas to the development of nutraceutical products.

References

Amado, D. A. V., A. M. Detoni, S. L. C. De Carvalho, A. S. Torquato, C. A. Martin, T. S. Tiuman, C. M. Aguiar, and S. M. Cottica. 2019. "Tocopherol and fatty acids content and proximal composition of four avocado cultivars (*Persea americana* Mill.)." *Acta Alimentaria* 48 (1):47-55. doi: 10.1556/066.2019.48.1.6.

Anwar, F., and M. I. Bhanger. 2003. "Analytical characterization of *Moringa oleifera* seed oil grown in temperate regions of Pakistan." *Journal of Agricultural and Food Chemistry* 51 (22):6558-6563. doi: 10.1021/jf0209894.

Araujo, R. G., R. M. Rodríguez-Jasso, H. A. Ruíz, M. Govea-Salas, M. Pintado, and C. N. Aguilar. 2021. "Recovery of bioactive components from avocado peels using microwave-assisted extraction." *Food and Bioproducts Processing* 127:152-161. doi: 10.1016/j.fbp.2021.02.015.

Araújo, R. G., R. M. Rodriguez-Jasso, H. A. Ruiz, M. Govea-Salas, M. E. Pintado, and C. N. Aguilar. 2020. "Process optimization of microwave-assisted extraction of bioactive molecules from avocado seeds." *Industrial Crops and Products* 154:112623. doi: 10.1016/j.indcrop.2020.112623.

Araújo, R. G., R. M. Rodriguez-Jasso, H. A. Ruiz, M. M. E. Pintado, and C. N. Aguilar. 2018. "Avocado by-products: Nutritional and functional properties." *Trends in Food Science & Technology* 80:51-60. doi: 10.1016/j.tifs.2018.07.027.

Awolu, O. O., E. T. Fole, O. A. Oladeji, H. N. Ayo-Omogie, and A. I. Olagunju. 2022. "Microencapsulation of avocado pear seed (*Persea Americana* Mill.) bioactive-rich extracts and evaluation of its antioxidants, in vitro starch digestibility and storage stability." *Bulletin of the National Research Centre* 46 (1):1-11. doi: 10.1186/s42269-022-00714-2.

Aysu, T., and H. Durak. 2015. "Assessment of avocado seeds (*Persea americana*) to produce bio-oil through supercritical liquefaction." *Biofuels, Bioproducts and Biorefining* 9 (3):231-257. doi: 10.1002/bbb.1535.

Barbosa-Martín, E., L. Chel-Guerrero, E. González-Mondragón, and D. Betancur-Ancona. 2016. "Chemical and technological properties of avocado (*Persea americana* Mill.) seed fibrous residues." *Food and Bioproducts Processing* 100:457-463. doi: 10.1016/j.fbp.2016.09.006.

Barrientos-Priego, A. F., and L. López-López. 2000. "*Historia y genética del aguacate* [*History and genetics of avocado*]." Fundación Salvador Sánchez Colín-CICTEMEX S.C., México, February 2002.

Bora, P. S., N. Narain, R. V. M. Rocha, and M. Queiroz Paulo. 2001. "Characterization of the oils from the pulp and seeds of avocado (cultivar: Fuerte) fruits." *Grasas y Aceites* 52 (3-4):4. doi: 10.3989/gya.2001.v52.i3-4.353.

Brai, B. I. C., J. A. Falode, R. A. Adisa, and A. A. Odetola. 2020. "Effects of aqueous leaf extract of avocado (*Persea americana*) on total cholesterol, triacylglycerols, protein and haematological parameters in CCl_4-intoxicated rats." *Clinical Phytoscience* 6 (1):1-6. doi: 10.1186/s40816-020-00159-y.

Brenes, C. H., G. D. J. G. Rivas, G. T. Amione, and D. G. R. Sanchez. 2019. Acetogenin molecules having antiplatelet and/or antithrombic activities, and methods and compositions thereof. edited by United States Patent. USA.

Bressani, R., B. Rodas, and A. S. Ruiz. 2009. *La composición química, capacidad antioxidativa y valor nutritivo de la semilla de variedades de aguacate* [*Chemical composition, antioxidant capacity and nutritive value of the seed of avocado varieties*]. Guatemala: Consejo Nacional de Ciencia y Tecnología-CONCYT, Secretaría Nacional de Ciencia y Tecnología-SENACYT, Fondo Nacional de Ciencia y Tecnología-FONACYT, Universidad del Valle de Guatemala-UVG.

Calderón-Oliver, M., H. B. Escalona-Buendía, O. N. Medina-Campos, J. Pedraza-Chaverri, R. Pedroza-Islas, and E. Ponce-Alquicira. 2016. "Optimization of the antioxidant and antimicrobial response of the combined effect of nisin and avocado by-products." *LWT-Food Science and Technology* 65:46-52. doi: 10.1016/j.lwt.2015.07.048.

Calderón-Oliver, M., R. Pedroza-Islas, H. B. Escalona-Buendía, J. Pedraza-Chaverri, and E. Ponce-Alquicira. 2017. "Comparative study of the microencapsulation by complex coacervation of nisin in combination with an avocado antioxidant extract." *Food Hydrocolloids* 62:49-57. doi: 10.1016/j.foodhyd.2016.07.028.

Calderón-Oliver, M., H. B. Escalona-Buendía, and E. Ponce-Alquicira. 2020. "Effect of the addition of microcapsules with avocado peel extract and nisin on the quality of ground beef." *Food Science & Nutrition* 8 (3):1325-1334. doi: 10.1002/fsn3.1359.

Ceballos, A., and S. Montoya. 2013. "Evaluación química de la fibra en semilla, pulpa y cáscara de tres variedades de aguacate [Chemical evaluation of the fiber in seed, pulp and peel of three varieties of avocado]." *Biotecnología en el Sector Agropecuario y Agroindustrial* 11 (1):103-112.

Chanderbali, A. S., D. E. Soltis, P. S. Soltis, and B. N. Wolstenholme. 2013. "Taxonomy and botany." In *The avocado: Botany, production and uses*, edited by Bruce A Schaffer, B Nigel Wolstenholme and Antony William Whiley, 31-50. Oxfordshire UK: CAB International.

Cheikhyoussef, N., and A. Cheikhyoussef. 2022. "Avocado (*Persea americana*) wastes: Chemical composition, biological activities and industrial applications." In *Mediterranean fruits bio-wastes*, edited by Mohamed Fawzy Ramadan and Farag A. Mohamed, 699-719. Cham, Switzerland: Springer.

Colombo, R., and A. Papetti. 2019. "Avocado (*Persea americana* Mill.) by-products and their impact: from bioactive compounds to biomass energy and sorbent material for removing contaminants. A review." *International Journal of Food Science & Technology* 54 (4):943-951. doi: 10.1111/ijfs.14143.

Cornelia, M., and A. Christianti. 2018. "Utilization of modified starch from avocado (*Persea americana* Mill.) seed in cream soup production." *IOP conference series: Earth and environmental science* 102 (1):012074. doi: 10.1088/1755-1315/102/1/012074.

Dabas, D., R. J. Elias, J. D. Lambert, and G. R. Ziegler. 2011. "A colored avocado seed extract as a potential natural colorant." *Journal of Food Science* 76 (9):C1335-C1341. doi: 10.1111/j.1750-3841.2011.02415.x.

Dabas, D., R. M. Shegog, G. R. Ziegler, and J. D. Lambert. 2013. "Avocado (*Persea americana*) seed as a source of bioactive phytochemicals." *Current Pharmaceutical Design* 19 (34):6133-6140.

Dávila, J. A., M. Rosenberg, E. Castro, and C. A. Cardona. 2017. "A model biorefinery for avocado (*Persea americana* Mill.) processing." *Bioresource Technology* 243:17-29. doi: 10.1016/j.biortech.2017.06.063.

De Almeida, A. P., M. M. F. S. Miranda, I. C. Simoni, M. D. Wigg, M. H. C. Lagrota, and S. S. Costa. 1998. "Flavonol monoglycosides isolated from the antiviral fractions of *Persea americana (Lauraceae)* leaf infusion." *Phytotherapy Research* 12 (8):562-567. doi: 10.1002/(SICI)1099-1573(199812)12:8%3C562::AID-PTR356%3E3.0.CO;2-6.

De Sousa Galvão, M., N. Narain, and N. Nigam. 2014. "Influence of different cultivars on oil quality and chemical characteristics of avocado fruit." *Food Science and Technology* 34:539-546. doi: 10.1590/1678-457x.6388.

FAO. 2022. "*Producción mundial de aguacate 2020* [*Avocado world production 2020*]." Last Modified June 28, 2022. https://www.fao.org/faostat/en/#data/QCL.

Ferreira da Vinha, A., J. Moreira, and S. Barreira. 2013. "Physicochemical parameters, phytochemical composition and antioxidant activity of the algarvian avocado (*Persea americana* Mill.)." *Journal of Agricultural Science* 5 (12):100-109. doi: 10.5539/jas.v5n12p100.

Ferreira, S. M., Z. Falé, and L. Santos. 2022. "Sustainability in skin care: Incorporation of avocado peel extracts in topical formulations." *Molecules* 27 (6):1782. doi: 10.3390/molecules27061782.

Figueroa, J. G., I. Borrás-Linares, R. Del Pino-García, J. A. Curiel, J. Lozano-Sánchez, and A. Segura-Carretero. 2021. "Functional ingredient from avocado peel: Microwave-assisted extraction, characterization and potential applications for the food industry." *Food Chemistry* 352:129300. doi: 10.1016/j.foodchem.2021.129300.

Figueroa, J. G., I. Borrás-Linares, J. Lozano-Sánchez, and A. Segura-Carretero. 2018a. "Comprehensive characterization of phenolic and other polar compounds in the seed and seed coat of avocado by HPLC-DAD-ESI-QTOF-MS." *Food Research International* 105:752-763. doi: 10.1016/j.foodres.2017.11.082.

Figueroa, J. G., I. Borrás-Linares, J. Lozano-Sánchez, and A. Segura-Carretero. 2018b. "Comprehensive identification of bioactive compounds of avocado peel by liquid chromatography coupled to ultra-high-definition accurate-mass Q-TOF." *Food Chemistry* 245:707-716. doi: 10.1016/j.foodchem.2017.12.011.

Flores, M., J. Ortiz-Viedma, A. Curaqueo, A. Rodriguez, G. Dovale-Rosabal, F. Magaña, C. Vega, M. Toro, L. López, and R. Ferreyra. 2019. "Preliminary studies of chemical and physical properties of two varieties of avocado seeds grown in Chile." *Journal of Food Quality* 2019. doi: 10.1155/2019/3563750.

García-Fajardo, J. A., M. d. R. Ramos-Godínez, and J. Mora-Galindo. 1999. "Estructura de la semilla de aguacate y cuantificación de la grasa extraída por diferentes técnicas [Structure of the avocado seed and quantification of the fat extracted by different techniques]." *Revista Chapingo Serie Horticultura* 5:123-128. doi: 10.5154/r.rchsh.1999.07.056.

Ge, Y., X. Si, B. Wu, X. Dong, Z. Xu, and W. Ma. 2018. "Oil content and fatty acid composition of the seeds of 16 avocado (*Persea americana*) accessions collected from southern China and their application in a soap bar." *The Journal of Agricultural Sciencies* 10:69-78. doi: 10.5539/jas.v10n11p69.

Ge, Y., X. Y. Si, J. Q. Cao, Z. X. Zhou, W. L. Wang, and W. H. Ma. 2017. "Morphological characteristics, nutritional quality, and bioactive constituents in fruits of two avocado (*Persea americana*) varieties from Hainan province, China." *The Journal of Agricultural Sciences* 9 (2). doi: 10.5539/jas.v9n2p8.

Gustavsson, J., C. Cederberg, U. Sonesson, and A. Emanuelsson. 2013. The methodology of the FAO study: "Global food losses and food waste–extent, causes and prevention"-FAO, 2011. In *Save food: An initiative on food loss and waste reduction*, edited by Sik. Sweden.

Hatzakis, E., E. P. Mazzola, R. M. Shegog, G. R. Ziegler, and J. D. Lambert. 2019. "Perseorangin: A natural pigment from avocado (*Persea americana*) seed." *Food Chemistry* 293:15-22. doi: 10.1016/j.foodchem. 2019.04.064.

IICA. 2022. "*Catálogo bioeconomía y desarrollo productivo inovación y tecnología* [*Catalog bioeconomy and productive development, innovation and technology*]." accessed april 13th 2022. https://catalogo-bioeconomia.iica.int/es/node/142.

Jensen, L. B. 1951. Process for extraction of antibiotic material. edited by United States Patent Office. USA.

Knapp, H. R., and M. A. Melly. 1986. "Bactericidal effects of polyunsaturated fatty acids." *Journal of Infectious Diseases* 154 (1):84-94. doi: 10.1093/infdis/154.1.84.

Kosińska, A., M. Karamać, I. Estrella, T. Hernández, B. Bartolomé, and G. A. Dykes. 2012. "Phenolic compound profiles and antioxidant capacity of *Persea americana* Mill. peels and seeds of two varieties." *Journal of Agricultural and Food Chemistry* 60 (18):4613-4619. doi: 10.1021/jf300090p.

Kringel, D. H., A. R. G. Dias, E. d. R. Zavareze, and E. A. Gandra. 2020. "Fruit wastes as promising sources of starch: Extraction, properties, and applications." *Starch-Stärke* 72 (3-4):1900200. doi: 10.1002/star. 201900200.

Krokida, M. K., and C. Philippopoulos. 2005. "Rehydration of dehydrated foods." *Drying Technology* 23 (4):799-830. doi: 10.1081/DRT-200054201.

Lara-Márquez, M., P. A. Spagnuolo, R. Salgado-Garciglia, A. Ochoa-Zarzosa, and J. E. López-Meza. 2018. "Cytotoxic mechanism of long-chain lipids extracted from Mexican native avocado seed (*Persea americana var. drymifolia*) on colon cancer cells." *The FASEB Journal* 32:804.32-804.32. doi: 10.1096/fasebj.2018.32.1_supplement.804.32.

Lawal, I. O., M. B. Olaniyi, S. O. Rufai, and A. O. Aremu. 2021. "Comparative assessment of the foliar micromorphology, phytochemicals and elemental composition of two cultivars of *Persea americana* Mill. leaves." *Scientific African* 14:e01034. doi: 10.1016/j.sciaf.2021.e01034.

Leite, J. J. G., É. H. S. Brito, R. A. Cordeiro, R. S. N. Brilhante, J. J. C. Sidrim, L. M. Bertini, S. M. d. Morais, and M. F. G. Rocha. 2009. "Chemical composition, toxicity and larvicidal and antifungal activities of *Persea americana* (avocado) seed extracts." *Revista da Sociedade Brasileira de Medicina Tropical* 42 (2):110-113. doi: 10.1590/S0037-86822009000200003.

Loh, Z. H., and Y. Y. Lim. 2018. "Drying effects on antioxidant activity, enzyme activity, and phytochemicals of avocado (*Persea americana*) leaves." *Journal of Food Processing and Preservation* 42 (10):e13667. doi: 10.1111/jfpp.13667.

López-Cobo, A., A. M. Gómez-Caravaca, F. Pasini, M. F. Caboni, A. Segura-Carretero, and A. Fernández-Gutiérrez. 2016. "HPLC-DAD-ESI-QTOF-MS and HPLC-FLD-MS as valuable tools for the determination of phenolic and other polar compounds in the edible part and by-products of avocado." *LWT-Food Science and Technology* 73:505-513. doi: 10.1016/j.lwt.2016.06.049.

Lourith, N., M. Kanlayavattanakul, K. Mongkonpaibool, T. Butsaratrakool, and T. Chinmuang. 2016. "Rambutan seed as a new promising unconventional source of specialty fat for cosmetics." *Industrial Crops and Products* 83:149-154. doi: 10.1016/j.indcrop.2015.12.045.

Mazyan, W. I., E. O'connor, E. Martin, A. Vogt, E. Charter, and A. Ahmadi. 2021. "Effects of temperature and extraction time on avocado flesh (*Persea americana*) total phenolic yields using subcritical water extraction." *Processes* 9 (1):159. doi: 10.3390/pr9010159.

McClements, D. J. 2018. "Encapsulation, protection, and delivery of bioactive proteins and peptides using nanoparticle and microparticle systems: A review." *Advances in Colloid and Interface Science* 253:1-22. doi: 10.1016/j.cis.2018.02.002.

Monzón, L., G. Becerra, E. Aguirre, G. Rodríguez, and E. Villanueva. 2021. "Ultrasound-assisted extraction of polyphenols from avocado residues: Modeling and optimization using response surface methodology and artificial neural networks." *Scientia Agropecuaria* 12 (1):33-40. doi: 10.17268/sci.agropecu.2021. 004.

Morais, D. R., E. M. Rotta, S. C. Sargi, E. G. Bonafe, R. M. Suzuki, N. E. Souza, M. Matsushita, and J. V. Visentainer. 2017. "Proximate composition, mineral contents and fatty acid composition of the different parts and dried peels of tropical fruits cultivated in Brazil." *Journal of the Brazilian Chemical Society* 28:308-318. doi: 10.5935/0103-5053.20160178.

Moure, A., J. M. Cruz, D. Franco, J. M. Domínguez, J. Sineiro, H. Domínguez, M. a. J. Núñez, and J. C. Parajó. 2001. "Natural antioxidants from residual sources." *Food Chemistry* 72 (2):145-171. doi: 10.1016/S0308-8146(00)00223-5.

Nallal, V. U. M., K. Prabha, and M. Razia. 2022. "Avocado-waste mediated synthesis of Silver nanoparticles and exploration of their potential biological properties." *Materials Today: Proceedings*. doi: 10.1016/j.matpr.2022.05.537.

Neeman, I., A. Lifshitz, and Y. Kashman. 1970. "New antibacterial agent isolated from the avocado pear." *Applied Microbiology* 19 (3):470-473. doi: 10.1128/am.19.3.470-473.1970.

Nguyen, T. T. T., T. V. A. Le, N. N. Dang, D. C. Nguyen, P. T. N. Nguyen, T. T. Tran, Q. V. Nguyen, L. G. Bach, and T. D. Thuy Nguyen Pham. 2021. "Microencapsulation of essential oils by spray-drying and influencing factors." *Journal of Food Quality* 2021. doi: 10.1155/2021/5525879.

Oboh, G., T. I. Adelusi, and A. J. Akinyemi. 2013. "Inhibitory effect of phenolic extract from leaf and fruit of avocado pear (*Persea Americana*) on Fe2 induced lipid peroxidation in rats' pancreas in vitro." *FUTA Journal of Research in Science* 2:276-286.

Oliveira, É. R., R. F. Silva, P. R. Santos, and F. Queiroz. 2019. "Potential of alternative solvents to extract biologically active compounds from green coffee beans and its residue from the oil industry." *Food and Bioproducts Processing* 115:47-58. doi: 10.1016/j.fbp.2019.02.005.

Owolabi, M. A., H. A. B. Coker, and S. I. Jaja. 2010. "Bioactivity of the phytoconstituents of the leaves of *Persea americana*." *Journal of Medicinal Plants Research* 4 (12):1130-1135. doi: 10.5897/JMPR09.429.

Ozolua, R., O. Anaka, S. Okpo, and S. Idogun. 2009. "Acute and sub-acute toxicological assessment of the aqueous seed extract of *Persea americana* Mill *(Lauraceae)* in rats." *African Journal of Traditional, Complementary and Alternative Medicines* 6 (4). doi: 10.4314/ajtcam.v6i4.57214.

Pacheco, A., D. G. Rodríguez-Sánchez, R. Villarreal-Lara, J. M. Navarro-Silva, C. Senés-Guerrero, and C. Hernández-Brenes. 2017. "Stability of the antimicrobial activity of acetogenins from avocado seed, under common food processing conditions, against *Clostridium sporogenes* vegetative cell growth and endospore germination." *International Journal of Food Science & Technology* 52 (11):2311-2323. doi: 10.1111/ijfs.13513.

Pahua-Ramos, M. E., A. Ortiz-Moreno, G. Chamorro-Cevallos, M. D. Hernández-Navarro, L. Garduño-Siciliano, H. Necoechea-Mondragón, and M. Hernández-Ortega. 2012. "Hypolipidemic effect of avocado (*Persea americana* Mill) seed in a hypercholesterolemic mouse model." *Plant Foods for Human Nutrition* 67 (1):10-16.

Permal, R., W. L. Chang, T. Chen, B. Seale, N. Hamid, and R. Kam. 2020. "Optimising the spray drying of avocado wastewater and use of the powder as a food preservative for preventing lipid peroxidation." *Foods* 9 (9):1187. doi: 10.3390/foods9091187.

Permal, R., W. L. Chang, B. Seale, N. Hamid, and R. Kam. 2020. "Converting industrial organic waste from the cold-pressed avocado oil production line into a potential food preservative." *Food Chemistry* 306:125635. doi: 10.1016/j.foodchem.2019.125635.

Plazola-Jacinto, C. P., V. Pérez-Pérez, S. C. Pereyra-Castro, L. Alamilla-Beltrán, and A. Ortiz-Moreno. 2019. "Microencapsulation of biocompounds from avocado leaves oily extracts." *Revista Mexicana de Ingeniería Química* 18 (3):1261-1276. doi: 10.24275/uam/izt/dcbi/revmexingquim/2019v18n3/Plazola.

Putra, I. M. W. A., K. A. Sukesi, and N. P. E. Sulistyadewi. 2020. "Antioxidant capacity and Î±-Amilase inhibition by avocado (*Persea americana* Mill) peel and red ginger (*Zingiber officinale var. Rubrum*) based functional drink." *Acta Chimica Asiana* 3 (1):135-142. doi: 10.29303/aca.v3i1.22.

Rabadán, A., J. E. Pardo, R. Gómez, A. Alvarruiz, and M. Álvarez-Ortí. 2017. "Usefulness of physical parameters for pistachio cultivar differentiation." *Scientia Horticulturae* 222:7-11. doi: 10.1016/j.scienta.2017.04.034.

Ramírez-Anaya, J. d. P., A. Cavazos-Garduño, Y. King-Loeza, M. C. Castañeda-Saucedo, E. H. Valdez-Miramontes, and P.-L. J. Elizabeth. 2021. "Avocado: Composition and processing technologies." In *Super and nutraceutical foods. Composition and processing technology*, edited by Sandra Teresita Martín-del-Campo, Ramírez-Anaya, Jessica del Pilar, Cardador-Martínez, Anaberta, 413. New York: Nova Science Publishers.

Raymond Chia, T. W., and G. A. Dykes. 2010. "Antimicrobial activity of crude epicarp and seed extracts from mature avocado fruit (*Persea americana*) of three cultivars." *Pharmaceutical Biology* 48 (7):753-756. doi: 10.3109/13880200903273922.

Regar, M. N., Y. F. Nuningtyas, and M. H. Natsir. 2022. "The different encapsulants of lileum leaves (*Clerodendrum minahassae* L.) and avocado seed (*Persea americana* Mill.) extraction on physical quality and microscopic atructure." *E3S Web of Conferences* 335:00040. doi: 10.1051/e3sconf/202233500040.

Rezig, L., M. Chouaibi, K. Msaada, and S. Hamdi. 2012. "Chemical composition and profile characterisation of pumpkin (*Cucurbita maxima*) seed oil." *Industrial Crops and Products* 37 (1):82-87. doi: 10.1016/j.indcrop.2011.12.004.

Rguibi, O. 2021. "*Food loss and waste in the food supply chain: current situation and potential solutions.*" Masters Tesis, School of Business and Economics, Nova University.

Rodríguez-Carpena, J. G., D. Morcuende, M. J. Andrade, P. Kylli, and M. Estévez. 2011. "Avocado (*Persea americana* Mill.) phenolics, in vitro antioxidant and antimicrobial activities, and inhibition of lipid and protein oxidation in porcine patties." *Journal of Agricultural and Food Chemistry* 59 (10):5625-5635. doi: 10.1021/jf1048832.

Rodríguez-López, C. E., C. Hernández-Brenes, and R. I. D. de la Garza. 2015. "A targeted metabolomics approach to characterize acetogenin profiles in avocado fruit (*Persea americana* Mill.)." *RSC advances* 5 (128):106019-106029. doi: 10.1039/C5RA22854A.

Rodríguez-Martínez, B., P. Ferreira-Santos, B. Gullón, J. A. Teixeira, C. M. Botelho, and R. Yáñez. 2021. "Exploiting the potential of bioactive molecules extracted by ultrasounds from avocado peels—food and nutraceutical applications." *Antioxidants* 10 (9):1475. doi: 10.3390/antiox10091475.

Rodríguez-Sánchez, D. G., A. Pacheco, M. I. García-Cruz, J. A. Gutiérrez Uribe, J. A. Benavides-Lozano, and C. Hernández-Brenes. 2013. "Isolation and structure elucidation of avocado seed (*Persea americana*) lipid derivatives that inhibit *Clostridium sporogenes* endospore germination." *Journal of Agricultural and Food Chemistry* 61 (30):7403-7411. doi: 10.1021/jf401407s.

Rodriguez Saona, C., and J. T. Trumble. 2000. "Biologically active aliphatic acetogenins from specialized idioblast oil cells." *Current Organic Chemistry* 4 (12):1249-1260. doi: 10.2174/1385272003375789.

Rosenblat, G., S. Meretski, J. Segal, M. Tarshis, A. Schroeder, A. Zanin-Zhorov, G. Lion, A. Ingber, and M. Hochberg. 2011. "Polyhydroxylated fatty alcohols derived from avocado suppress inflammatory response and provide non-sunscreen protection against UV-induced damage in skin cells." *Archives of Dermatological Research* 303 (4):239-246. doi: 10.1007/s00403-010-1088-6.

Rosero, J. C., S. Cruz, C. Osorio, and N. Hurtado. 2019. "Analysis of phenolic composition of by-products (seeds and peels) of avocado (*Persea americana* Mill.) cultivated in Colombia." *Molecules* 24 (17):3209. doi: 10.3390/molecules24173209.

Rotta, E. M., D. R. de Morais, P. B. F. Biondo, V. J. dos Santos, M. Matsushita, and J. V. Visentainer. 2016. "Use of avocado peel (*Persea americana*) in tea formulation: a functional product containing phenolic compounds with antioxidant activity." *Acta Scientiarum. Technology* 38 (1):23-29. doi: 10.4025/actascitechnol.v38i1.27397.

Saavedra, J., A. Córdova, R. Navarro, P. Díaz-Calderón, C. Fuentealba, C. Astudillo-Castro, L. Toledo, J. Enrione, and L. Galvez. 2017. "Industrial avocado waste: Functional compounds preservation by convective drying process." *Journal of Food Engineering* 198:81-90. doi: 10.1016/j.jfoodeng.2016.11.018.

Salinas-Salazar, C., C. Hernández-Brenes, D. G. Rodríguez-Sánchez, E. C. Castillo, J. M. Navarro-Silva, and A. Pacheco. 2017. "Inhibitory activity of avocado seed fatty acid derivatives (acetogenins) against *Listeria monocytogenes*." *Journal of Food Science* 82 (1):134-144. doi: 10.1111/1750-3841.13553.

Sarabandi, K., P. Gharehbeglou, and S. M. Jafari. 2019. "Spray-drying encapsulation of protein hydrolysates and bioactive peptides: Opportunities and challenges." *Drying Technology*:1-19.

Schaffer, B. A., B. N. Wolstenholme, and A. W. Whiley. 2013. "Introduction." In *The avocado: Botany, production and uses*, edited by Bruce A. Schaffer, B. Nigel Wolstenholme and Antony William Whiley, 560. Oxfordshire, UK: CABI.

Secretaría de Economía. 2012. *Monografía del sector aguacate en México: Situación actual y oportunidades de mercado.* Edited by Secretaría de Eonomía. Mexico. [*Monograph on the avocado sector in Mexico: Current situation and market opportunities*. Edited by the Economy Ministry].

Segovia Gómez, F., S. Peiró Sánchez, M. G. Gallego Iradi, N. A. Mohd Azman, and M. P. Almajano. 2014. "Avocado seeds: extraction optimization and possible use as antioxidant in food." *Antioxidants* 3 (2):439-454. doi: 10.3390/antiox3020439.

Silva, I. R. A., M. Magnani, F. S. M. de Albuquerque, K. S. Batista, J. d. S. Aquino, and V. Queiroga-Neto. 2017. "Characterization of the chemical and structural properties of native and acetylated starches from avocado (*Persea americana* Mill.) seeds." *International Journal of Food Properties* 20 (sup1):S279-S289. doi: 10.1080/10942912.2017.1295259.

Sivanathan, S., and N. K. B. Adikaram. 1989. "Biological activity of four antifungal compounds in immature avocado." *Journal of Phytopathology* 125 (2):97-109. doi: 10.1111/j.1439-0434.1989.tb00644.x.

Soong, Y. Y., and P. J. Barlow. 2004. "Antioxidant activity and phenolic content of selected fruit seeds." *Food Chemistry* 88 (3):411-417. doi: https://doi.org/10.1016/j.foodchem.2004.02.003.

Takenaga, F., K. Matsuyama, S. Abe, Y. Torii, and S. Itoh. 2008. "Lipid and fatty acid composition of mesocarp and seed of avocado fruits harvested at northern range in Japan." *Journal of Oleo Science* 57 (11):591-597. doi: 10.5650/jos.57.591.

Téliz-Ortiz, D., and F. J. Marroquín-Pimentel. 2007. "Importancia histórica y socioeconómica del aguacate [Historical and socioeconomic importance of the avocado] " In *El aguacate y su manejo integrado* [*The avocado and its integrated manegement*], edited by Daniel Téliz-Ortiz and Antonio Mora, 321. México: Editorial Mundi Prensa.

Tochihuitl-Martiñón, A., S. H. Chávez-Franco, C. Saucedo-Veloz, J. Suarez-Espinosa, and D. Guerra-Ramírez. 2018. "Extracts of *Persea americana* Mill. that delay ripening n avocado fruits." *Revista Mexicana de Ciencias Agrícolas* 9 (8):1639-1650. doi: 10.29312/remexca.v9i8.1720.

Tremocoldi, M. A., P. L. Rosalen, M. Franchin, A. P. Massarioli, C. Denny, É. R. Daiuto, J. A. R. Paschoal, P. S. Melo, and S. M. d. Alencar. 2018. "Exploration of avocado by-products as natural sources of bioactive compounds." *PloS one* 13 (2):e0192577. doi: 10.1371/journal.pone.0192577.

Trujillo-Mayol, I., M. M. C. Sobral, O. Viegas, S. C. Cunha, J. Alarcón-Enos, O. Pinho, and I. M. P. L. V. O. Ferreira. 2021. "Incorporation of avocado peel extract to reduce cooking-induced hazards in beef and soy burgers: A clean label ingredient." *Food Research International* 147:110434. doi: 10.1016/j.foodres. 2021.110434.

USDA. 2021. "*Food Data Central. Egg albumin.*", Last Modified April 17, 2021. https://fdc.nal.usda.gov/fdc-app.html#/food-details/173573/nutrients.

Vargas-Torrico, M. F., E. von Borries-Medrano, S. Valle-Guadarrama, and M. A. Aguilar-Méndez. 2022. "Development of gelatin-carboxymethylcellulose coatings incorporated with avocado epicarp and coconut endocarp extracts to control fungal growth in strawberries for shelf-life extension." *CyTA-Journal of Food* 20 (1):27-38. doi: 10.1080/19476337.2021.2024607.

Velderrain-Rodríguez, G. R., L. Salvia-Trujillo, G. A. González-Aguilar, and O. Martín-Belloso. 2021. "Interfacial activity of phenolic-rich extracts from avocado fruit waste: Influence on the colloidal and oxidative stability of emulsions and nanoemulsions." *Innovative Food Science & Emerging Technologies* 69:102665. doi: 10.1016/j.ifset.2021.102665.

Villarreal-Lara, R., D. G. Rodríguez-Sánchez, R. I. Díaz De La Garza, M. I. García-Cruz, A. Castillo, A. Pacheco, and C. Hernández-Brenes. 2019. "Purified avocado seed acetogenins: Antimicrobial spectrum and complete inhibition of *Listeria monocytogenes* in a refrigerated food matrix." *CyTA-Journal of Food* 17 (1):228-239. doi: 10.1080/19476337.2019.1575908.

Wang, W., T. R. Bostic, and L. Gu. 2010. "Antioxidant capacities, procyanidins and pigments in avocados of different strains and cultivars." *Food Chemistry* 122 (4):1193-1198. doi: https://doi.org/10.1016/ j.foodchem.2010.03.114.

Weremfo, A., F. Adulley, and M. Adarkwah-Yiadom. 2020. "Simultaneous optimization of microwave-assisted extraction of phenolic compounds and antioxidant activity of avocado (*Persea americana* Mill.) seeds using response surface methodology." *Journal of Analytical Methods in Chemistry* 2020:7541927. doi: 10.1155/2020/7541927.

Wong, M., L. Eyres, and L. Ravetti. 2013. "Modern aqueous oil extraction—centrifugation systems for olive and avocado oils." In *Green Vegetable Oil Processing*, edited by Walter E. Farr and Proctor Andrew, 19-51. Urbana Illinois: AOCS Press.

Wu, S., M. Tokuda, A. Kashiwagi, A. Henmi, Y. Okada, S. Tachibana, and M. Nomura. 2015. "Evaluation of the fatty acid composition of the seeds of *Mangifera indica L.* and their application." *Journal of Oleo Science* 64 (5):479-484. doi: 10.5650/jos.ess14238.

Yamassaki, F. T. 2013. "*Caracterização química, atividade imunomoduladora e antioxidante de extratos aquosos e hidroalcoólicos da goma de exsudato vegetal de Anacardium occidentale L. E de folhas de Persea americana Mill* [*Chemical characterization, immunomudatori and antioxidant activity of aqueous and hydroalcoholic extracts of Anacardium occidentale L plant exudate gum and Persea americana Mill leaves*]." Mestre, Universidade Federal do Paraná.

Yamassaki, F. T., L. H. Campestrini, S. F. Zawadzki-Baggio, and J. B. B. Maurer. 2017. "Avocado leaves: Influence of drying process, thermal incubation, and storage conditions on preservation of polyphenolic compounds and antioxidant activity." *International Journal of Food Properties* 20 (sup2):2280-2293. doi: 10.1080/10942912.2017.1369105.

Yamassaki, F. T., L. H. Campestrini, S. F. Zawadzki-Baggio, and J. B. B. Maurer. 2018. "Chemical characterization and complement modulating activities of an arabinogalactan-protein-rich fraction from an aqueous extract of avocado leaves." *International Journal of Biological Macromolecules* 120:513-521. doi: 10.1016/j.ijbiomac.2018.08.072.

Zavala-Guerrero, B., A. Hernandez-Garcia, R. Torres-Martinez, E. Melendez-Herrera, P. Rios-Chavez, A. Ochoa-Zarzosa, J. E. López-Meza, A. Saavedra-Molina, and R. Salgado-Garciglia. 2020. "Antioxidant and anti-inflammatory activities of methanolic fraction from native Mexican avocado seed oil." *The FASEB Journal* 34 (S1):1-1. doi: 10.1096/fasebj.2020.34.s1.07300.

Chapter 2

Berry Byproducts

Elia Herminia Valdés Miramontes*
Zyanya Reyes Castillo
Norma – Piña Contreras
Lourdes Barajas García
Ana Cristina Espinoza Gallardo
and Patricia Martín del Campo – Rayas
Behavioral Feeding and Nutrition Research Institute,
University Center of the University of Guadalajara, Jalisco, México

Abstract

The great demand for berries worldwide has increased, not only because of these fruits attractive color and flavor or the multiple byproducts that can be obtained from them, but also because of the bioactive compounds that contribute to health. The blueberry, blackberry, raspberry, and strawberry are among the most prominent berries. The main berry-producing countries in 2020 were the United States of America, with 269,257 tons of blueberries, Mexico, with 298,024 tons of blackberries, Russia, with 164,602 tons of raspberries, and China, with 3,801,865 tons of strawberries; quantities that generate an important economic flow after exports. Jams, juices, liqueurs, frozen and dehydrated are among the main products obtained from berries, made mainly from a harvest that does not meet quality standards for marketing fresh. The waste that constitutes an important source of bioactive compounds that present potential applications in the nutraceutical industry, such as the extraction and commercialization of ketones extracted from raspberries, which are used for dyslipidemias treatment, since scientific evidence indicates that these compounds can reduce blood's triglyceride and cholesterol levels. In the same way, scientific studies indicate that anthocyanins, have hypoglycemic, antioxidant, and lipid-lowering capacity after their chronic consumption, which represents an important adjuvant in prevention and treatment for chronic degenerative diseases. Obtaining extracts with bioactive properties from berries that do not meet requirements for fresh marketing or those derived from the industrialization of these fruits is a very promising study area, since it is also involved taking advantage of otherwise wasted berry fruits, often disposed of without any use.

---

* Corresponding Author's Email: eliav@cusur.udg.mx.

In: Food Byproducts
Editors: S. Teresita Martín-del-Campo, A. Cardador-Martínez and J. del Pilar Ramírez-Anaya
ISBN: 979-8-88697-612-0

Keywords: berries byproducts, bioactive molecules, compounds recovery, berries processing, plant-based colorants

Byproducts Generated in Processing and/or Products Discarded during Harvest, and Transportation Due to Non-compliance with Quality Standards

In berries, quality can be defined by the presence of suitable characteristics in the fruit, these can be grouped into visual, organoleptic and nutritional quality. Visual quality refers to the appearance of the fruit; that it has a uniform color, it does not present defects of mechanical damage and decay, a uniform shape, adequate size and characteristic firmness. The organoleptic and nutritional quality, this is determined by a balanced content of sugars, acids and volatile compounds responsible for each fruit's characteristic aroma. Therefore, all pre-harvest and postharvest processes must aim to obtain a quality product that maintains adequate characteristics until the consumption stage (Madrid, 2021).

Main Causes of Quality Loss

Various causes can damage berries' quality, a) inadequate fruit harvesting; exert too much pressure during harvesting, b) delays in cooling; doing so in less than one hour after harvest is recommended for optimal conservation point, any delay in cooling can cause fungi appearance and reduce shelf life, c) damage by the sun; berries must be protected from direct sunlight exposure, mainly, keeping the fruit under shade after harvesting and while waiting for transport to a refrigeration warehouse, additionally, transport must always be covered, d) collection of wet berries either by rain or dew; these fruits should always be collected dry, e) poor health in the field; the level of fungal inoculum must be kept as low as possible, it is necessary to remove damaged or rotten fruits from the soil or the plant, as well as dead, fallen or fungal leaves, f) it must be fumigated only with authorized fungicides, respecting re-entry safety periods and agrochemical products labels. These practices help postharvest disease incidence reduction (Madrid, 2021).

As mentioned, in berries production, extreme care must be taken to preserve fresh fruit's good quality; however, it is common that not all meet these requirements, to avoid waste, various commercial alternatives add economic value to the products. Berries cannot be kept fresh for a long as they are delicate fruits, so a large quantity has to be processed for better use, so their production and marketing require coordination between the supply chain and production to avoid large losses (Oludemi & Akanbi, 2013).

These fruit's consumption is increasing due to the multiple investigations supporting their health benefits; additionally, an increase in sales has been noted since SARS-CoV-2's appearance. However, preserving shelf life is a challenge, since they are susceptible fruits to physical and microbial deterioration considering them as perishable, so their transformation into different products is important to facilitate consumption, and producing food with added value and guaranteeing year round accessibility (Mugdal, 2016; Jiménez-Ortega, Sánchez-Valdez, & Heredira, 2021).

Both nationally and internationally, around 50% of production is subjected to some type of processing. Berries are utilized by freezing and dehydrating the fruit and making jams, jellies, extracts and/or nutraceuticals and juices. Other industries take advantage of its sensory qualities to extract their fiber and anthocyanins, the latter responsible for the intense color of the berries, as well as aldehydes and ketones that give the aroma. These compounds are usually included in processed products such as dairy or confectionery and food supplements' production (Aryee & Boye, 2015).

Various byproducts and waste obtained from food processing contain components with great market value, like the phytochemicals derived from fruits such as berries. These byproducts can be used in cosmetics, pharmaceuticals, and nutraceuticals' production. This is why there's a great interest in taking advantage of these components that add value to other industry products (De Feria, 2018).

Products Made from Berries

"Plant-Based" Colorants

Current food trends focus on the production and consumption of foods of plant origin, which in addition to having attractive sensory qualities, contribute to improving the health of consumers (Sloan, 2020). Therefore, the food and pharmaceutical industries aim to replace artificial additives with natural ones, which, additionally to supplying technological function, provide bioactive molecules that help improve consumers' health (Jiménez-Ortega et al., 2021).

Berries are a source of natural colorants due to compounds like anthocyanidins and anthocyanins (cyanidin, delphinidin, malvidinin, peonidin, and petudinidine). Their color ranges from red to blue, passing through violet and pink, depending on pH. Pigments extracted from berries are approved to be used in foods with the code E163. Work to apply new techniques is being performed for pigments production, such as freeze-drying, which allows these fruits' color intensity and antioxidants' preservation. Pigments could be extracted from fruits damaged in pre and postharvest, which would help reduce an economic loss (Neves, Silva & Meireles, 2021; Rózyto, 2020; Teoli et al., 2016).

Jams

Jams' production is a preservation method that consists on reducing fruits' water activity by adding sugars such as sucrose, glucose and fructose. The most consumed jam on the market is made from strawberries, so a large part of the primary production is used to make this product. The preparation of berry jams such as blueberries, strawberries, blackberries, raspberries and blueberries is important because, in addition to preserving the fruits and increasing their shelf life considerably, their consumption is beneficial due to their nutritional qualities, phytochemicals, and antioxidant content (Diaconeasa et al., 2019).

Frozen

Most of the berries (blueberry, raspberry, and blackberry) are kept frozen because this extends their shelf life for several months, preserving some sensory and nutritional properties. In addition, it is an easy way to transport them for sale (Jiménez- Ortega et al., 2021). Frozen blueberries export has increased in Mexico; in 2019 it was the main form of consumption (Coriolis, 2020). Similarly, in the United States, frozen berries account for approximately 50% of frozen fruit consumed, mainly destined for industries, food services, restaurants, hospitals, schools, convenience stores and supermarkets (Universidad de Chile, 2008).

Ultra-fast freezing is an optimal method for preserving berries since it preserves their sensory qualities, achieving consumers' acceptance. It also allows significant antioxidants, anthocyanidins and anthocyanins maintenance so that the colors remain intense and attractive to the consumer (Giusti & Jing, 2008; Zhao, 2008).

Dehydrated and Freeze-Dried

Dehydration and lyophilization methods allow to significantly preserve fruit's nutritional compounds and some of its organoleptic qualities. In the United States, dehydrated and freeze-dried berries represent the main consumption forms. In freeze-drying, the fruits are previously ultra-frozen so that water sublimation is efficient, and the components are preserved, avoiding the formation of large crystals that can break the cellular vacuoles. This type of preservation is used in the industry for food supplements' formulation, as well as nutraceuticals and beverages such as "smoothies," milkshakes, dairy products and desserts. In dehydration, water activity is reduced through the use of heat; although there is some compounds lost by this method, most of them remain, the flavor is usually intensified, color, aroma, and texture are also modified. However, it is considered a good way to extend these fruits' shelf life. Conserving them for months or years, the time is longer the less water they retain and they are perfectly preserved in closed containers; additionally, it's highly preferred because it reduces storage, handling and transport space. Small growers employ the sunlight drying technique, where the process is generally slower and exposed to microbial contamination if not cared for properly. In the same way as freeze-dried berries, dehydrated ones are often used to make "snacks," cereal products, ice cream, and confectionery products (Zhao 2008; Figuerola, 2008).

Juices and Functional Drinks

Juices and beverages are an excellent way to consume functional foods because their intake is easy and fast, allowing their nutritional and organoleptic conditions to be maintained without significant changes. The United States is the country with the highest number of sales for juices and beverages made from berries, followed by European countries (Goel-Lal, 2015).

It is increasingly common for consumers to be concerned about their health status and demand vegetable foods, looking for organic or "plant-based" foods. These products are characterized by being produced without using preservatives, colorants and/or artificial flavors. The target market for these drinks is people between 18-34 years old, who usually consume energy and sports drinks (Goel-Lal, 2015). Berry extracts favor products in this industry because they are rich in ketones and aromatic aldehydes, and also provide a pleasant flavor as well as functional compounds to the consumer (Jiménez-Ortega et al., 2021).

Undoubtedly, functional foods have great versatility and offer the opportunity to add nutritional and functional ingredients to innovative beverages however, berry extracts can be added as color and flavor substitutes to another traditional products (Jiménez-Ortega et al., 2021).

Nutraceuticals

The nutraceutical ingredients industry has grown rapidly in recent years, mainly in developed countries. Japan is the largest consumer and has a wide variety of functional foods found in solid foods, beverages and food supplements (Jiménez-Ortega et al., 2021).

Some nutraceuticals are enriched with phytochemicals that are present in berries that contain phenolic compounds such as hydroxybenzoic acids and flavonoids, particularly anthocyanins; other compounds that are often used for this industry are polyunsaturated fatty acids, dietary fiber and antioxidants (Gosh, Das and Sen, 2008). Most berries used in this industry come from Mexico and Latin America (Ministerio de Agricultura de Chile, 2009).

The market for nutraceutical products has increased in recent years due to growing demand in the market. Berries are added to products such as juices, extracts and encapsulated food supplements, which is why it is considered one of the most productive market segments and profitable since the majority of berry production is used in this way (Aryee & Boye, 2015).

However, research and development cost for these products is high, so their cost is higher compared to conventional products (Singh & Orsat, 2015). In the case of berries, once processed, the byproducts obtained from them are usually profitable because there are bioactive compounds that can be subjected to purification processes in the leaves and bagasse (Jiménez-Ortega et al., 2021).

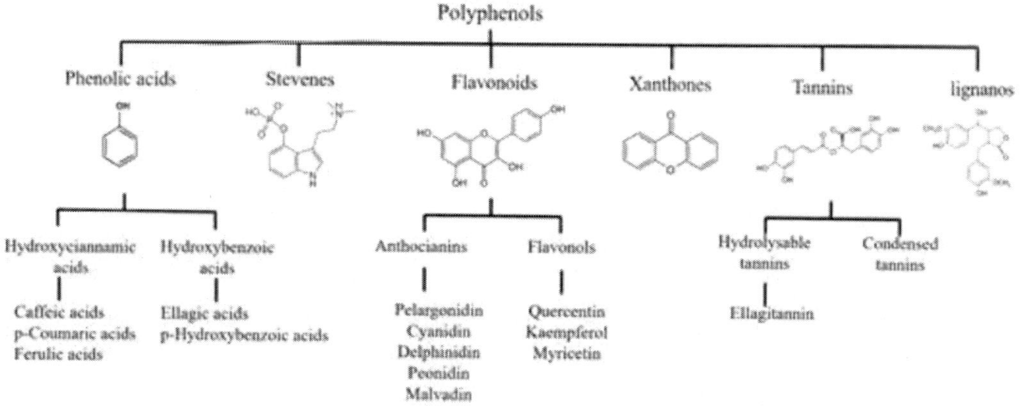

Figure 1. Bioactive components in berries.

Bioactive Molecules Present or That Could Be Generated from Byproducts

In recent decades there has been a steady increase in popularity and research interest in berries as they are one of the best dietary sources of bioactive compounds (Halvorsen, et al., 2002;

Slatnar et al., 2017). Species from the Rosaceae families: strawberry (*Fragaria vesca*), raspberry (*Rubus idaeus*), blackberry (*Rubus ulmifolius*) and the Ericaceae family: lingonberry (*Vaccinium vitis-idaea*), blueberry (*Vaccinium corymbosum*) contain higher amounts of bioactive compounds, extracts of these isolated compounds are used as ingredients in the preparation of functional foods or as nutraceuticals (Skrovankova et al., 2015).

These fruits are characterized by a high content and wide diversity of phenolic compounds, which differ in terms of structure and molecular weight, represented by phenolic acids (derived from benzoic and cinnamic acid), tannins, stilbenes and flavonoids such as anthocyanins, flavonols, isoflavones, lignans, vitamins, lutein, phytosterols and triterpenic esters. Figure 1 shows the berry fruits'main polyphenols classification and chemical structure.

Numerous studies have evaluated the functional properties of bioactive compounds present in different berries in situ, in vitro and in vivo, these health beneficial effects are related to their antioxidant, antimutagenic, anticancer, anti-inflammatory, antimicrobial, cardio protective, hypoglycemic, lipid-lowering, and, antihypertensive effects, among others (see Figure 2 and Table 1).

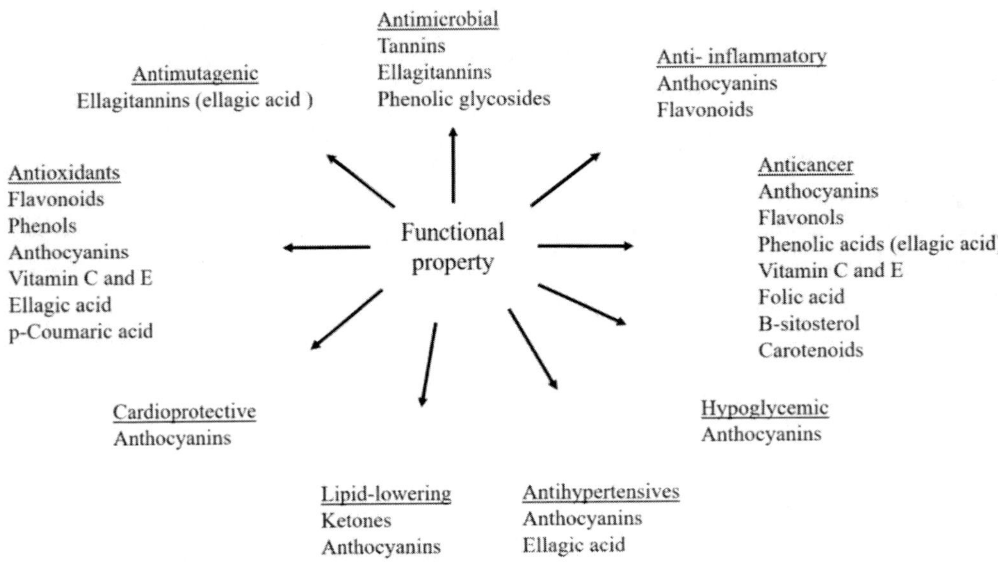

Figure 2. Functional properties of bioactive compounds present in berries.

Composition and Factors That Influence Bioactive Compounds Concentration in Berries

Berries are frequently used in the human diet. As mentioned above, they can be consumed as fresh or frozen fruit or as processed and derived products, including dried and canned fruits, beverages, jams and jellies (Nile and Park, 2014). In recent years, there has been an increased berry extracts use as a functional food ingredient and dietary supplement. The bioactive compounds in berries are of great interest to both, nutritionists and food technologists. Next, the bioactive compounds of some berries and the factors that intervene in their concentration are described.

Table 1. Bioactive compounds of berries and their effect on diabetes, dyslipidemia, oxidative stress and inflammation prevention or treatment

Compound	Food source		References
Vitamins			
A, C, E and B complex vitamin	Blackcurrants, strawberries and raspberries	It improves cardiac function in subjects with diabetes and prevents infections due to dry skin lacerations and diabetic foot complications. It also regulates LDL cholesterol levels in dyslipidemias.	Jacamna and Alfaro (2017); Scalzo et al., (2018)
Polyphenols (flavonoids)			
Anthocyanins	Strawberry, raspberry, blackberry, cranberry, black currant	Anthocyanins contribute to lower cardiac inflammation in mouse models of induced diabetes. In other studies, a decrease in cholesterol and low-density lipoprotein concentrations is reported, as well as an increase in high-density lipoprotein. Antioxidant capacity.	Jacamna and Alfaro (2017); Li, Zhang, Liu, Sun, and Xia (2015); Chen et al., (2016); Lopes da-Silva (2007)
Flavonols	Blackberries	In subjects with diabetes, decreases pro-inflammatory enzymes and molecules, reduces reactive oxygen species concentrations and regulates glucose homeostasis.	Jacamna and Alfaro (2017); Leo and Woodman (2015)
Flavonoids	Blueberries, black and red raspberries, blackberries, currants	Inhibits low-density lipoproteins (LDL) oxidation.	Nile and Park (2014)
Polyphenols (not flavonoids)			
Hydrolyzable tannins	Blueberry, blackberries, strawberries and raspberries.	They increase the production of insulin, contributes to the repair of the diabetic mitochondria caused by free radicals.	Laddha and Kulkarni (2019); Sieniawska and Baj (2017)
Stilbenes	Blueberries, blackberries	Stilbenes help reduce free radicals in diabetic individuals, providing protection against endothelial damage that commonly occurs due to excessive blood glucose concentrations, which generates increased oxidative stress and chronic inflammation.	Rodríguez-Bonilla, Gandía-Herrero, Matencio, García-Carmona and López-Nicolás (2017); Reinisalo, Kårlund, Koskela, Kaarniranta and Karjalainen (2015)
Phenolic acids	Strawberry, blackberry, blueberry, kiwi, grape, banana, coffee, orange, cherry, apple, cinnamon.	They promote insulin secretion by increasing glucose absorption in the small intestine and its uptake in insulin-sensitive tissues. It inhibits cell proliferation and cell death in prostate cancer cells.	Vinayagam, Jayachandran and Xu (2015)
Ketones	Red raspberries	Anti-obesity activity, cardio/hepatoprotection and anti-inflammatory properties. Hypolipidemic.	Hameed, Galli, Adamska-Patruno, Kretowski and Ciborowski (2020); Rao, et al., (2021)

Strawberries

The most abundant phenolic compounds in strawberries are flavonoids such as Kaempferol glycosides, quercetin glycosides, anthocyanins such as cyanidin glycosides, pelargonidine glycosides and peonidin glycosides. Some phenolic acids and hydrolyzable tannins such as ellagic acid, ellagitannins, gallic acid, gallotannins and caffeic acid are also present (Giampieri, et al., 2014). These compounds' variability and concentration in strawberries depend on many factors, including genetic characteristics, maturity stage, growing conditions, type and storage conditions. The total phenol content of strawberries is similar to the values presented by raspberries and blackberries but lower than in blueberries (Fredes, et al., 2014).

Strawberries are generally consumed fresh; however, they are highly perishable, and their quality deteriorates rapidly, which is why jams, jellies, purees and juices are made from these fruits. Freshly produced juices have a higher content of total polyphenols, anthocyanins and proanthocyanidins than those stored for six months at 4 and 30 °C; this means that storage generates total phenolic compounds content loss (Oszmiański and Wojdyło, 2009).

The main phenolic compounds in strawberries are anthocyanins, but the content is lower than in other berries such as blueberries, blackberries and raspberries (Fredes, et al., 2014). Its presence is influenced by the crop type, environmental factors such as light, temperature and agricultural methods. Anthocyanins' degradation depends on temperature and storage time; the degradation is greater at 0°C than at 30°C in a one-week period. Anthocyanins' stability also depends on oxygen presence; the strict absence during strawberry processing seems to have a stabilizing effect (Howard, Brownmiller and Prior, 2014). pH can influence anthocyanins' content; a 2.5 pH for a few months maintains the polyphenol content of these fruits compared to higher pH values (Oliveira, et al., 2015).

The total antioxidant capacity in strawberries can be influenced by storage time and temperature; the antioxidant capacity during one-week storage increases on average 1.5 times at a 10 to 20 °C temperature. Total antioxidant capacity of these fruits increases with ripening (Jin, et al., 2011; Mandave, et al., 2014).

Red Raspberries

The main phenolic compounds in raspberries are flavonols and flavons, including Kaempferol glycosides, quercetin glycosides and epigenins, as well as anthocyanins such as cyanidin glycosides and pelargonidine glycosides. They also present significant amounts of phenolic acids and hydrolysable tannins (Borges, et al., 2010; Strik, 2007). Factors that influence these compounds' variation are: genotype, variety, cultivation location, cultivation technique (conventional, organic), cultivation condition (light, maturation), and storage (time, material, atmosphere) (Wang, Chen and Wang, 2009; Hassani, et al., 2015).

The total phenol content of raspberries is similar to that of strawberries, approximately half that of blackberries and four times less than blueberries (Benvenuti, et al., 2004). Raspberries' phenolic compounds increase approximately 1.5 times after one week of storage, and the content of anthocyanins can increase 2.5 times after one week of storage at 20 °C (Kalt, Forney, Martin, and Prior, 1999). The anthocyanin content in raspberry is higher than in strawberries but lower than that in blackberries, blackcurrants and blueberries (Benvenuti, et al., 2004).

The total antioxidant capacity of raspberries is similar to that of strawberries but three times less than in blueberries. Storage at temperatures above 0°C could increase the total antioxidant capacity in this fruit (Kalt, Forney, Martin, and Prior, 1999).

Byproducts Quality Evaluation and Their Implications (Economic, Environmental and Social)

Production Statistics

Currently, berries' production is not limited to a single season, but production strategies have been developed to maintain availability for practically the entire year; in the same way, berries are produced in numerous countries, being the main producers: Russia, United States, Poland, and, Mexico, among others. World production of berries reached 11 million tons in 2013, representing an increase of 5.3% compared to 2012. The main producers of berries in 2013 were China (27.3%), the United States (19%), Russia (6.8%), Poland (5.4%) and Mexico (4.9%). These countries contributed 63.5% of world production (FIRA, 2016). In the last years the main producers of berries; blueberries, raspberries, blackberries, and strawberries worldwide are the United States, Canada, Mexico, Chile, Serbia, China, and Poland (Jiménez-Ortega, Sánchez-Valdez, & Heredira, 2021). Mexico has a large number of factors that favor production and marketing of these fruits, such as its geographical position with the US, a suitable climate, technological-scientific knowledge, as well as marketing and development strategies (González-Ramírez et al., 2020). Concerning to the worldwide commerce of strawberry, Spain, Vietnam and Thailand were positioned as the main countries in strawberry exports worldwide, and the main countries for import were China, the United States and Vietnam (González-Ramírez, Santoyo-Cortes, Arana-Coronado, and Muñoz-Rodríguez, 2020).

Blueberries' world production is estimated at 552,505 tons per year, with the United States being the main producer, followed by Canada and Mexico. Its estimated that blackberry production is around 778,405 tons per year, with Russia being the largest producer in the world with 164,602 tons per year, followed by the United States and the United Kingdom (FIRA, 2016).

The increase in world production of berries, coupled with the accelerated ripening and decomposition processes of these fruits, has also made it necessary to implement biotechnological processes for the generation of byproducts from berries (Luzón-Quintana, Castro & Durán-Guerrero, 2021). Among the main technified byproducts of berries are jams and jellies, juices and concentrates, frozen and dehydrated fruits, as well as syrups and liqueurs, while the main processed berries are strawberries, raspberries, blackberries and blackberries blue, although it is also possible to find less commercialized berry byproducts on the market, such as currant, lingonberry, goji berry, among others (Salo et al., 2021).

Byproducts Manufacture

Production processes of berries byproducts must be developed in accordance with regulations that specify evaluations to ensure they meet requirements such as adequate nutritional value for product type, adequate additives or preservatives content, product safety, and increase of product' shelf life, among others (Bartkiene et al., 2019). Berries byproducts is a growing industry and an excellent business opportunity because these generate profit, as they are easily marketed, presenting an added value for consumers. Blackberry and raspberry juices are an excellent source of phenolic compounds, depending on the preparation type and conservation treatment. Additionally, one byproduct can be made from another, such is the case of blackberry or raspberry liqueur, from the fruits' juice (Luna-Ramírez, Arellano- Cárdenas, García-Pinilla, & Mazón-Cornejo, 2017), Figure 3 shows juice production from pasteurized berries.

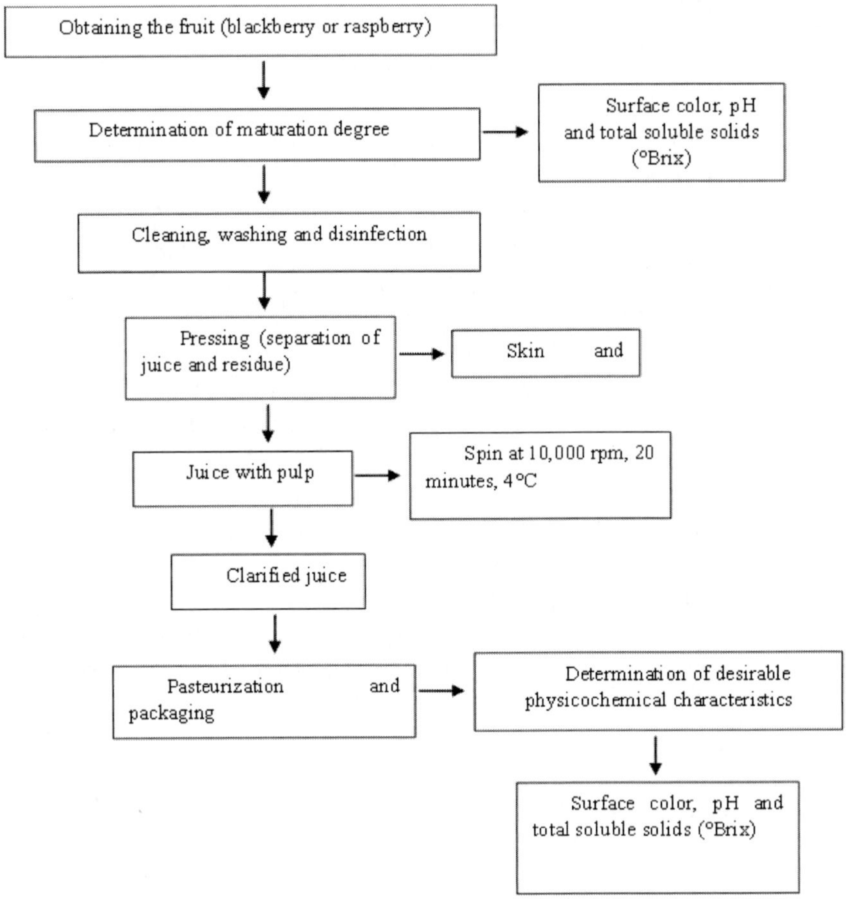

Figure 3. Preparation and quality evaluation of berry juice.

On the other hand, beverages production from blackberries and raspberries generates residues that, like juice and liquor, contain abundant phenolic compounds, as well as dietary fiber and pectins. Such residues are used in other byproducts production, thus avoiding losses and achieving the greatest economic benefit from the fruits (Dupas de Matos et al., 2017). Such residues are used by the industry for the extraction, isolation and commercialization of bioactive compounds such as ketones, anthocyanins and many other phenolic compounds that can be used as antioxidants or for various chronic diseases treatment, a great contribution to public health problem surrounding an incidence and prevalence increase in chronic degenerative diseases worldwide (Kshatriya, Hao, Li, & Bello, 2020). On the other hand, the dietary fiber obtained after juice manufacture is used in the production of products for gastrointestinal disorders' treatment, such as constipation; as well as for the treatment of other disorders such as hypercholesterolemia (Jurevičiūtė, Milda, Loreta, Daiva, & Jasutienė, 2022). While obtaining pectin from juices production's waste, is used in the jam and jellies production industry (Villegas, Handford, Alcalde, & Pérez-Donoso, 2016) (see figure 3).

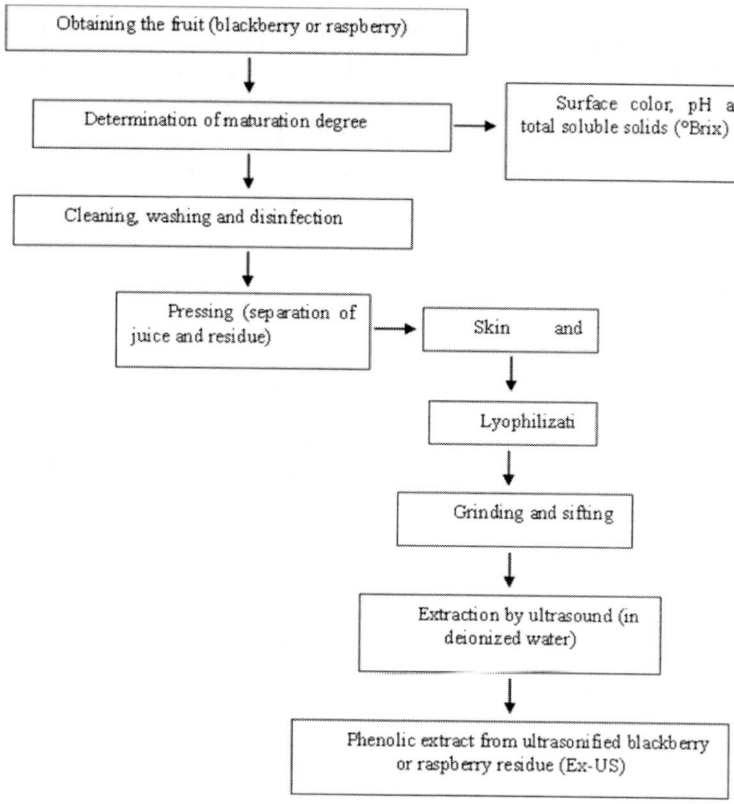

Figure 4. Extraction of phenols from residues obtained in the production of berry juice.

Berry jams, mainly strawberry jams, are another byproduct that represents an alternative for these fruits' conservation and marketing, lengthening their shelf life, facilitating distribution and access even in regions where, due to climatic conditions, it is not possible to obtain fresh fruit. However, berries' phenolic compounds tend to decrease during the processing and storage of jams and jellies; the loss of bioactive compounds of this byproduct, whether from strawberries or other berries, can vary depending on the severity of the processing applied; elevated temperature or a long exposure period, sugar and pectin content, storage conditions such as humidity, temperature and the period before consumption. Strawberry jam loses between 70 and 78% of total polyphenol content during preparation and storage; in the same way, blueberry jam can lose up to 80% of the total polyphenol content in fresh fruit, while the polyphenols in a raspberry jam can rise to 13%, contrary to other berry jams (see Figure 5) (Elídio et al., 2020).

Recovery Systems and Uses

Processing and Industrialization

Three main trends are currently guiding food development: sustainability, health and convenience. Sustainability implies promoting awareness of environmental effects caused by conventional agricultural practices and redirecting them towards organic agricultural practices to reduce the negative effects on the environment (Grunert, 2013). On the other hand, trends towards health are mainly influenced by consumers' growing interest maintaining good health, decreasing unhealthy foods' consumption and increasing those that provide a benefit in diseases prevention with diet, as well as a healthier lifestyle (Rodríguez-Mateos et al., 2014). Likewise, the trend of convenience is related to having easy-to-consume foods; ready to eat/drink, to be transported and characterized by having a long shelf life (frozen, dehydrated, etc.) (Asiolli et al., 2017).

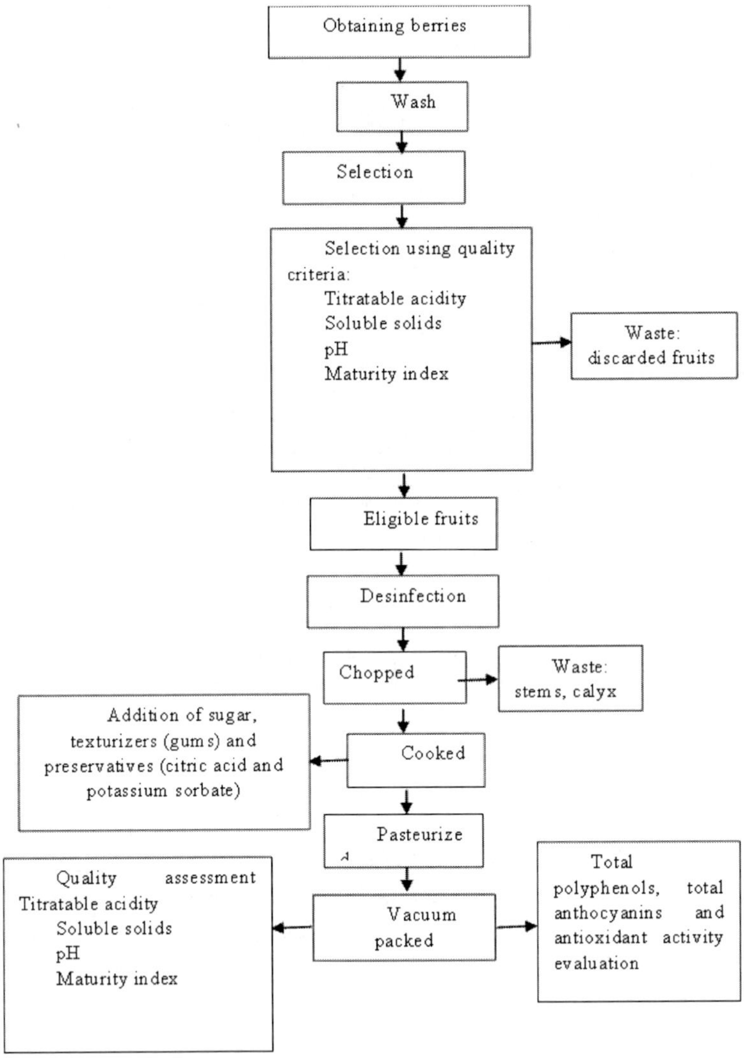

Figure 5. Diagram of berry jam production and quality evaluation.

In this regard, the incorporation of berries or their extracts in food products is a relatively easy way to add bioactive compounds, increase their commercial value and prolong their shelf life. Various technological developments have been implemented in these fruits' processing in order to give them these characteristics, such is the case of traditional technologies (pasteurized UHT, HTST, frozen, dehydrated, etc.) and emerging technologies (spray drying, high pressure, cold plasma, lyophilization and ultrasound, among others).

Growing evidence on the health benefits has promoted consumption of berries, suggesting the need to develop new products with these fruits and their extracts, using new technologies and thus increasing the consumption of different varieties of berries (Burton-Freeman et al., 2018). There is a wide variety of products developed with berries as their main component or, being fortified, and/or added with these fruits. In relation to the previous idea, there is a large amount of work focused on berry food products (Wang et al., 2019).

Emerging Technologies for Berries Processing and Industrialization

Emerging technologies arise as procedures with the ability to obtain more natural foods, with minimal deterioration on characteristics like color, texture, flavor and nutrient content with the fewest possible changes in the food matrix, allowing to provide sufficient shelf life for transport from their production place to the consumer (Contigiani et al., 2019). Some of the emerging techniques used in the treatment of berries are described below.

Spray Drying

The spray-drying technique has been widely applied on the production of functional foods and pharmaceutical products, since it improves the fruits' manipulation properties, as well as oxidation stability by protecting the organic compounds content and their functional properties (Mohammed et al., 2020). The process consists on removing moisture from a fluid material that can be a solution, dispersion or paste, by breaking it into small droplets in the presence of hot air, in order to obtain a dry powder (Pallicer et al., 2019).

In a study carried out by Leyva-Porras et al., (2019), the effect of the spray-drying process conditions used for cranberry juice, with different mixtures of maltodextrin, on the content and retention of resveratrol was analyzed. Quantitatively, the results showed that maltodextrin concentration was the variable that most influenced resveratrol content.

Lyophilization

One of the most effective methods of drying is freeze-drying, a great loss of water is generated, allowing the preservation of fresh foods' quality characteristics as it inhibits microbial deterioration and enzymatic activity. It is a process that basically consists of freezing the product, followed by sublimation of frozen water to reduced pressure (De Bruijn et al., 2016; Prosapio et al., 2017).

In one study, the quality of dehydrated organic strawberries, obtained by convection drying, freeze-drying, and spray-drying methods, was compared. Changes in polyphenol content, antioxidant activity, sensory quality and their physical-chemical properties were taken into account. Strawberries dehydrated by freeze-drying were found to have the highest vitamin C and polyphenols content, with values very close to those found in fresh strawberries after conversion to dry matter (Sadowska et al., 2020).

Due to the fact that lyophilization is a long drying process that generates a high energy consumption, it is currently used by applying some pretreatments; the case of its application in berries is not the exception. In this sense, Zhang et al., (2020) investigated the effects of high hydrostatic pressure pretreatment on water mobility and distribution, and freeze-drying duration, finding that this pretreatment caused a significant increase in water mobility of the strawberry slices, which resulted in a reduction of freeze-drying in the duration from 9 to 24%.

Cold Plasma

Cold plasma consists in the application of ionized gases through an electric field. Although the specific interaction mechanisms of cold plasma with food are not yet fully understood, it is known that some conditions must be met for exposure to food, like: the food must be in imbalance with the temperature of the gas at approximately 20 to 50 °C, in order to reduce or prevent vitamins, proteins and other biological compounds degradation, and the treatment gas contains a high concentration of agents that initiate bacterial inactivation (Bogdanov et al., 2018). Therefore, to achieve the best bactericidal results with minimal feed quality deterioration during plasma treatment, low temperatures and short treatment times are recommended (Won et al., 2017).

High Pressures

In a study by Li et al., (2018) anthocyanins were extracted from *Lonicera caerulea* berries using ultra-high pressure; To do this, they optimized the process through the response surface design, with which it was possible to determine that the optimal conditions for anthocyanin extraction were an extraction time of 7 minutes 43 seconds. Likewise, the authors reported a maximum total anthocyanin content of 323.85 mg/100 g of berries. Chromatographic analysis by HPLC-DAD-MS showed that the compounds to which the high antioxidant capacity can be attributed were 5-methylpyranocyanidin-3-hexoside and cyanidin-3-hexoxide-ethyl-catechin. The authors attribute a better yield of anthocyanins through the high pressure method because it allows a greater and faster infiltration through the cell walls, which makes mass transfer more efficient.

Ultrasound

Ultrasound-assisted extraction has been used to obtain compounds of interest. Some of the advantages it has over other extraction methods is that it is a fast, simple, low-cost, and low-energy (Bamba et al., 2018; Vazquez-Espinosa et al., 2019). Ultrasound consists of very high frequency pressure waves, which cause the materials' contraction and expansion, achieving energy transmission through them. These ultrasonic signals generate physical and chemical changes that culminate in causing cavitation, which breaks the cell wall and favors solvent permeability and the subsequent release of metabolites of interest (Vázquez-Espinosa et al., 2019).

Another advantage this method offers is that allows higher compounds of interest yields to be obtained in less time. In this sense, a study published by Vázquez Espinosa et al., (2019), reported an efficient method to extract compounds of biological interest from maqui berries. Optimal variables were an extraction time of 15 minutes to obtain phenolic compounds and 5 minutes for anthocyanins. Furthermore, the authors showed that the ultrasound method was better than the microwave method for the anthocyanins' extraction.

On the other hand, ultrasonic processing of berries can also improve the obtained compounds' bioavailability. For example, Zafra-Rojas et al., (2020) demonstrated this in blackberry residue extracts (Rubus fruticosus) obtained by ultrasound-assisted extraction, evidenced by a higher content of total phenols and anthocyanins than in extracts without ultrasound treatment, with around a higher yield of 70 and 51%, respectively.

Conclusion

Strategies for the agro industrial use of berries have great impact on trade, as they are highly perishable fruits with a short shelf life, which is why research into new forms of industrial transformation and conservation has been considered important, developing nutraceutical and functional foods, which provide greater benefits to consumers. Industries are currently developing preservatives from berry extracts to use them in meat products, preventing oxidation and increasing their shelf life (Lorenzo et al., 2018). Products made with berry extracts represent a high potential for the bakery, confectionery, non-alcoholic beverage, dairy, meat, food supplements and, herbal remedies industries, just to name a few. In addition, research continues for the development of new products such as "smoothies," liqueurs, fermented products, vinegars, syrups, pulps, and, extracts; as well as the use of innovative production and preservation technologies such as high hydrostatic pressure, supercritical fluid extraction and sonication, among others (Jiménez-Ortega et al., 2021).

References

Alberta, N. A. and Joyce Boye. 2015. "Current and Emerging Trends in the Formulations and Manufacture of Nutraceuticals and Funcional Food Products." On " *Nutraceutical and Funcional Food Processing Technology*," Joyce Boye. Ed. West Sussex, UK: John Wiley & Sons, Ltd.

Asioli, Daniele, Jessica Aschemann-Witzel, Vicenzina Caputo, Riccardo Vecchio, Azzurra Annunziata, Tormod Naes and Paula Varela. (2017). Making sense of the "clean label" trends: "A review of consumer food choice behavior and discussion of industry implications." *Food Research International*, 99 (1). pp. 58-71. ISSN 0963-9969 doi: 10.1016/j.foodres.2017.07.022.

Bamba, Bio Sigui Bruno, John Shi, Carole C. Tranchant, Sophia Jun Xue, Charles F. Forney and Loong-Tak Lim. (2018). "Influence of extraction conditions on ultrasound-assisted recovery of bioactive phenolics from blueberry pomace and their antioxidant activity." *Molecules*, 23(7). doi.org/10.3390/molecules23 071685.

Bartkiene, Elena, Vita Lele, Vytaute Sakiene, Paulina Zavistanaviciute, Modestas Ruzauskas, Jurga Bernatoniene, Valdas Jakstas, Pranas Viskelis, Daiva Zadeike and Grazina Juodeikiene. 2019. "Improvement of the antimicrobial activity of lactic acid bacteria in combination with berries/fruits and dairy industry by-products." *Society of Chemical Industry* 99:3992-4002. Accessed March 02, 2021. doi:10.1002/jsfa.9625.

Benvenuti, S., F. Pellati, M. Melegari and D. Bertelli. 2004. "Polyphenols, anthocyanins, ascorbic acid, and radical scavenging activity of Rubus, Ribes, and Aronia." *J. Food Sci.* 69:164–169. Accessed February 28, 2022.

Bogdanov, Todor, Ivan Tsonev, Plamena Marinova, Evgenia Benova, Krasimir Rusanov, Mila Rusanova, Ivan Atanassov, Zdenka Kozáková and František Krčma. "Microwave Plasma Torch Generated in Argon for Small Berries Surface Treatment." *Applied Sciences* (2018) doi:10.3390/app8101870.

Borges, Gina, Alexandra Degeneve, William Mullen, Alan Crozier. 2010. "Identification of flavonoid and phenolic antioxidants in black currants, blueberries, raspberries, red currants, and cranberries." *J. Agric. Food Chem.* 58:3901–3909. Accessed March 03, 2022.

Burton-Freeman, Britt M., Patricia M. Guenther, Miyoung Oh, David Stuart and Helen H. Jensen (2018) "Assessing the consumption of berries and associated factors in the United States using the National Health and Nutrition Examination Survey (NHANES), 2007-2012." *Food Function*, 9(2):1009-1016. doi: 10.1039/c7fo01650f.

Chen, Yu-Feng, Shibu, Marthandam Asokan, Ming-Jen Fan, Ming-Cheng Chen, Vijaya Padma Viswanadha, Yi.Lin Lin and Chih-Yang Huang, C. Y. 2016. "Purple rice anthocyanin extract protects cardiac function in STZ-induced diabetes rat hearts by inhibiting cardiac hypertrophy and fibrosis." *The Journal of Nutritional Biochemistry* 31:98-105. doi:10.1016/j.jnutbio.2015.12.020.

Cieślik, Ewa, Anna Gręda and Wiktor Adamus. 2006. "Contents of polyphenols in fruit and vegetables." *Food Chem.* 94:135–142. doi:10.1016/j. foodchem.2004.11.015.

Contigiani, Eunice Valentina, Stella Maris, Alzamora and Paula Luisina Gómez. (2019). "*Desarrollo de estrategias alternativas para la conservación post-cosecha de frutillas.*" ["*Developement of alternitive strategies for the postharvest cnservation of strawberries*"] (Doctor), Universidad de Buenos Aires, Buenos Aires, Argentina.

Coriolis. 2020. *Opportunities in the New Zealand Blueberry Industry.* https://www.mbie.govt.nz/dmsdocument/11670-opportunities-in-the-new-zealand-blueberries-industry.

De Bruijn, Johannes, Fernando Rivas, Yeannina Rodriguez, Cristina Loyola, Adan Flores, Pedro Melin, and Rodrgio Bórquez. (2016). "Effect of Vacuum Microwave Drying on the Quality and Storage Stability of Strawberries." *Journal of Food Processing and Preservation, 40(5)*, 1104-1115. doi:10.1111/jfpp.12691.

De Feria, Francisco, Rossier, Francisco and Sepúlveda, Abigail. 2018. "*Cero Pérdida de Materia Prima en la Agroindustria: Informe de Etapa 1 (Diagnóstico).*" ["*Zero loss of raw material in agroindustry: Stage 1 Report (Diagnosis)*"] Accessed March 02. https://www.academia.edu/41179322Transformaalimentos_cero_perdidas_en_la_agroindustria_informe_de_diagnóstico.

Diaconeasa, Zorita, Cristian I. Luhas, Huseyin Ayvaz, Dumitrita Rugina, Andreea Stanila, Fransisc Dulf, Andrea Bunea, Sonia Ancuta Socaci, Carmen Socaciu, and Adela Pintea. 2019. "Phytochemical Characterization of Comercial Processed Blueberry, Blackberry, Blackcurrant, Cranberry, and Raspberry and Their Antioxidant Activity." *Antioxidants.* 8:1-15. Accessed March 02 2021. doi:https://doi.org/10.3390/antiox8110540.

Dupas de Matos, Amanda, Andrea Curioni, Alan T. Bakalinsky, Matteo Marangon, Gabriella Pasini and Simone Vincenzi. 2017. "Chemical and sensory analysis of verjuice: An acidic food ingredient obtained from unripe grape berries." *Innovative Food Science and Emerging Technologies.* Accessed March 01, 2021. doi:10.1016/j.ifset.2017.09.014.

Fideicomisos Instituidos en Relación con la Agricultura (FIRA). *Panorama Agroalimentario* [*Agri-food Panorama*]. (2016). https://www.gob.mx/cms/uploads/attachment/file/61952/Panorama_Agroalimentario_Ma_z_2015.pd. Acces: 13 de octubre de 2022.

Figuerola, Fernando E. 2008. "Dehydration of berries." *Berry Fruit Value-Added Products for Health Promotion* by Zhao, Y. March 01 2021.

Fredes, Carola, Gloria Montenegro, Juan Pablo Zoffoli, Francisca Santander and Paz Robert. 2014. "Comparison of the total phenolic content, total anthocyanin content and antioxidant activity of polyphenol-rich fruits grown in Chile." *Cienc. Inv.* Agr 41:49–60. Accessed March 05, 2022.

Ghosh, Nandini, Amidativa Das, Chandan k. Sen. 2008. "Nutritional Supplements and Funcional Foods: Functional Significance and Global Regulations." *Nutraceutical and Functional Food Regulations in the United States and Around the World* by Bagchi. March 01 2021.

Giampieri, Francesca, José M. Alvarez-Suarez, Luca Mazzoni, Tamara Y. Forbes-Hernandez, Massimiliano Gasparrini, Ana P. Gonzàlez-Paramàs, Celestino Santos-Buelga, José L. Quiles, Stefano Bompadre, Bruno Mezzetti and Maurizio Battino. 2014. "An anthocyanin-rich strawberry extract protects against oxidative stress damage and improves mitochondrial functionality in human dermal fibroblasts exposed to an oxidizing agent." *Food Funct.* 5:1939–1948. Accessed February 25, 2022.

Giusti, M. Monica and Pu Jing. 2008. "Natural Pigments of berries: Functionality and applications." *Review of the Berry Fuit Value-Added Products for Health Promotion* by Zhao, Y. March 01 2021.

Goel-Lal, Garima. 2015. "Processing of Beverages for the Health Food Market Consumer." *Nutraceutical and Functional Food Processing Technology* by Boye, J.l. March 01 2021.

González-Ramírez María, Vinicio Santoyo-Cortes, José Arana-Coronado and Manrrubio Muñoz-Rodríguez. 2020. "The insertion of Mexico into the global value chain of berries." *World Develoment Perspectives* 20:1-11. Accessed March 02, 2021. doi:https://doi.org/10.1016/j.wdp.2020.100240.

Grunert, Klaus G. (2013). Trends in food choice and nutrition. *Consumer attitudes to food quality products: Empashis on Southem Europe* (133). 23-30. doi: doi.org/10.3920/978-90-8686-762-2_2.

Halvorsen, Bente, Kari Holte, Mari Myhrstad, Ingrid Barikmo, Erlend Hvattum, Siv Remberg, Anne-Britt Wold, Karin Haffner, Halvard Baugerød, Lene Andersen, Jan Moskaug, David Jacobs and Rune Blomhof. 2002. "A systematic screening of total antioxidants in dietary plants." *Journal of Nutrition* 132:461–471. Accessed March 01, 2021. doi:10.1093/jn/132.3.461.

Hameed, Ahsan., Mauro Galli, Edyta Adamska-Patruno, Adam Krętowski and Michal Ciborowski. 2020. "Select Polyphenol-Rich Berry Consumption to Defer or Deter Diabetes and Diabetes-Related Complications." *Nutrients* 12:2538. Accessed February 28, 2022. doi:10.3390/nu12092538.

Hassani, Shokoufeh, Marjan Shariatpanahi, Fatemeh Tavakoli, Amir Nili-Ahmadabadi and Mohammad Abdollahi. 2015. "The changes of bioactive ingredients and antioxidant properties in various berries during jam processing." *Int. J. Biosci.* 6:172–179. Accessed February 25, 2022.

Howard, Luke R., Cindi Brownmiller and Ronald L. Prior. 2014. "Improved color and anthocyanin retention in strawberry puree by oxygen exclusion." *J. Berry Res.* 4:107–116. Accessed February 23, 2022.

Jacamna, Nicodemo and Sarela Alfaro. 2017. "*Antioxidantes en los alimentos.*" ["*Antioxidants in food*"] PhD diss., National University of Barranca.

Jiménez-Ortega, Luis., Eduardo Sánchez-Valdez and J. Basilio Heredira. 2021. Production, market and commercialization of processed products and derivatives based on berries. Martín eds. 2021. *Cultivation, post-harvest, processing and trade of berries.* https://issuu.com/horticulturaposcosecha/docs/cultivo_poscosecha_procesado_y_comercio_de_berries.

Jin, Peng, Shiow Y. Wang, Chien Y. Wang and Yonghua Zheng, Y. 2011. "Effect of cultural system and storage temperature on antioxidant capacity and phenolic compounds in strawberries." *Food Chem.* 124:262–270. Accessed March 05, 2022.

Jurevičiūtė, Ieva., Milda, Keršienė., Loreta, Bašinskienė., Daiva, Leskauskaitė., and Ina Jasutienė. 2022. "Characterization of Berry Pomace Powders as Dietary Fiber-Rich Food Ingredients with Functional Properties." *Foods* 11, no. 5: 716. Accessed March 01, 2021. doi:https://doi.org/10.3390/foods11050716.

Kalt, Whilhelmina., Charles Forney, Anotnio Martin, A., and Ronald Prior. 1999. "Antioxidant Capacity, Vitamin C, Phenolics, and Anthocyanins after Fresh Storage of Small Fruits." *J. Agric. Food Chem.* 47:4638–4644. Accessed February 28, 2022.

Kshatriya, Dushyant., Lihong Hao, Xianyi Li and Nicholas Bello. 2020. "Raspberry Ketone [4-(4-Hydroxyphenyl)-2-Butanone] Differentially Effects Meal Patterns and Cardiovascular Parameters in Mice." *Nutrients* 1754:1-16. Accessed March 01, 2021. doi:10.3390/nu12061754.

Laddha, Ankit and Yogesh Kulkarni. 2019. "Tannins and vascular complications of Diabetes: An update." *Phytomedicine* 56:229-245. Accessed March 02, 2022. doi:https://doi.org/10.1016/j.phymed.2018.10.026.

Landim Neves, María Isabel., Eric Keven Silva, Maria Angela A. Meireles. 2021. "Natural blue food colorants: consumer acceptance, current alternatives, trends, challenges, and future strategies." *Trends in Food Science & Technology* 112:163-173. Accessed March 02 2021. doi:https://doi.org/10.1016/j.tifs.2021.03.023.

Leo, Chen-Huei and Owen Woodman. 2015. "Flavonols in the Prevention of Diabetes-induced Vascular Dysfunction." *Journal of Cardiovascular Pharmacology* 65:532-544. Accessed March 03, 2022. doi:https://doi.org/10.1097/FJC.0000000000000180.

Leyva-Porras, C.; Saavedra-Leos, M. Z.;Cervantes-González, E.; Aguirre-Bañuelos, P.; Silva-Cázarez, M. B.; Álvarez-Salas, C. (2019). "Spray Drying of Blueberry Juice-Maltodextrin Mixtures: Evaluation of Processing Conditions on Content of Resveratrol." *Antioxidants* (Basel, Switzerland), 8(10), 437. https://doi.org/10.3390/antiox8100437.

Li, Dan., Yuhua Zhang, Yan Liu, Ruifang Sun and Mia Xia. 2015. "Purified Anthocyanin Supplementation Reduces Dyslipidemia, Enhances Antioxidant Capacity, and Prevents Insulin Resistance in Diabetic Patients." *The Journal of Nutrition* 145:742-748. Accessed March 05, 2022. doi:https://doi.org/10.3945/jn.114.205674.

Li, Xinyuan., Xinyuan Meng, Hui Tan, Qingqi Ren, Dognan Li and Bin Li. (2018). "Ultra-High Pressure Extraction of Anthocyanins from Lonicera caerulea and its Antioxidant Activity Compared with Ultrasound-assisted Extraction." *International Journal of Agriculture and Biology*, 20 (10). doi: 10.17957/IJAB/15.0775.

Lopes da-Silva, Fatima., María Teresa Escribano-Bailon, José Joaquin Perez-Alonso, Julián C. Rivas-Gonzalo and Celestino Santos-Buelga. 2007. "Anthocyanin pigments in strawberry." *Food Science Technology* 40:374–82. Accessed March 01, 2021. doi: https://doi.org/10.1016/j.lwt.2005.09.018.

Luna-Ramírez, K., Arellano-Cárdenas, S., García-Pinilla, M., and Mazón-Cornejo. 2017. "Kinetic analysis of the stability of antioxidants in blackberry (*Rubus fruticosus* l.) liquor." *Mexican Journal of Chemical Engineering* 1:121-130. Accessed March 01,2021.

Luzón-Quintana, Luis María., Remedios Castro and Enrique Durán-Guerrero. 2021. "Biotechnological Processes in Fruit Vinegar Production." *Foods* 10, 945:2-23. Accessed March 01, 2021. doi:https://doi.org/10.3390/foods10050945.

Madrid, M., 2021. Postharvest recommendations in arrears. Martín, L eds. 2021. *Cultivation, post-harvest, processing and trade of berries*. https://issuu.com/horticulturaposcosecha/docs/cultivo_poscosecha_procesado_y_comercio_de_berries.

Mandave, Pallavi., Pankaj Pawar, Prabhakar Ranjekar, Nitin Mantri and Aniket Kuvalekar. 2014. "Comprehensive evaluation of in vitro antioxidant activity, total phenols and chemical profiles of two commercially important strawberry varieties." *Sci. Hortic.* 172:124–134. Accessed March 06, 2022.

Maurício Elídio, Patrícia da Silva, Luciana Affonso, Esperanca Baptista, Jaime Vilela, Elisangela Nunes and Eduardo de Barros. 2020. " Impact of processing, packages, and storage on quality of mangaba (*Hancornia speciosa* Gomes) jelly." *Journal of Foo of Processing and Preservation* 1-13. Accessed March 01, 2021. https://doi.org/10.1111/jfpp.14814.

Ministerio de Agricultura de Chile. 2009. "*Resultados y Lecciones en productos a Base de Berries Nativos*." ["*Results and lessons in products based on native berries*."] Accessed March 03 2021. http://bibliotecadigital.fia.cl/handle/20.500.11944/2081.

Mohammed Nameer Khairulah, Chin Ping Tan, Yazid Abd Manap, Belal J. Muhialdin, Anis Shobirin Hussin. (2020). "Spray Drying for the Encapsulation of Oils-A Review." *Molecules*. 25(17). doi: 10.3390/molecules25173873.

Mugdal, Samriddh. 2016. "*Technologies for Proccessing Value-Added Berry Fruit Products*." Accessed Febrary 20, 2021. http://chonbds.com/pdf/jbt-white-paper-berry-processing.pdf.

Nile, Shivraj Hariram and Se Won Park. 2014. "Edible berries: Bioactive components and their effect on human health." *Journal of Nutrition* 30:134-144. Accessed February 19, 2021. doi:10.1016/j.nut.2013.04.007.

Oliveira, Ana., Maria Helena Gomes, Elisabete Alexandre, Fátima Poças, Domingos Almeida and Manuela Pintado. 2015. "Phytochemicals preservation in strawberry as affected by pH modulation." *Food Chem.* 170:74–83. Accessed March 02, 2022.

Oludemi, Femi O. and Charles T. Akanbi. 2013. "Chemical, antioxidant and sensory properties of tomato-watermain-pineapple blends, and changes in their total antioxidant capacity during storage." *International Journal of Food Science & Technology* 48:1416-1425. Accessed March 02, 2021. doi:https://doi.org/10.1111/ijfs.12104.

Oszmiański, Jan and Aneta Wojdyło. 2009. "Comparative study of phenolic content and antioxidant activity of strawberry puree, clear, and cloudy juices." *Eur. Food Res. Technol.* 228:623–631. Accessed February 22, 2022.

Pellicer, José Antonio., María Isabel Fortea, Juan Trabal, María Isabel Rodríguez-López, José Antonio Gabaldón and Estrella Núñez-Delicado. (2019). "Stability of microencapsulated strawberry flavour by spray drying, freeze drying and fluid bed." *Powder Technology*, 347, 179-185. https://doi.org/10.1016/j.powtec.2019.03.010.

Rao, Shailaja., Mallesh Kurakula, Nagarjuna Mamidipalli, Papireddy Tiyyagura, Bhaumik Patel and Ravi Manne. 2021. "Pharmacological Exploration of Phenolic Compound: Raspberry Ketone—Update 2020." *Plants* 10:1323. Accessed March 06, 2022. doi:10.3390/plants10071323.

Reinisalo, Mika., Anna Kårlund, Ali Koskela, Kai Kaarniranta and Reijo O. Karjalainen. 2015. "Polyphenol Stilbenes: Molecular Mechanisms of Defence against Oxidative Stress and Aging-Related Diseases." *Oxidative Medicine and Cellular Longevity* 1-24. Accessed February 20, 2022. doi:https://doi.org/10.1155/2015/340520.

Rodríguez-Bonilla, Pilar., Fernando Gandía-Herrero, Adrián Matencio, Francisco García-Carmona and José Manuel López-Nicolás. 2017. "Comparative Study of the Antioxidant Capacity of Four Stilbenes Using ORAC, ABTS+, and FRAP Techniques." *Food Analytical Methods* 10:2994-3000. Accessed February 18, 2022. doi:https://doi.org/10.1007/s12161-017-0871-9.

Rodriguez-Mateos, Ana., Christian Heiss, Gina Borges and Alan Crozier. (2014) "Berry (Poly) phenols and Cardiovascular Health." *Journal of Agricultural and Food Chemistry*, 62 (18), 3842-3851 doi: 10.1021/jf403757g.

Rózyto, Renata. 2020. "Recent trends in methods used to obtain natural food colorants by freeze-drying." *Trends in Food Science & Technology* 102:39-50. Accessed March 02 2021 doi:https://doi.org/10.1016/j.tifs.2020.06.005.

Sadowska, Anna., Franciszek Świderski and Ewelina Hallmann. (2020). "Bioactive, physicochemical and sensory properties as well as microstructure of organic strawberry powders obtained by various drying methods." *Applied Sciences*, 10(14), doi.org/10.3390/app10144706.

Salo, Heikki., Nga Nguyen, Emmi Alakärppä, Linards Klavins, Anne Linn, Hykkerud, Katja Karppinen, Laura Jaakola, Maris Klavins and Hely Häggman. 2021. "Authentication of berries and berry-based food products." *Comprehensive Reviews in Food Science and Food Safety* 20:5197-5225. Accessed March 01, 2021. doi:10.1111/1541-4337.12811.

Scalzo, Rebecca L., Timothy A. Bauer, Kylie Harrall, Kerrie Moreau, Cemal Ozemek, Leah Herlache, Shawna McMillin, Amy G. Huebschmann, Jennifer Dorosz, Jane E. B. Reusch and Judith G. Regensteiner. 2018. "Acute vitamin C improves cardiac function, not exercise capacity, in adults with type 2 diabetes." *Diabetology & Metabolic Syndrome* Accessed February 19, 2022. doi:10.1186/s13098-018-0306-9.

Sieniawska, E., and T. Baj. 2017. "Tannins." *Pharmacognosy* 99-232. Accessed March 02, 2022. doi: https://doi.org/10.1016/B978-0-12-802104-0.00010-X.

Singh, Ashutosh and Valérie Orsat. 2015. "Key Considerations in the Selection of Ingredients and Processing Technologies for Funcional Foods and Nutraceutical Products." *Nutraceutical and Functional Food Processing Technology* by Boye, J. March 01 2021. Doi: doi.org/10.1002/9781118504956.ch3.

Skrovankova, Sona., Daniela Sumczynski, Jiri Mlcek, Tunde Jurikova and Jiri Sochor. 2015. "Bioactive Compounds and Antioxidant Activity in Different Types of Berries." *International Journal of Molecular Sciences* 10:24673-24706. Accessed February 20, 2021. doi:https://doi.org/10.3390/ijms161024673.

Slatnar, Ana., Jerneja Jakopic, Franci Stampar, Robert Veberic and Polona Jamnik. 2017. "The effect of bioactive compounds on in vitro and in vivo antioxidant activity of different berry juices." *PLoS ONE* 7:e47880. Accessed February 28, 2021. doi:10.1371/journal.pone.0047880.

Sloan, Elizabeth. 2020. *"The Top 10 Functional Food trends."* Accessed March 02. https://www.ift.org/news-and-publications/food-technology-magazine/issues/2018/april/features/top-10-functional-food-trends-2018#:~:text=Functional%20mushrooms%20%28e.g.%2C%20chaga%2C%20reishi%2C%20and%20cordyceps%29%20are,source%20of%20quick%20energy%20are%20also%20getting%20attention.

Strik, Bernadine C. 2007. "Berry crops: Worldwide area and production systems". In *Berry Fruit: Value-Added. Review of The Products for Health Promotion* by Zhao, Y., Ed.; CRC Press: Boca Raton, FL, USA, pp. 3–51.

Teoli, F., Lucioli, S., Nota, P., Frattarelli, A., Matteocci, F., Di Carlo, A., Caboni, E., and Forni, C. 2016. "Role of pH and pigment concentration for natural dye-sensitized solar cells treated with anthocyanin extracts of common fruits." *Journal of Photochemistry and Photobiology A: Chemistry*. 316:24-30. Accessed March 02 2021. doi:https://doi.org/10.1016/j.jphotochem.2015.10.009.

Universidad de Chile. 2009. "*Technical and economic evaluation of agroindustrial products rich in antioxidants based on native berries (Aristotelia chilensis, Ugni molinae and Fragaria chiloensis).*" Accessed March 01 2021. http://bibliotecadigital.fia.cl/handle/20.500.11944/145657.

Vázquez-Espinosa, Mercedes., Ana V. González de Peredo, Marta Ferreiro-González, Ceferino Carrera, Miguel Palma, Gerardo F. Barbero and Estrella Espada-Bellido. (2019). "Assessment of ultrasound assisted extraction as an alternative method for the extraction of anthocyanins and total phenolic compounds from maqui berries (*Aristotelia chilensis* (Mol.) Stuntz)." *Agronomy- Basel*, 9(3), 17. doi.org/10.3390/agronomy9030148.

Villegas, Daniel., Michael Handford, José Antonio Alcalde and Alonso Pérez-Donoso. 2016. "Exogenous application of pectin-derived oligosaccharides to grape berries modifies anthocyanin accumulation, composition and gene expression." *Plant physiology and Biochemistry* 104:125-133. Accessed March 01, 2021. doi:https://doi.org/10.1016/j.plaphy.2016.03.020.

Vinayagam, Ramachandran., Muthukumaran Jayachandran and Baojun Xu. 2015. "Antidiabetic Effects of Simple Phenolic Acids: A Comprehensive Review." *Phytotherapy Research* 30:184-199. Accessed February 25, 2022. doi:10.1002/ptr.5528.

Wang, Jin., Jun Wang, Yinghua Ye, Sai Kranthi Vanga and Vijaya Raghavan. (2019). "Influence of high-intensity ultrasound on bioactive compounds of strawberry juice: Profiles of ascorbic acid, phenolics, antioxidant activity and microstructure." *Food Control*, 96, 128-136. https://doi.org/10.1016/j.foodcont.2018.09.007.

Wang, Shiow Y., Chi T-sun Chen, and Chien Y. Wang. 2009. "The influence of light and maturity on fruit quality and flavonoid content of red raspberries." *Food Chem* 112:676–684. Accessed March 02, 2022.

Won, Mee Yeon., Seung jo Lee and Sea C. Min. (2017). "Mandarin preservation by microwave-powered cold plasma treatment." *Innovative Food Science & Emerging technologies*. 39, 25-32. doi: 10.1016/j.ifset.2016.10.021.

Zafra-Rojas, Quinatzin Yadira., Blanca Edelia González-Martínez, Nelly del Socorro Cruz-Cansino, Manuel López-Cabanillas, Ángela Suárez-Jacobo, Alicia Cervantes-Elizarrarás and Esther Ramírez-Moreno. (2020). "Effect of ultrasound on in vitro bioaccessibility of phenolic compounds and antioxidant capacity of blackberry (*Rubus fruticosus*). Residues cv. Tupy." *Plant Foods for Human Nutrition*, 75(4), 608-613.

Zhang, Lihui., Yu Qiao, Chao Wang, Li Liao, Defang Shi, Kejing An, Jianzhong Hu, Jun Wang and Liu Shi. 2020. "Influence of high hydrostatic pressure pretreatment on properties of vacuum-freeze dried strawberry slices." *Food Chemistry*, 331, 127203. doi.org/10.1016/j.foodchem.2020.127203.

Zhao, Yanyun. 2008. "Freezing process of berries." *Berry Fuit Value-Added Products for Health Promotion* by Zhao, Y. March 01 2021.

Chapter 3

Cacao Byproducts: Their Generation and Valorization Strategies

Martina Klinsky-Del Castillo
Karol Michelle Doniz-Allende
and Blanca Isabel Maldonado-Guevara*
Escuela de Ingeniería y Ciencias, Tecnológico de Monterrey, Querétaro, Querétaro

Abstract

Cacao byproducts are generated along the cacao production chain: the production of dry beans leads to mainly three fractions (i) pod husks (CPH), (ii) bean shells (CBS) and (iii) mucilage (CM); while harvesting, storing and transporting stages result in cacao beans that non-accomplishment the quality standards due to mechanical damage and presence of insects among others. Of these byproducts, CPH is the one that is generated in the highest proportion (70%).

When the byproducts are considered waste, they cause environmental problems, especially when they are left to rot on the cacao plantation generating unpleasant odors and constituting an inoculum of fungal diseases for cacao crops, among other effects.

Processing cacao byproducts has led to its valorization and, at the same time, has diminished its impact on the environment. Byproducts have been used as fertilizers and soil amendments; in producing organic acids like lactic, succinic, and citric; and in the generation of biogas, biofilms, and biopolymers, among many other applications.

Cacao beans that do non-accomplishment the quality parameters due to physical damage or that are affected by diseases such as black rot and monilla; as well as grains that are moldy, smoky, over fermented or damaged by insects; are used for the production of refined and deodorized cacao butter for the cosmetic, pharmaceutical and food industry.

Cacao byproducts are also a source of bioactive molecules like dietary fiber and polyphenolic compounds, with antioxidant and antimutagenic activity. Currently, cacao byproducts constitute an important source of raw materials for different industries.

Keywords: cacao byproducts, cacao pod husk, cacao bean shells, cacao mucilage, biomolecules, cacao biomass

* Corresponding Author's Email: bimaldon@tec.mx.

In: Food Byproducts
Editors: S. Teresita Martín-del-Campo, A. Cardador-Martínez and J. del Pilar Ramírez-Anaya
ISBN: 979-8-88697-612-0

Introduction

Cacao (*Theobroma cacao L*) is a tree native to Central and South America, but nowadays, it is also cultivated in other countries that meet the conditions of humidity and temperature (above 27°C) for its optimal growth. Its cultivation is important for many countries, especially developing countries that base their economy on this crop. Seven countries produce up to 90% of the cacao worldwide: Cóte d'Ivoire, Ghana, Indonesia, Cameroon, Nigeria, Ecuador, and Brazil.

According to the International Cacao Organization (ICCO) (ICCO 2022), in 2018/2019, African countries were responsible for 80% of the total global production of cacao beans, being the main producers: Cóte d'Ivoire (2,154 thousand tons) and Ghana (812 thousand tons). The major producer of América is Ecuador, with 322 thousand tons (Figure 1).

Cacao is an important source of economic profit for the producer countries and those participating in the value chain. This value chain is one of the most important worldwide due to the logistics necessary for the production, transportation, manufacturing, distribution and commercialization of the raw material and derived products (Vásquez et al. 2019).

Given the economic importance of cacao production, its social and ecological impact is also considerable and has its greatest effect, especially in producing countries. Cacao production is familiar-based cultivation; hence, price fluctuation and speculation directly impact the producers. Nowadays, guaranteeing a decent and fair livelihood for cacao farmers remains a priority (Vásquez et al. 2019).

According to the Global Market Report (2019), 40 to 50 million people all over the world earn an income from cacao production, of which 5 million are agricultural households and 70% are smallholder farms (Voora, Bermúdez, and Larrea 2019).

The environmental aspect of the cacao industry also presents relevant issues like:

(a) better administration of natural resources (land and water),
(b) regulation and use of agrochemicals to reduce negative effects on producers' health and soils without affecting production;
(c) conservation of local ecosystems and biodiversity;
(d) improving the capacity to respond to climate changes (Vásquez et al. 2019) and
(e) management and use of byproducts.

Byproducts Generated during Harvesting, Processing, and Transport of Cacao Beans

Cacao beans are mainly used to produce confectionery, functional food, and beverages; nevertheless, their use in cosmetics and the pharmaceutical industry has been growing in recent years due to their functional properties and health benefits.

The transformation of cacao beans into final products requires a series of steps that can be grouped on pre-processing and processing practices. Pre-processing practices are developed on the field and involve pod harvesting, pod opening, bean recovery, fermentation, and drying of the beans. After those steps, cacao beans are transported to the industrial plant.

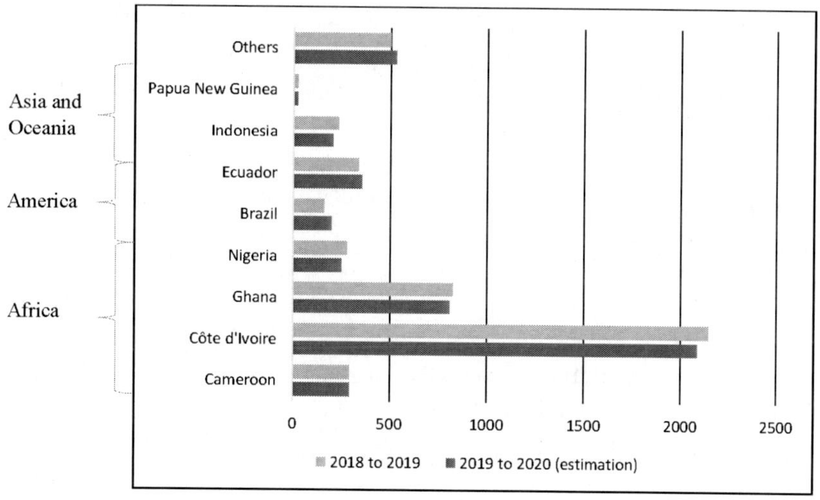

Figure 1. World cacao production (10^3 tons), 2018-2019 and 2019-2020 (Voora, Bermúdez, and Larrea 2019).

During pre-processing practices, an important number of byproducts is generated (Figure 2).

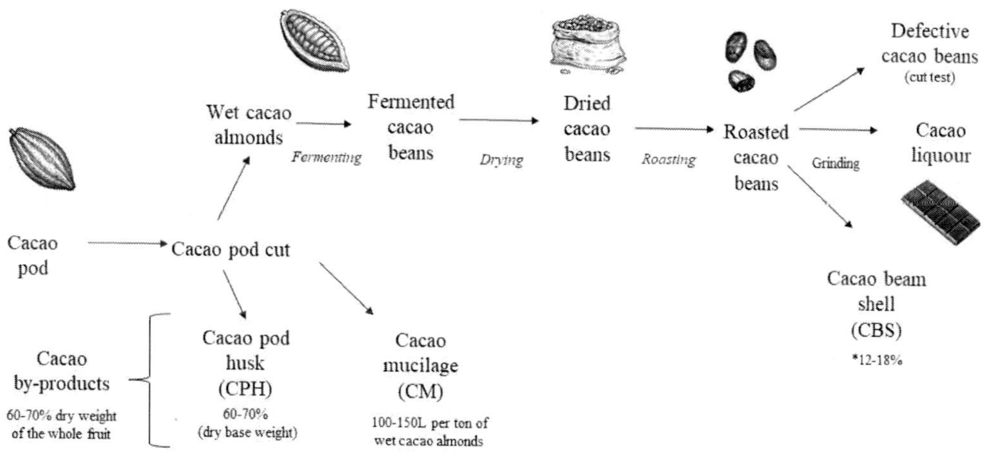

Figure 2. Byproducts generated during cacao processing.

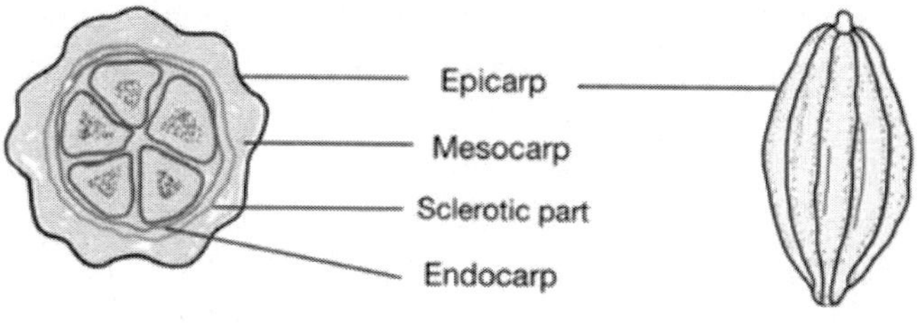

Figure 3. Cacao pod husk parts (CPH).

Cacao pod husk (CPH) and cacao mucilage (CM) are the main byproducts obtained during pre-processing practices. CPH is the by-product generated in a major proportion, constitutes 60-70% of the cacao pod and comprises the epicarp, mesocarp, sclerotic part and endocarp (Lu et al. 2018) (Figure 3). CM is the second in quantity; 100-150L are produced per ton of wet cacao beans. Defective practices in this pre-processing phase, especially during fermentation, produce cacao beans that do not meet the quality standards for commercialization and constitute part of the generated biomass.

After the pre-processing step, cacao beans are transported to the processing plants mainly located outside the cacao-producing countries. Complicated logistic procedures are necessary to collect the cacao from the plantations and send it, generally by ship to the processing plants. During the stage of transportation, especially the phase of primary recollection on farms, inadequate practices will be associated with the generation of byproducts like broken or contaminated beans (physical, chemical, and microbiological), which diminish their quality or even produce partial or total loose of the cacao beans.

Besides the damage to the cacao beans associated with its transportation, the different stages of buying and selling cacao by intermediaries are also related to biomass generation due to the lack of compliance with quality standards. Cacao beans are classified into different categories that determine their final destination.

The processing, commercialization, and consumer use stages are not exempt from biomass generation. The quantity generated depends on the plant production size and technology, the distribution channels, and the type of final user.

It is estimated that 45% of fruits and vegetable products are loosed during the entire process, being the highest amount of food loose and waste compared with other food and agricultural products: 20% for dairy, meat, and poultry products; 30% for cereal products, and 35% of fish and seafood products. The amount of cacao waste produced constitutes an important issue with implications beyond the surface of losing resources. Food waste gets more expensive the further the value chain travels, both; economically and environmentally. Rescuing food at the earlier stages is more impactful; however, according to the United States Environmental Protection Agency (EPA), it is preferred to rescue food in the last stages.

The cacao industry requires cacao beans that meet quality requirements to be commercialized. Those requirements have been grouped into three main areas: economic, quality, and wholesomeness (Lima et al. 2011). Economic is related to the content of useful material, which depends on parameters like bean size, fat content, or the presence of germinated and infested beans. Quality is related to the flavor of the cacao beans, including the absence of off-flavors, presence of desired flavors (e.g., floral, spicy, fruity), and physical properties such as cacao butter hardness, melting and solidification behavior. Wholesomeness is related to food safety (mycotoxins, pesticides, heavy metals or foreign materials). The limits of those materials are regulated by the country's current legislation where the factory is located or through regulatory bodies like the European Union (EU) and the Food and Drug Administration (FDA) in the United States.

All along the cacao production chain are generated cacao beans that non-accomplishment the quality standard required for the industry. The total losses of cacao beans due to lack of quality are about 20% to 30% of the annual production (Lima et al. 2011), which constitutes an important mass of byproducts generated. Table 1 shows the quantity (tons) of cacao beans and cacao biomass as CPH and CBS produced by the different countries in 2020 and biomass as total (tons) produced worldwide this year. Reported amounts of CPH and CBS were roughly calculated using variables like bean index (dry bean weight), pod index (number of seeds per pod), and proportion of the CPH and CBS weight about the total pod, respectively. Estimated data reported varies depending on the method used for calculation.

Table 1. Calculated amount of tones of biomass as CPH and CBS generated during 2020 by different countries

Area	Source of information*	Biomass 2020 (tones)*		
		Cacao Production 2020	CPH (tones)	CBS (tones)
Suriname	4	5	13.2	3.40
Fiji	2	9	23.8	6.13
Saint Lucia	2	18	47.6	12.25
Micronesia (Federated States of)	1	31	82.1	21.10
Comoros	1	42	111.2	28.59
Gabon	3	100	264.7	68.06
Belize	2	103	272.6	70.10
Thailand	4	112	296.5	76.23
Central African Republic	1	131	346.7	89.16
Timor-Leste	4	171	452.6	116.39
Saint Vincent and the Grenadines	4	228	603.5	155.18
Jamaica	2	268	709.4	182.41
El Salvador	2	371	982.0	252.51
Angola	1	380	1005.8	258.64
Costa Rica	2	500	1323.4	340.31
Guyana	1	431	1140.8	293.35
Samoa	4	479	1267.9	326.02
Trinidad and Tobago	3	480	1270.5	326.70
Panama	2	706	1868.7	480.52
Grenada	3	800	2117.5	544.50
Malaysia	2	926	2451.0	630.26
Cuba	2	1000	2646.9	680.63
Dominica	3	1000	2646.9	680.63
Equatorial Guinea	3	1000	2646.9	680.63
Honduras	2	1455	3851.2	990.31
Vanuatu	3	1577	4174.1	1073.35
Sri Lanka	2	2000	5293.8	1361.25
Sao Tome and Principe	2	3041	8049.1	2069.78
Congo	2	4500	11910.9	3062.81
Solomon Islands	3	5875	15550.4	3998.67
Bolivia	2	7000	18528.1	4764.38
Nicaragua	2	7000	18528.1	4764.38
Togo	3	7508	19872.7	5110.13
Philippines	2	9341	24724.5	6357.72
Liberia	3	10000	26468.8	6806.25
Guatemala	1	12000	31762.5	8167.50
Madagascar	2	12000	31762.5	8167.50
Democratic Republic of the Congo	3	12564	33255.3	8551.37
United Republic of Tanzania	3	12608	33371.8	8581.32
Haiti	1	14000	37056.3	9528.75
India	2	14942	39549.6	10169.90
Guinea	2	20000	52937.5	13612.50

Table 1. (Continued)

Area	Source of information*	Biomass 2020 (tones)*		
		Cacao Production 2020	CPH (tones)	CBS (tones)
Venezuela	2	26000	68818.8	17696.25
Mexico	2	29429	77894.9	20030.11
Papua New Guinea	3	35000	92640.6	23821.88
Uganda	3	38000	100581.3	25863.75
Sierra Leone	2	63416	167854.2	43162.52
Dominican Republic	2	77681	205611.9	52871.63
Colombia	2	160289	424264.9	109096.70
Peru	2	193156	511259.8	131466.80
Ecuador	2	269731	713944.2	183585.66
Brazil	2	290000	767593.8	197381.25
Cameroon	3	327903	867918.3	223178.98
Nigeria	4	340163	900368.9	231523.44
Indonesia	2	739483	1957319.1	503310.62
Ghana	3	800000	2117500.0	544500.00
Côte d'Ivoire	2	2200000	5823125.0	1497375.00
Total		5756921	15029718.6	3864784.8

Source*(FAO 2022): based on imputation methodology (1), Official data (2), Unofficial Figure (3), FAO data (4).
#Calculated

The increment in cacao production worldwide due to increasing demand for cacao products also means an important generation of biomass that urges the recovery of byproducts to reduce their environmental and economic burden.

Cacao bean quality, defined by the cacao industry, is the sum of good agriculture practices and good post-harvest processing practices. Factors associated with agricultural practices are plant population and genotype, soil and climate, plant phytosanitary state, and fruit maturation and harvest. In contrast, the factors associated with post-harvest practices are related to pod opening, fermentation, drying, storage stages, etc. (Figure 4). The way those practices are carried out defines the final cacao quality and the amount of biomass generated by the process. A defective process on one or more of those practices modifies the throughput of the whole chain. An example is the harvest process of cacao pods, which is usually done using subjective factors, such as thorough color observation or shine loss. If those practices flail, unripe pods are harvested; it being difficult to pulp the beans, which will contain fewer sugars, fats, and compounds that contribute to the potential flavor than mature pods. The harvest process will determine the quality, price, percentage of organic waste, and the cacao bean's final destination.

Plant population and genotype are important factors for by-product generation, e.g., differences in resistance to cacao black pod disease caused by different *Phytophthora* species have been observed (Barreto et al. 2015, Luz et al. 2006, Iwaro, Sreenivasan, and Umaharan 1997). *Phytophthora* spp. has caused losses of 40% of the worldwide production of cacao crops, but the production of new resistant cacao clones has reduced these losses considerably. Similar examples of diseases like witches' broom, watery pod rot, and swollen shoot are reported.

Cacao is a perennial tropical crop with higher mineral nutrient requirements than other tropical crops. Cacao is a very demanding crop that requires high to medium fertility and no physical limitations. Soil organic matter is one of the most important parameters for selecting suitable cacao soils. All major elements and micronutrients are essential for cacao trees. (Cabala

Rosand, Santana, and De Santana 2019). Recent studies have suggested that climate change could severely impact cacao production (Lahive, Hadley, and Daymond 2019). Nutritional deficiency on soil and stress by climate conditions during cacao growing; cause decreasing trees' yield and produces lower-quality cacao beans. As in the case of resistance to diseases, the climate change effect depends on the genotypic variation of the tree.

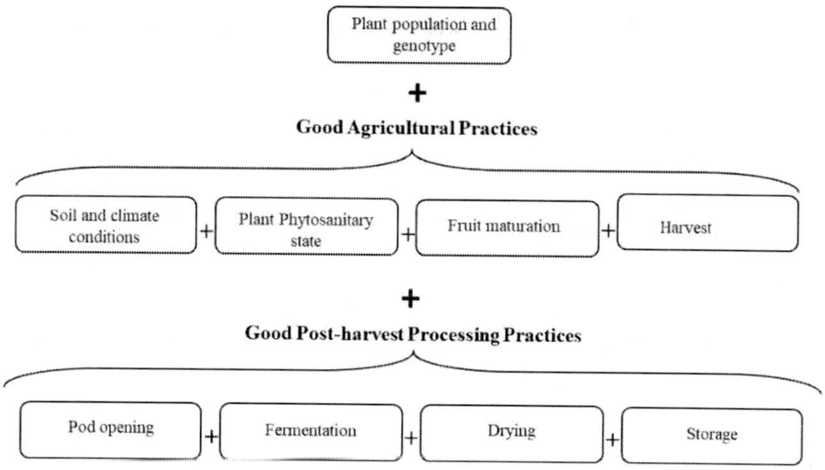

Figure 4. Cacao quality is the sum of good agricultural practices and post-harvest processing practices.

During the post-harvest process, pets like rats, squirrels, and monkeys are responsible for cacao waste generation and loss of product due to consuming significant quantities of ripe pods. Nevertheless, when losses are associated with post-harvest processing practices, the fermentation stage is of core importance because cacao beans under fermented or over-fermented produce defective grains with less economic o nonvalue for the cacao industry.

Drying and storage are also post-harvesting operations that compromise the cacao quality. Sun drying is common when beans are exposed to physical and microbiological contamination.

An important concern about microbiological contamination is Ochratoxin A (OTA), a mycotoxin produced mainly by species of *Aspergillus* and *Penicillium*. Food contamination with OTA is a major consumer health hazard, being a neurotoxic, immunosuppressive, genotoxic, carcinogenic and teratogenic mycotoxin. Mycotoxins result from storing cacao grains at high temperatures (24°C) and humidity (95%-99%), conditions that can be reached during deficient post-harvesting operations. The OTA content has been quantified during post-harvesting stages for cacao from Côte d'Ivoire: pod opening (0.25 ± 0.02mg/kg), fermentation (0.275 ± 0.2 mg/kg), drying (0.56 ± 0.015mg/kg) and storage (0.558 ± 0.04 mg/kg) (Dano et al. 2013).

The aflatoxins (AF) are highly associated with OTA, which are mycotoxins considered carcinogenic. *Aspergillus flavus*, *A. parasiticus* and *A. nomius* are the main producers of AF, with relevance to the B1, B2, G1 and G2 analogs. It has been observed that OTA and AF concentrations are significantly high at the end of the drying process. It is estimated that up to 70% of cacao is contaminated with OTA and AF. Roasting does not remove toxins, and those remain in the final product. The presence of high concentrations of OTA/AF causes total crop loss. FAO/WHO Expert Committee on Food Additives established a maximum tolerable weekly intake value of OTA of 100 ng/kg of body weight, equivalent to a daily intake of 14 ng/kg of body weight. A lethal dose of AF is 50 (DL50) (Ravelo Abreu et al. 2011).

Besides physical and microbiológical contamination, some of the most common defects of cacao beans related to post-harvest production stages are shown in Table 2 (Lima et al. 2011).

High-quality cacao beans must have the variety's appearance and odor characteristics and be fully ripened and correctly fermented of a brown to dark red color; having a fragile kernel with a pure, bitter flavor (TIS 2022). The quality of the cacao beans is evaluated by a grading process that varies across producing and consuming countries. Over the years, the physical market has developed standard practices set out by the main international cacao trade associations: Federation of Cacao Commerce Ltd (FCC) and the Cacao Merchants'Association of America, Inc. (CMAA) (ICCO n.d.) FCC distinguishes three grades; Good Fermented (GF), Fair Fermented (FF) and Fair Average Quality (FAQ). Cacao described as either Good or Fair Fermented is defined by specific quality parameters (not more than 5% defective or 5% slate, not more than 10% defective or 10% slate, respectively). At the same time, FAQ is used when GF or FF are not customarily applicable to the origin described (ICCO n.d.).

To determine the percent of defective beans, experts evaluate the aroma and appearance of the whole grain; and the color and appearance of the cut grain ("cut test").

Table 2. Common quality defects of cacao beans associated with the post-harvest process

Quality defect	Cause	Stage of production
Bitter	underfermentation (strong violet color)	fermentation
Salty and firm kernel		
Astringent		
Dull	overfermentation (purple color)	
Putrid		
Ham-like odor		
Lack of cocoa flavor		
Salty	beans that were dried without undergoing fermentation	
Germinated	lack of temperature and humidity control	
Flat	bad practices	
Moldy	non-adequate drying, high moisture	drying
Smoky	beans contaminated by smoke during artificial drying	
Acid	fast (artificial) drying and other not yet well-known causes	
Small	unripe pods	harvesting
Contaminated	beans that have absorbed off-flavors from other products, like rubber or oil-based fuels	production chain
Insect-damaged	bad quality assurance practices	
Broken	defective logistic practices	

The cut test consists of cutting lengthwise 50 cacao beans to infer, based on its interior coloration and compartmentalizing, which are the characteristics related to the chemical composition and fermented stages of the bean that later will determine the flavor of the chocolate (Santos, Palmeira, and Jesus 2019). Separate counts are made of the number of moldy, slaty, insect damaged, germinated, or flat beans (ICCO n.d.). Cacao beans that non-accomplishment by the international trade standards will be part of the residual biomass of the cacao chain production (Figure 5) (ISCQF n.d.).

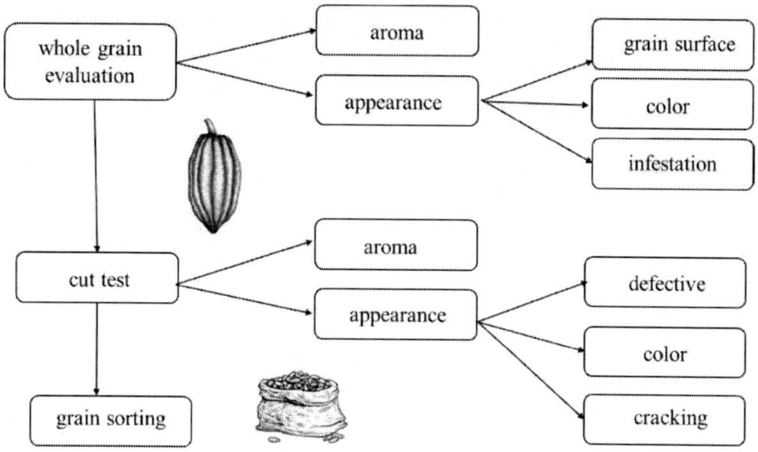

Figure 5. Diagram of the parameters evaluated on the whole and cut cacao bean during the cut test.

Byproducts Generated during the Supply Chain

The organic material produced for human consumption, discarded, lost or degraded primarily at the manufacturing and retail stages is called supply chain waste (FSCW). The food supply chain (FSC) of cacao has been recognized as inefficient by producing large and accumulative quantities of waste (Pfaltzgraff et al. 2013). FSCW is divided into pre-consumer FSCW and post-consumer FSCW. Table 3. FAO reported 32% of losses due to pre-consumer FSCW and 60% to post-consumer FSCW.

Table 3. Simplified food waste chain

Pre-consumer FSCW	*Food production* Agrigultural waste
	Food processing Food processing residues
	Food distribution Mixed domestic food waste production & waste packaging (e.g., damaged food products)
Post-consumer FSCW	*Food consumption* Mixed domestic food waste production & waste packaging (e.g., meal preparation waste)

Pre-consumer FSCW was addressed before; hence in this section, the main objective is to describe the waste biomass generated during food distribution since it is difficult to determine the amount of waste cacao-derived associated with food consumption.

When discussing the cacao distribution process, it isn't easy to estimate the number of byproducts generated due to the variability of the marketing channels, which differ from region to region within the same producing country and across producing countries. At one extreme of the spectrum, the marketing channel between cacao farmers and exporters comprises at least two intermediaries. Small traders buy cacao beans directly from farmers, visiting them individually. In a second phase, small buyers sell the beans to wholesalers, which will re-sell them to exporters. At the other extreme of the spectrum, cacao beans are sold directly to exporters by farmers' cooperatives or even directly exported by the cooperative (ICCO n.d.).

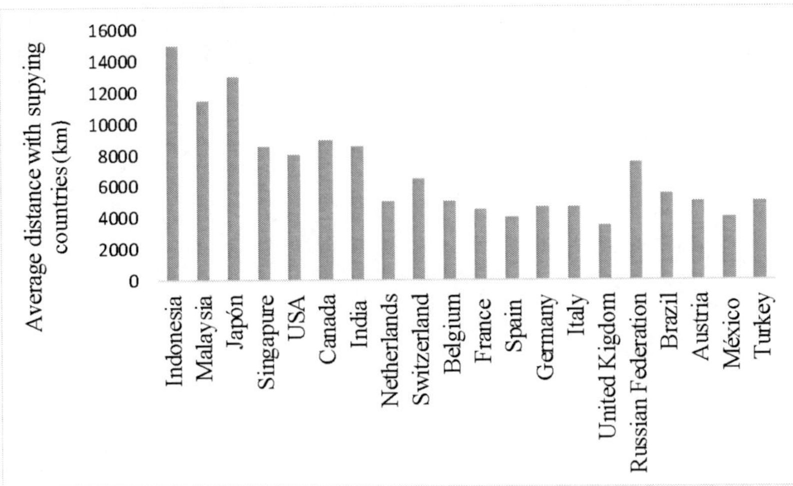

Figure 6. Average distance with supplying countries.

Finally, cacao is transported to its last destination, mainly North America and Europe, mostly through sea transportation. The average distance that cacao beans, whole or broken, raw or roasted, travel from supplying countries to importing countries varies from 3000 km to almost 15,000 km (ITC n.d.) (Figure 6).

The more distance the cacao is transported, the more the opportunity to get damaged during the cacao bean supply chain. Cacao's logistic chain involves internal and international transportation. The percentage of waste biomass generated during transportation varies considerably depending on the supplying country. African countries generally have the highest rate of waste produced during internal transportation (Rodríguez-Bonilla, Enriquez-Martínez, and Esteves-Ceballos 2020).

Raw cacao belongs in the "most sensitive" cargo category. It requires particular temperature, humidity/moisture, and ventilation conditions during transportation to maintain its quality. Cacao is vulnerable to external factors such as temperature fluctuation: day to night or en-route delays, and temperature drops between the loading and unloading process (almost 15-20°C in the hottest seasons of the year). Cacao is hygroscopic, meaning it can absorb moisture from the environment or release water when it is not properly dried or fermented. Cacao water content has been observed to fall by 1-3% during extended voyages.

Due to its high-fat content, raw cacao must not be stowed near heat sources. Rancidity and overfermentation readily occur at temperatures greater than 25°C. Cacao cargo tends to self-heate, and there is a risk of fire on contact with flammable substances like oil.

The recommended conditions for cacao transportation are: relative humidity -70-75%; water content - 6-8%; critical water content - 8.5%, maximum equilibrium moisture content: 65%.

The sensitivity of cacao requires great care during transportation to ensure its optimum quality when arriving with the final user. Otherwise, it will be directed to a different industrial process like the production of cacao butter or even will constitute a residue depending on the magnitude and type of damage caused.

Cacao Byproducts and Their Implications: Economic, Environmental and Social

Ghana and Indonesia are two of the most important suppliers of cacao beans globally, but this achievement is not balanced with the welfare of the farmers. Farmer's income is below the poverty line. Cacao farmers made hard labor, and their income doesn't fulfill their needs, so they have to do extra work to cover their basic needs. It is calculated that only 6 cents end up in the farmer's pocket for each dollar spent on a chocolate bar. For this reason, so many farmers undertake land conversion. Cacao history is linked to slavery history in African countries, which partly explains the low prices traditionally paid to the farmers and their working conditions; but these conditions are not exclusive to African countries but are shared by Latin American countries.

At present, the producer country's name is not linked to the reputation for high-quality chocolate, and big monopolies control the cacao value chain and the highest percentage of their incomes.

Another important factor is the environmental factor, especially when it comes to byproducts; that are not necessarily managed adequately.

The three pillars of sustainability are economic, environmental, and social. Studies of sustainability in the cacao supply chain agroindustry found that the sustainable index for all dimensions was below 50% or less sustainable. Social dimension: 46.42%, economic dimension: 39.85%, and environmental dimension: 49.19% (Sriwana et al. 2017).

The most important elements stated for the pillars were: (Sriwana et al. 2017)

Environmental Dimension

1. Doses of fertilizers
2. Good agriculture practices
3. Wastes at the farm level
4. Environmental pollution

Economic Dimension

1. Fair profit distribution
2. Profit for farmers
3. Cacao productivity
4. Additional income from cacao wastes

Social Dimension

1. Institution (creation)
2. Ability to send children to higher education
3. Farmer's awareness on occupational health and safety
4. Workforce absorption

The challenge is not minor for the cacao value chain, but the use and valorization of byproducts will impact favorably on the three dimensions of sustainability.

Bioactive Molecules Present in Cacao Byproducts

Phytochemicals are compounds derived from plants that act as a natural defense, protecting them from microbes and infections while giving them color, aroma, and flavor. Phytochemicals are not considered essential for the organism; however, they have beneficial properties for health. There are more than 2000 phytochemicals in plants grouped according to their chemical structure into four groups: terpenoids, phenolic compounds, nitrogen compounds, and sulfur compounds (Martillanes, Rocha-Pimienta, and Delgado-Adámez 2018).

Phenolic compounds are an important group of phytochemicals with a molecular structure characterized by one or more phenolic rings. Plants can synthesize phenolic compounds in large quantities as products of their secondary metabolism. Some phenolic compounds are essential for plant physiological functions, defense functions in stressful situations, and other stimuli (water, light, etc.). (Quiñones, Miguely, and Aleixandre 2012).

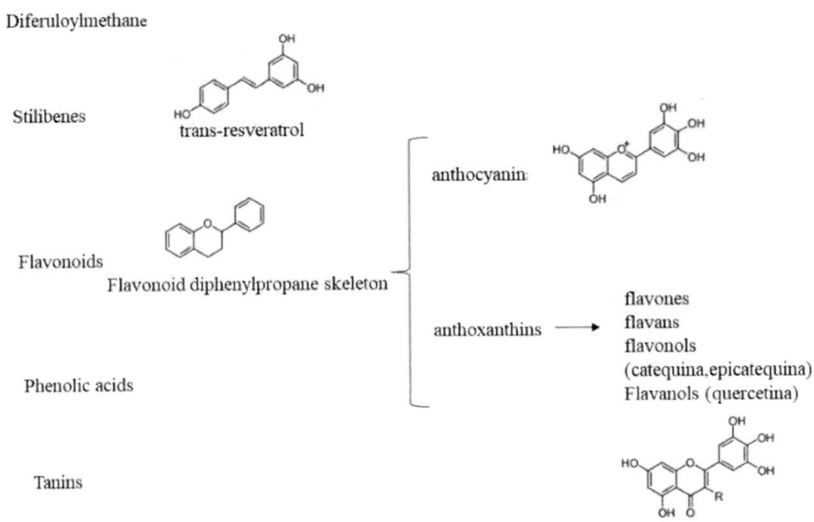

Figure 7. Phenolic compound classification.

Polyphenols are a group of non-caloric substances that modulate the activity of different enzymes and interfere with signaling mechanisms and different cellular processes. Polyphenols participate in different reactions in the cellular metabolism of oxidation-reduction. Polyphenols act as a chemopreventive agent against oxidative damage preventing free radicals from causing degenerative diseases such as cancer, cerebrovascular diseases and heart diseases. When polyphenols are included in the diet, they positively affect health, reducing the incidence of cardiovascular diseases, among other positive effects.

There are different types of polyphenols (Figure 7). One of the most important groups is flavonoids; this group contains the anthoxanthins, compounds that are the most common flavonoids in cacao and cacao byproducts (Han, Shen, and Lou 2007).

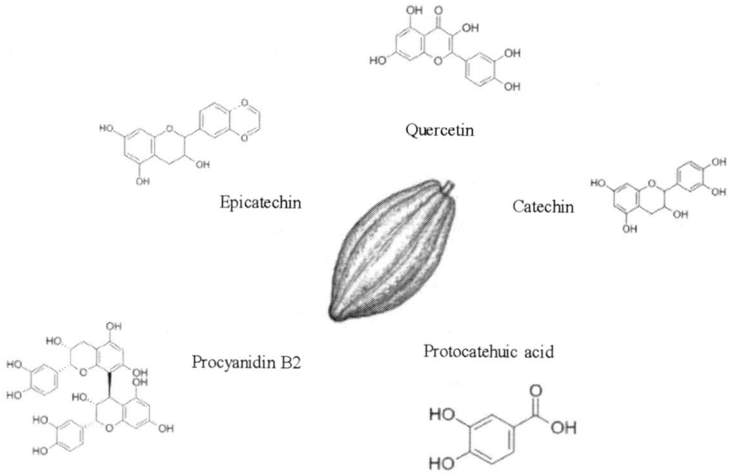

Figure 8. Major polyphenols found in cacao and cacao.

Cacao contains numerous phytochemicals, of which polyphenols are the most abundant group. Polyphenols are found mainly inside the seed and have numerous biological properties, such as antioxidant, antiproliferative, antiapoptotic, anti-inflammatory and anti-cancer activity. Cacao biomolecules have been investigated to treat health conditions, including heart diseases, dyspepsia, nervous system diseases, and circulation problems. Some of the major cacao flavonoids belong to the anthoxanthins, especially the flavonols y flavanols groups (Cinar et al. 2021) (Figure 8).

Given the number of byproducts generated along the value chain of cacao, it is important to evaluate the type and concentration of biomolecules on them for its valorization.

Cacao byproducts are an important source of biomolecules like flavonoids, polyphenols, procyanidins, catechins, polypeptides, oligopeptides, theobromine, proteins, lipids and dictary fiber. Cacao biomolecules have demonstrated their value in being incorporated into foods and different products. The type and number of biomolecules present on cacao byproducts depend on the type of by-product and are subjected to several factors like cacao variety, climatic and growing conditions, and also depends on processing (fermentation, drying, roasting), and transportation conditions (temperature, humidity, etc.)

The extraction of biomolecules from cacao byproducts is considered a challenge because they are susceptible to oxidation since the pod is open, which causes its deterioration.

Cacao

Pod Husk (CPH)

Once the beans are extracted from the fruit, the first cacao by-product is generated: the cacao pod husk. CPH is the principal by-product generated during harvesting and cacao production, representing 67-80% of the fruit weight (Porto de Souza Vandenberghe et al. 2022); in other words, 10 tons of wet CPH for each ton of dry beans is obtained (Campos-Vega, Nieto-Figueroa, and Oomah 2018). CPH is lignocellulosic biomass and is rich in theobromine, fiber (principally lignin, cellulose, hemicellulose, and pectin), protein, antioxidants, such as phenolic acids (catechin, quercetin, epicatechin, gallic acid, coumaric acid, and protocatechuic acid), and also contains minerals (potassium, calcium, and phosphorous) (Ouattara et al. 2021)

The CPH comprises three layers: endocarp, mesocarp, and pericarp, which constitute the inner, middle, and outer layers, respectively. Each layer aims to protect the fruit and contains different amounts of biomolecules. The main biomolecules found in the pericarp are hemicellulose and minerals, like potassium, calcium, and phosphorous, while the mesocarp is rich in crude fiber and cellulose. The endocarp is rich in protein, fat and pectin (Campos-Vega, Nieto-Figueroa, and Oomah 2018). Table 4 shows the chemical composition of CPH.

Table 4. Chemical composition of CPH

Compound	Content (g/100g)
Lignin	19.5 to 25.5
Cellulose	26 to 32
Hemicellulose	15.2 to 21.2
Pectin	6 to 12
Ash	8.3 to 14.3
Protein	4.21 to 10.74
Phenolic acids*	4.60 to 6.90
Theobromine	0.34
Phosphorus	0.19-0.25
Potassium	3.26-3.32
Calcium	0.36-0.42
Magnesium	0.21-0.26
Iron	0.016-0.022
Manganese	0.018-0.024
Sodium	0.12-0.13
Zinc	0.02-0.03

*(Marsiglia et al. 2016, Sandesh, Shishir, and Rao 2020, Titiloye, Abu Bakar, and Odetoye 2013, Soares and Oliveira 2022, Antwi et al. 2019, Ouattara et al. 2021, Lu et al. 2018, Asiedu, Neba, and Addo 2019, Daud et al. 2014, Nazir et al. 2016, Vásquez et al. 2019, Djeke, Angui, and Kouadio 2011, Gyedu-Akoto et al. 2015)

Table 5. Chemical composition of CM

Compound		Content (g/100g dry weight)*
Lipids		1.91 to 3.54
Ash		3.70 to 7.68
Fiber	Pectin	0.82 to 6.61
	Cellulose	19.17 to 25.17
	Hemicellulose	4.42 to 10.42
	Lignin	4.4 to 10.4
Total organic acids		17.52
Protein		4.21 to 10.74
Phenolic acids*		4.60 to 6.90
Theobromine		0.34
Phosphorous		0.19 - 0.25
Potassium		3.26 - 3.32
Calcium		0.36 - 0.42
Magnesium		0.21 - 0.26
Iron		0.016 - 0.022
Manganese		0.018 - 0.024
Sodium		0.12 - 0.13
Zinc		0.02 - 0.03

*(Soares and Oliveira 2022, Endraiyani et al. 2017, Campos-Vega, Nieto-Figueroa, and Oomah 2018)

Mucilage (CM)

Cacao mucilage, also known as cacao pulp, is the second generated by-product. CM is a white mass that coats the cacao beans, and once CPH is opened, it begins to detach from the beans during the fermentation stage. During mucilage pretreatment and due to the action of microorganisms, mostly yeasts, and acidophilic bacteria, it releases a turbid liquid called cacao pulp juice. After fermentation, one ton of wet cacao beans produces 100 to 150 liters of cacao pulp juice (Soares and Oliveira 2022).

Cacao pulp has a low pH, which ranges from 3.0 to 3.80, mainly derived from the presence of organic acids, and has a content of total solids around 16.17°Bx. CM is also composed of water, sugars (sucrose, fructose, and glucose), organic acids (citric and ascorbic), proteins, free amino acids, vitamins, minerals, and pectin, among other compounds (Table 5). (Afoakwa et al. 2013, Endraiyani et al. 2017).

Bean Shell (CBS)

The third cacao by-product is the cacao bean shell. After the roasting and grinding process, CBS is obtained. Those processes facilitate CBS separation from the cacao beans. CBS represents 12% to 20% of the cacao bean and contains high nutritional benefits (Lee et al. 2020, Lozano Moreno 2020).

Table 6. Chemical composition of CM

Compound		Content (g/100g)*
Ash		5.96-11.42
Carbohydrates		7.85-70.25
Proteins		10.30-27.40
Fats		1.50-8.49
Dietary fiber	Soluble fiber	7.03-16.91
	Insoluble fiber	28.34-50.42
Methylxanthines	Theobromine	0.39-1.83
	Caffeine	0.04-0.42
Pectin		7.62-15.59
	Total phenolic	3.12–94.95#
	Total flavonoid	1.65–40.72#
Polyphenol	Total tannin	1.70–25.30#
	Epicatechin	0.21–34.97
	Catechin	0.18–4.50
Flavanols	Procyanidin B1	0.55–0.83
	Procyanidin B2	0.23–1.38
Volatile organic compounds (aromatics)		4.92–16.10&

*(Rojo-Poveda et al. 2020)
(mg/g of dried CBS), &μg/g

CBS contains proteins, carbohydrates, dietary fiber, phenolic compounds, methyl-xanthines, like theobromine and caffeine; phytosterols (e.g., campesterol, stigmasterol, and B-sitosterol) and fats (Cinar et al. 2021, Belwal et al. 2022). CBS is also rich in minerals, being sodium and potassium the major components (Osundahunsi, Bolade, and Akinbinu 2007). CBS has lower fat content than cacao beans, but it has the highest fiber content. CBS is also rich in proteins, fats, and carbohydrates compared to other cacao byproducts, like CPH. (Rojo-Poveda et al. 2020)

CBS contains the same phenolic compounds as cacao beans because components migrate during fermentation, roasting, and alkalizing processes. During these processes, there is an enrichment of CBS in polyphenols (catechins, epicatechins, procyanidins B2, and protocatechuic); and a reduction of those components in cacao beans (Panak Balentić et al. 2018). Table 6.

CBS is also a rich source of phytochemicals and has been extensively studied. Table 7.

Table 7. Phytochemicals present on CBS

Compound	Concentration (μg/g extract)
Hydroxybenzoic acids	
Gallic acid	19.2 ± 0.4
Protocatechuic acid	765.5 ± 47.6
4-hydroxybenzoic acid	70.1 ± 9.3
Salicylic acid	3.3 ± 0.2
Hydroxycinnamic acids	
Caffeic acid	1.9 ± 0.2
p-coumaric acid	4.2 ± 0.6
Mandelic acids	
Mandelic acid	11.18 ± 1.2
Phenylacetic acids	
3,4-dihydroxyphenylacetic acid	25.9 ± 2.4
4-hydroxyphenylacetic acid	48.5 ± 4.3
Flavan-3-ols: monomers	
(-)-epicatechin	221 ± 13.8
(+)-catechin	200.8 ± 16.9
Flavan-3-ols: dimers	
Procyanidin B1	83.6 ± 7.8
Procyanidin B2	219.9 ± 11.4
Flavonols	
Quercetin-3-O-galactoside	9.3 ± 0.4
Quercetin-3-O-glucoside	11.12 ± 0.77
Alkaloids	
Caffeine	2433.5 ± 7.8
Theobromine	10,035.0 ± 4.5

*(Rebollo-Hernanz et al. 2019)

Cacao Biomolecules Extraction

CPH, CS, and CM have been used as raw materials to extract biomolecules, such as pectin, fibers, polyphenols, antioxidants, methylxanthines and phytosterols for further applications. Several studies have been carried out to study the impact of different kinds of extractions. Conventional methods comprising organic solvents are very time-consuming and demand high quantities of solvents using temperature. Therefore, the novel extraction techniques are environmentally friendly, and high-quality extracts and higher yields of biomolecules can be obtained. Table 8 shows different methods studied to extract biomolecules present in cacao byproducts.

The effect of the type and conditions of the extraction process is directly related to the yield of the extracted phytochemical. Significant differences have been observed in the quantities obtained by the different methods, especially in the case of total phenols. Table 9.

Table 8. Extraction methods for cacao biomolecules

Biomolecule	Extraction method*
Polyphenols and methylxanthines	Pulsed electric field assisted
Phenolic compounds	Combination of supercritical CO2 and pressurized liquid extraction with ethanol
Theobromine	Supercritical CO2. Solvent extraction (70% ethanol
Pectins	Hot acid extraction (pH of 2.5, per 1 h) Citric acid (pH 2.5, at 95°C, per 3 h) Subcritical water extraction.
Polyphenol and polysaccharide	Microwave-assisted extraction
Phenolic compounds	Water extraction
Fat and methylxanthines	Supercritical CO2
Dietary fiber	High voltage electric discharge conditions
Flavonoids	Ultrasound-assisted extraction
Alkaloids	Hydrated alcohol

*(Panak Balentić et al. 2018, Belwal et al. 2022)

Table 9. Extraction performance of CBS biomolecules by different methods*

Extraction method (mg/g)	MetOH-HOH	EtOH-acid HOH	HOH	MetOH-acid HOH	Acid HOH	Pulsed Electric Field	Supercritical CO2
Total phenols	14.64	49.46	5.77	20.39	9.4	24.9- 2.3	18.2
Theobromine	10.20	11.00	8.47	11.62	6.6	4.6- 10.92	12.9
Caffeine	n.d.	n.d.	n.d.	n.d.	n.d.	1.6 - 4.21	n.d.
Catechin	1.02	1.97	1.65	4.00	6.16	n.d.	n.d.
Epicatechin	15.84	9.00	6.93	17.7	7.04	0.21-2.12	n.d.

*(Hernández-Hernández et al. 2018, Panak Balentić et al. 2018)

Recovery and Uses Systems

Cacao byproducts have received growing attention because of the increasing shortage of natural resources and environmental problems (Nieto Figueroa, Mendoza García, and Campos Vega 2020). The main goal is to convert those byproducts into other value-added applications. Each by-product has different applications in agronomic areas, as feedstuff, for energy production, as raw material for the extraction of biomolecules, and in food uses, among many other applications. The valorization of the byproducts implies the use of different technologies that vary within a wide range of specialization; for example, CPH has been used to make soap in Ghana and Nigeria, and CPS has been used for obtaining biopolymers, biofilms and nano-cellulose.

Table 10. Main applications of cacao byproducts reported in the literature

Application	CPH	CBS	CBP
Agronomic	Fertilizer	Fertilizer	
	Soil organic matter	Biofungicide	
	Soil amendments		
	Biomass		
	Composting to suppress weed	Agricultural composting	
		Wastewater treatments	
Feed	feedstuff	Cow diets	
		Fish and Pig diets	
	Coal		
	Briquettes		
Energy	3G Biofuels	Biofuels	
	1G/2G Biofuel	Chemical industry	
	Biogas		
	Biodiesel		
	Bioethanol		
Biomolecules	Lactic Acid	Lactic Acid	
	Succinic Acid		
	Polylactic acid		
	Citric Acid		
	Xylitol		
	Pectin		
	Cellulose-Nanocrystalline		
	Polyhydroxyalk-anoate		
	Propionic acid		
	Methylxanthine-rich extract	Methylxanthine-rich extract	
	Propionic acid		
	Methylxanthine-rich extract	Methylxanthine-rich extract	
	Fiber-rich extract	Fiber-rich extract	
	Theobromine		
Food applications	Gums	Gums	Cacao juice
	Cacao pigment	Nutritional fortification agent, snack)	Alcoholic products
	Production of functional beverage	Antioxidant agent	Marmalade
	Bakery products	Bakery_applications (addition_of polyphenols)	Jelly
	Chocolates (addition of dietary fiber and rheological modifier)	Tea	
Miscellaneous applications	Biopolymers	Papermaking	Fermented products (kefir)
	Biofilms (cellulose: fiber)/Plastics	Growth regulator	
	Soap	Soap	
	Nano cellulose	Cacao pigment	
	Activated carbon	Activated carbon	
	Sheets (electromagnet_tic wave	Absorbent	
	protection)	Fabric and cotton drying	
		Antimicrobial activity Dermatologic applications (UV protection)	

CPH: Cacao pod husk, CBS: Cacao ben shell; CBP: Cacao bean pulp

(Nieto Figueroa, Mendoza García, and Campos Vega 2020, Peñaloza Albarracín et al. 2021, Figueira et al. 1992, Unten et al. 1991, Tran et al. 2017, Titiloye, Abu Bakar, and Odetoye 2013, Sandesh, Shishir, and Rao 2020, Panak Balentić et al. 2018, Osundahunsi, Bolade, and Akinbinu 2007, Cinar et al. 2021).

In recent years, lucrative companies have been developed around the utilization of cacao byproducts, and the options of usage have become more specialized so that new industries benefit from the materials or biomolecules obtained from byproducts that were originally considered waste. Many studies have been reported focused on developing new and more sophisticated alternatives to recovery cacao byproducts. From 1980 to 2019, 1042 scientific articles were generated around new applications of cacao byproducts; and in the period from 2000 to 2020, 3717 patent families were also reported (Peñaloza Albarracín et al. 2021). Those numbers reveal the attention generated regarding the use of cacao byproducts (Figure 9).

The number of published articles shows an increasing trend over the years (9.37% annual) and, in the case of patents, a growth of 26.01% (Peñaloza Albarracín et al. 2021). The countries with more publications on the topic are in order of importance: Brazil, Nigeria, the United States, and Malaysia (Peñaloza Albarracín et al. 2021). According to the number of publications, the studied matrices in order of importance are 147 for CBS, 146 for CPH, and 12 for CM (Peñaloza Albarracín et al. 2021).

The principal studied applications for cacao byproducts depend on the countries: Brazil, Nigeria, and The United Stated focused their research efforts on feed uses, biomolecules extraction, activated charcoal production, and biomass production. Colombia is focused on extracting biomolecules, while Mexico is mainly interested in producing activated charcoal (Peñaloza Albarracín et al. 2021). Table 10. List the main applications of cacao byproducts found in the literature.

Currently, the cacao industry only takes advantage of 10% of the fresh fruit. While wasting important resources, this situation causes environmental problems that directly affect the cacao value chain, like deterioration of landscapes, problems with disposition, and the appearance of foil odors, among many other consequences.

In recent years, the use of cacao byproducts has gone further and brought various applications in different areas of industry, benefiting society and, above all, the environment. Some of these applications are discussed.

Antimicrobial Activity

It was found that cacao residues have inhibitor and antimicrobial activity: an extract of CBS was used as an antibacterial agent against Bacillus cereus and is proved to have inhibitory activity against S. agalactiae. Cacao mucilage has been used to prepare fermented mixtures containing weak acids like citric acid used against some phytopathogens (Moreno et al. 2021). After applying the mixture to the cacao tree, it was observed that it was protected from insects or fungi such as Moniliophthora roreri. (Moreno et al. 2021).

Food Industry Applications

Cacao byproducts have been more frequently used in food applications; due to the food industry is experiencing a constantly growing demand for new ingredients from natural sources (Nieto Figueroa, Mendoza García, and Campos Vega 2020). CBS has been proved to be useful for producing nutritionally fortified shacks (Panak Balentić et al. 2018). CBS also can be used as a raw material to produce extracts rich in fibers, polyphenols, antioxidants, etc. Some cacao byproducts derived have found applications in the bakery and meat industries.

Cacao gums from CPH have been evaluated as a replacement for gum karaya (Berbert 1972, Figueira et al. 1992, Marsiglia et al. 2016), and as alkalinizing agent on cacao products (Osundahunsi, Bolade, and Akinbinu 2007).

Feedstuff Applications

CBS has been used as feedstuff. The nutritional value of CBS varies depending on the animal. Horses and rabbits consume this by-product more efficiently than pigs (Nieto Figueroa, Mendoza García, and Campos Vega 2020). The protein digestibility of CBS is 83g/kg (Nieto Figueroa, Mendoza García, and Campos Vega 2020). The main concern about using CBS as feedstuff is theobromine's presence, which may negatively affect some species (Nieto Figueroa, Mendoza García, and Campos Vega 2020). Treatments such as boiling CBS for 15 minutes were used on hens without influencing egg production and feed conversion.

Agriculture Applications

CPH has been used to suppress weeds in fruit corps gardens, urban landscapes, and crops in organic production systems (Nieto Figueroa, Mendoza García, and Campos Vega 2020). CPH has also been used as fertilizer, soil organic matter and soil amendment.

Pharmaceutical Applications

CPH extract, which is a mixture of flavonoids, has been shown to inhibit the cytopathic effects of HIV in cell culture. The anti-HIV activity was attributable to interference with the virus adsorption rather than inhibition of the virus replication after adsorption (Unten et al. 1991).

Uses as Biofuel

New alternatives to fossil fuels are needed. Ethanol production from lignocellulosic biomass like agricultural residue is one important alternative. CBS has proved to be useful in ethanol production by acid hydrolysis and Saccharomyces cerevisiae (Panak Balentić et al. 2018). The use of CBS for energy purposes showed its utility as an additive for thermal emission.

Uses as Adsorbent

Some studies have proved the utility of modified CBS as a low-cost absorbent to entrap pollutants like metal ions, gases, or industrial dye. CBS has also proved to have good potential for treating agro-industrial wastewaters (Panak Balentić et al. 2018). One of the most common uses of CBS is converting it into activated carbon, which has the potential to be used as an absorbent for 4-nitrophenol and methylene blue dye in water or wastewater treatments (Ahmad et al. 2012).

Uses as Dye

Biofilaments based on CS and biodegradable poly Ɛ-caprolactonethey (PCL) were prepared using a single-screw extruder. This solvent-free fabrication technique produced biofilaments for the 3D printing of objects with potential household and biomedical applications (Tran et al. 2017).

Besides the principal byproducts (CPH, CBS, and CM), an important amount of biomass is generated derived from the non-accomplishment of the quality standards of cacao beans due to its size, and defective process of fermentation; or derived to physical or microbiological damage. Those rejected grains and the grains damaged in transport or affected by poor storage conditions; are processed to obtain refined and deodorized cacao butter, perfectly suitable for use in cosmetics, pharmaceuticals, and even in chocolate production. This industry can process grains affected by diseases such as black rot monilla and grains that are moldy, smoked, over fermented, damaged by insects, etc. Removing those beans from plantations has advantages like avoiding infections or pets and minimizing the cost of waste removal from plantations. It is common practice that insurance companies have contracts with butter producers to derive the beans damaged during transportation.

All the usage alternatives for cacao byproducts constitute opportunities to diminish the environmental impact and open new possibilities for industries around the cacao value chain.

Conclusion

All along the cacao value chain, an important amount of byproducts are generated. If those byproducts are considered waste, the proportion of useful material will be minimum and the process highly inefficient. New studies have opened the possibility of using those materials as a source of bioactive compounds. They have proved their utility for the production of high-valuable new products that are useful in applications in the food, cosmetic, and pharmaceutics industries. Cacao byproducts have also been useful in agriculture, producing biofuels and as feedstuff, among many other applications. The income generated from the new industries developed around the cacao byproducts can complement the income of farmers and the actors throughout the cacao value chain, including side industries like the transportation industry.

References

Afoakwa, E., Kongor, J., Takrama, J., and Budu, A. (2013). "Changes in acidification, sugars and mineral composition of cocoa pulp during fermentation of pulp pre-conditioned cocoa (*Theobroma cacao*) beans." *International Food Research Journal*, 20 (3), 1215-1222.

Ahmad, F., Daud, W. M. A. W., Ahmad, M. A., and Radzi, R. (2012). "Cocoa (*Theobroma cacao*) shell-based activated carbon by CO_2 activation in removing of Cationic dye from aqueous solution: Kinetics and equilibrium studies." *Chemical engineering research and design*, 90 (10), 1480-1490. doi: 10.1016/j.cherd.2012.01.017.

Antwi, E., Engler, N., Nelles, M., and Schüch, A. (2019). "Anaerobic digestion and the effect of hydrothermal pretreatment on the biogas yield of cocoa pods residues." *Waste Management*, 88, 131-140. doi: 10.1016/j.wasman.2019.03.034.

Asiedu, N. Y., Neba, F. A., and Addo, A. (2019). "Modeling the attainable regions for catalytic oxidation of renewable biomass to specialty chemicals: waste biomass to carboxylic acids." *South African Journal of Chemical Engineering*, 30, 1-14. doi: 10.1016/j.sajce.2019.07.003.

Barreto, M., Santos, J., Corrêa, R., Luz, E., Marelli, J., and Souza, A. (2015). "Detection of genetic resistance to cocoa black pod disease caused by three *Phytophthora* species." *Euphytica*, 206 (3), 677-687. doi: 10.1007/s10681-015-1490-4.

Belwal, T., Cravotto, C., Ramola, S., Thakur, M., Chemat, F., and Cravotto, G. (2022). "Bioactive Compounds from Cocoa Husk: Extraction, Analysis and Applications in Food Production Chain." *Foods*, 11 (6), 798. doi: 10.3390/foods11060798.

Berbert, P. (1972). "Estudo da pectina do mel e de casca do fruto de cacau [Pectin studies of cacao "sweatings" and cacao pod husks]." *Theobroma*, 49-51.

Cabala Rosand, P., Santana, M. B., and De Santana, C. J. (2019). "Cacao." In *Detecting mineral nutrient deficiencies in tropical and temperate crops*, edited by Donald L Plucknett, 409-425. CRC Press.

Campos-Vega, R., Nieto-Figueroa, K. H., and Oomah, B. D. (2018). "Cocoa (*Theobroma cacao* L.) pod husk: Renewable source of bioactive compounds." *Trends in Food Science & Technology*, 81, 172-184. doi: 10.1016/j.tifs.2018.09.022.

Cinar, Z. Ö., Atanassova, M., Tumer, T. B., Caruso, G., Antika, G., Sharma, S., Sharifi-Rad, J., and Pezzani, R. (2021). "Cocoa and cocoa bean shells role in human health: An updated review." *Journal of Food Composition and Analysis*, 103, 104115. doi: 10.1016/j.jfca.2021.104115.

Dano, S. D., Manda, P., Dembélé, A., Kouassi Abla, A. M. J., Bibaud, J. H., Gouet, J. Z., and Ze Maria Sika, C. B. (2013). "Influence of fermentation and drying materials on the contamination of cocoa beans by ochratoxin A." *Toxins*, 5 (12), 2310-2323. doi: 10.3390/toxins5122310.

Daud, Z., Awang, H., Mohd Kassim, A. S., Mohd Hatta, M. Z., and Mohd Aripin, A. (2014). "Cocoa pod husk and corn stalk: Alternative paper fibres study on chemical characterization and morphological structures." *Advanced Materials Research*, 911, 331-335. doi: 10.4028/www.scientific.net/AMR.911.331.

Djeke, M. D., Angui, P. K. T., and Kouadio, J. Y. (2011). "Décomposition des broyats de coques de cacao dans les sols ferrallitiques de la zone d'Oumé, centre-ouest de la Côte d'Ivoire: effets sur les caractéristiques chimiques des sols [Decomposition of shredded cocoa shells in ferrallitic soils in the Oumé area, center-west of Côte d'Ivoire: effects on the chemical characteristics of the soils]." *Biotechnology, Agronomy, Society and Environment.*

Endraiyani, V., Ludescher, R. D., Di, R., and Karwe, M. V. (2017). "Total phenolics and antioxidant capacity of cocoa pulp: processing and storage study." *Journal of Food Processing and Preservation*, 41 (4), e13029. doi: 10.1111/jfpp.13029.

FAO. (2022). *FAOSTAT*_data_2_17_2022.

Figueira, A., Janick, J., Yadav, M., and BeMiller, J. (1992). "Cacao gum: a potential new economic product." *Proceedings of the International Cocoa Conference Challenges in the 90s*.

Gyedu-Akoto, E., Yabani, D., Sefa, J., and Owusu, D. (2015). "Natural skin-care products: the case of soap made from cocoa pod husk potash." *Advances in Research*, 4 (6), 365-370. doi: 10.9734/AIR/2015/1702.

Han, X., Shen, T., and Lou, H. (2007). "Dietary polyphenols and their biological significance." *International journal of molecular sciences*, 8 (9), 950-988. doi: 10.3390/i8090950.

Hernández-Hernández, C., Viera-Alcaide, I., Morales-Sillero, A. M., Fernández-Bolaños, J., and Rodríguez-Gutiérrez, G. (2018). "Bioactive compounds in Mexican genotypes of cocoa cotyledon and husk." *Food chemistry*, 240, 831-839. doi: 10.1016/j.foodchem.2017.08.018.

ICCO. (2022). *Cocoa Market Report.* March 2022 (QBSC-XLVI-No-4). Retrieved 17_11_2022 from: icco.org/wp-content/uploads/Production-QBSC-XLVI-No-4.pdf.

ICCO. n.d. "*Trading and Shipping cocoa beans.*" International Cocoa Organization (ICCO), accessed 13/02/2022. https://www.icco.org/trading-shipping/.

ISCQF. n.d. "External Analysis and Cut Test of Cocoa Beans." *International Standards for the Assessment of Cocoa Quality and Flavour*, accessed 03/02/2022. https://www.cocoaqualitystandards.org/protocols-for-review/download.

ITC. n.d. "*Trade map*." International Trade Center, accessed 11/04/2022. https://www.trademap.org/Index.aspx.

Iwaro, A., Sreenivasan, T., and Umaharan, P. (1997). "*Phytophthora* resistance in cacao (*Theobroma cacao*): Influence of pod morphological characteristics." *Plant Pathology*, 46 (4), 557-565. doi: 10.1046/j.1365-3059.1997.d01-47.x.

Lahive, F., Hadley, P., and Daymond, A. J. (2019). "The physiological responses of cacao to the environment and the implications for climate change resilience. A review." *Agronomy for Sustainable Development*, 39 (1), 1-22. doi: 0.1007/s13593-018-0552-0.

Lee, J. K., Patel, S. K. S., Sung, B. H., and Kalia, V. C. (2020). "Biomolecules from municipal and food industry wastes: an overview." *Bioresource Technology*, 298, 122346. doi: 10.1016/j.biortech.2019.122346.

Lima, L. J., Almeida, M. H., Nout, M. R., and Zwietering, M. H. (2011). "*Theobroma cacao* L.," The food of the Gods": quality determinants of commercial cocoa beans, with particular reference to the impact of fermentation." *Critical reviews in food science and nutrition*, 51 (8), 731-761. doi: 10.1080/10408391003799913.

Lozano Moreno, M. S. (2020). "*Utilización de los subproductos del beneficio del cacao: una revisión* [*Utilization of cocoa processing by-products: a review*]." Bacherlos, Facultad de Ciencias Naturales e Ingeniería, Universidad de Bogotá Jorge Tadeo Lozano.

Lu, F., Rodriguez-Garcia, J., Van Damme, I., Westwood, N. J., Shaw, L., Robinson, J. S., Warren, G., Chatzifragkou, A., McQueen Mason, S., Gomez, L., LauraFaas, Balcombe, K., Srinivasan, C., Picchioni, F., Hadley, P., and Charalampopoulos, D. (2018). "Valorisation strategies for cocoa pod husk and its fractions." *Current Opinion in Green and Sustainable Chemistry*, 14, 80-88. doi: 10.1016/j.cogsc.2018.07.007.

Luz, E., Paim, M., Silva, S., Pires, J., Santos Filho, L., and Monteiro, W. (2006). "Correlations of cocoa clones' reactions to different species of *Phytophthora*." In *Global approaches to cocoa germplasm utilization and conservation*, edited by B. Eskes and Y. Efron, 132-135.

Marsiglia, D., Ojeda, K., Ramírez, M., and Sánchez, E. (2016). "Pectin extraction from cocoa pod husk (*Theobroma cacao* L.) by hydrolysis with citric and acetic acid." *International Journal of Chem Tech Research*, 9 (7), 497-507.

Martillanes, S., Rocha-Pimienta, J., and Delgado-Adámez, J. (2018). "Agrifood by-products as a source of phytochemical compounds." In *Descriptive food science*, 43-58. London: IntechOpen.

Moreno, S., Morán, E., Quijije, I., and Ochoa, D. (2021). "Mucílago de *Theobroma Cacao* L. como base para un bioantimicrobiano mezclado con dos ácidos débiles: alternativas ecológicas [Mucilage of *Theobroma Cacao* L. as a base for a bioantimicrobial mixed with two weak acids: ecological alternatives]." *Ecuadorian Science Journal*, 5 (4), 98-108. doi: 10.46480/esj.5.4.173.

Nazir, N., Juita, E., Amelia, C., and Fatli, R. (2016). "Optimization of pre-treatment process of cocoa pod husk using various chemical solvents." *International Journal on Advanced Science, Engineering and Information Technology*, 6 (3), 403-409.

Nieto Figueroa, K. H., Mendoza García, N. V., and Campos Vega, R. (2020). "Cocoa By-products." In *Food Wastes and By-products: Nutraceutical and Health Potential*, edited by B. Dave Oomah, Rocio Campos-Vega and Hayde Azeneth Vergara-Castaneda, 373-411. John Wiley & Sons.

Osundahunsi, O., Bolade, M., and Akinbinu, A. (2007). "Effect of cocoa shell ash as an alkalizing agent on cocoa products." *Journal of Applied Sciences*, 7, 1674-1678.

Ouattara, L. Y., Kouassi, E. K. A., Soro, D., Soro, Y., Yao, K. B., Adouby, K., Drogui, A. P., Tyagi, D. R., and Aina, P. M. (2021). "Cocoa pod husks as potential sources of renewable high-value-added products: A review of current valorizations and future prospects." *BioResources*, 16 (1), 1988-2020. doi: 10.15376/biores.16.1.Ouattara

Panak Balentić, J., Ačkar, Đ., Jokić, S., Jozinović, A., Babić, J., Miličević, B., Šubarić, D., and Pavlović, N. (2018). "Cocoa shell: A by-product with great potential for wide application." *Molecules*, 23 (6), 1404. doi: 10.3390/molecules23061404.

Peñaloza Albarracín, D. F., Laiton Daza, L. J., Caballero Yáñez, D. F., Blanco Tirado, T. d. S., Acevedo Argüello, C., and Cervantes Díaz, M. (2021). "Estudio cienciométrico de tendencias en el aprovechamiento de los subproductos del cacao (*Theobroma cacao* L.) [Scientometric study of trends in the use of cocoa by-products (*Theobroma cacao* L.)]." *Espacio I+ D, Innovación más desarrollo*, 10 (27), 83-94.

Pfaltzgraff, L. A., De bruyn, M., Cooper, E. C., Budarin, V., and Clark, J. H. (2013). "Food waste biomass: a resource for high-value chemicals." *Green Chemistry*, 15 (2). doi: 10.1039/c2gc36978h.

Porto de Souza Vandenberghe, L., Valladares-Diestra, K. K., Bittencourt, G. A., Murawski de Mello, A. F., Sarmiento Vásquez, Z., Zwiercheczewski de Oliveira, P., de Melo Pereira, G. V., and Soccol, C. R. (2022). "Added-value biomolecules' production from cocoa pod husks: A review." *Bioresource Technology*, 344:126252. doi: 10.1016/j.biortech.2021.126252.

Quiñones, M., Miguely, M., and Aleixandre, A. (2012). "Los polifenoles, compuestos de origen natural con efectos saludables sobre el sistema cardiovascular [Polyphenols, naturally occurring compounds with healthy effects on the cardiovascular system]." *Nutrición hospitalaria*, 27 (1), 76-89.

Ravelo Abreu, A., Rubio Armendáriz, C., Gutiérrez Fernández, A., and Hardisson De la Torre, A. (2011). "La ocratoxina A en alimentos de consumo humano: revisión [Ochratoxin A in food for human consumption: review]." *Nutrición Hospitalaria*, 26 (6), 1215-1226.

Rebollo-Hernanz, M., Zhang, Q., Aguilera, Y., Martín-Cabrejas, M. A., and Gonzalez de Mejia, E. (2019). "Relationship of the phytochemicals from coffee and cocoa by-products with their potential to modulate biomarkers of metabolic syndrome *in vitro*." *Antioxidants*, 8 (8), 279. doi: 10.3390/antiox8080279.

Rodríguez-Bonilla, C. G., Enriquez-Martínez, M. R., and Esteves-Ceballos, S. L. (2020). "Alternativa de Cadena logística para la exportación de cacao usando el modelo justo a tiempo [Logistics chain alternative for cocoa export using the just-in-time model]." *EmTHYMÓS, Revista de Estudios Empresariales* [*EmTHYMÓS, Business Studies Magazine*], 1 (3), 199-211.

Rojo-Poveda, O., Barbosa-Pereira, L., Zeppa, G., and Stévigny, C. (2020). "Cocoa bean shell—a by-product with nutritional properties and biofunctional potential." *Nutrients*, 12 (4), 1123. doi: 10.3390/nu12041123.

Sandesh, K., Shishir, R., and Rao, C. V. (2020). "Optimization and comparison of induction heating and LPG assisted acid pretreatment of cocoa pod for ABE fermentation." *Fuel*, 262, 116499. doi: 10.1016/j.fuel.2019.116499.

Santos, F. A., Palmeira, E. S., and Jesus, G. (2019). "An image dataset of cut-test-classified cocoa beans." *Data in brief*, 24, 103916. doi: 10.1016/j.dib.2019.103916.

Soares, T. F., and Oliveira, M. B. P. (2022). "Cocoa By-Products: Characterization of Bioactive Compounds and Beneficial Health Effects." *Molecules*, 27 (5), 1625. doi: 10.3390/molecules27051625.

Sriwana, I. K., Arkeman, Y., Syah, D., and Marimin. (2017). "Sustainability improvement in cacao supply chain agro-industry." *World Review of Science, Technology and Sustainable Development*, 13 (3), 256-275. doi: 10.1504/WRSTSD.2017.087154.

TIS. (2022). "*Cargo Information.*" German Insurance Association (GDV e.V.), accessed 1/03/2022. https://www.tis-gdv.de/tis_e/ware/genuss/kakao/kakao-htm/.

Titiloye, J. O., Abu Bakar, M. S., and Odetoye, T. E. (2013). "Thermochemical characterisation of agricultural wastes from West Africa." *Industrial crops and products*, 47, 199-203. doi: 10.1016/j.indcrop.2013.03.011.

Tran, T. N., Bayer, I. S., Heredia-Guerrero, J. A., Frugone, M., Lagomarsino, M., Maggio, F., and Athanassiou, A. (2017). "Cocoa shell waste biofilaments for 3D printing applications." *Macromolecular Materials and Engineering*, 302 (11), 1700219. doi: 10.1002/mame.201700219.

Unten, S., Ushijima, H., Shimizu, H., Tsuchie, H., Kitamura, T., Moritome, N., and Sakagami, H. (1991). "Effect of cacao husk extract on human immunodeficiency virus infection." *Letters in applied microbiology*, 13 (6), 251-254. doi: 10.1111/j.1472-765X.1991.tb00621.x.

Vásquez, Z. S., de Carvalho Neto, D. P., Pereira, G. V., Vandenberghe, L. P., de Oliveira, P. Z., Tiburcio, P. B., Rogez, H. L., Neto, A. G., and Soccol, C. R. (2019). "Biotechnological approaches for cocoa waste management: A review." *Waste management*, 90, 72-83. doi: 10.1016/j.wasman.2019.04.030.

Voora, V., Bermúdez, S., and Larrea, C. (2019). Global Market Report: Cocoa. In *Sustainable Commodities Marketplace Series 2019*: International Institute for Sustainable Development (IISD).

Chapter 4

Chili and Pepper Byproducts

Noé Rodríguez-Barajas[1]
Efigenia Montalvo-González[2]
Zuamí Villagrán[1]
Sughey González-Torres[1]
Angélica Villaruel-López[3]
and Luis M. Anaya Esparza[1,*]

[1]University Center of Los Altos, University of Guadalajara, Tepatitlán, Jalisco
[2]Integral Food Research Laboratory, Technological Institute of Tepic, Tepic, Nayarit
[3]University Center of Exact Sciences and Engineering, University of Guadalajara, Guadalajara, Jalisco

Abstract

The *Capsicum* genus belongs to the Solanaceae family and contains more than 30 species, where *Capsicum annum (C. annum)* is one of the most cultivated worldwide. *Capsicum* spp. includes hot and sweet peppers, depending on their pungency. They are commonly eaten as a fresh vegetable or used as a spice in many culinary dishes, enhancing their texture, appearance, and taste due to their unique flavor, color, and aroma. Furthermore, *Capsicum* spp. exhibited ethnopharmacological importance in treating toothache, rheumatism, ulcers, and alopecia.

On the other hand, various byproducts (leaves, peel, seeds, peduncles, and placenta) are generated during pepper harvesting and processing. Byproducts are usually discarded as food waste. These byproducts have been recognized as valuable raw materials for obtaining vitamins (A, C, and E) and bioactive compounds such as capsaicinoids, carotenoids, phenolic compounds, and flavonoids, as well as oleoresins. The most-reported biological activities *(in vitro* and *in vivo*) include antioxidant, anti-inflammatory, neuroprotective, and antimicrobial, attributed to bioactive compounds. Moreover, these byproducts can be used for diverse food applications for humans (nuggets, spaghetti, and bakery products) and animals (poultry production), bioremediation, textile coloring, biodiesel production, and as a stabilizing agent for the synthesis of nanoparticles. This chapter aims to describe and discuss the current scientific advances on the phytochemical profile of the byproducts of chilies and peppers, their related biological activities, and

[*] Corresponding Author's Email: luis.aesparza@academicos.udg.mx.

In: Food Byproducts
Editors: S. Teresita Martín-del-Campo, A. Cardador-Martínez and J. del Pilar Ramírez-Anaya
ISBN: 979-8-88697-612-0

technological uses as an alternative to revalorize losses and wastes from the harvesting and processing of hot and sweet peppers.

Keywords: *Capsicum* spp., byproducts, bioactive compounds, food waste, food revalori-zation, circular economy

Introduction

In recent years, food production has increased considerably, generating an exponential increase in food waste in the form of skin, seeds, pulp, leaves, shells, and stems, among others, which is a global concern with significant economic losses and environmental impact (Anaya-Esparza et al. 2021, Mármol et al. 2021). It is estimated that >200 million tons of food are wasted every year during production and in retail chains (FAO 2019). Traditionally, food waste has been used for composting, livestock feeding, landfilling, and energy production (biohydrogen, bioethanol, and biogas) or incineration; however, the majority of food waste can contain health-beneficial and functional compounds with potential technological uses in the pharmaceutical, food, and bioenergetics industries (Reguengo et al. 2022).

The European Commission (EC 2010) defined food waste as "raw or cooked food materials discarded at any point in the life cycle of food," and the term "byproduct" refers to any substance or compound resulting after the processing of primary foods. Thus, the term "food byproducts" indicates that food waste is a substrate for obtaining new value-added products for diverse applications. In this context, a growing trend exists for utilizing food waste as a natural source to get nutritional and functional compounds, including proteins, polysaccharides, vitamins, minerals, oils, and secondary metabolites such as polyphenols, flavonoids, anthocyanins, and carotenoids, among others, leading to a bioeconomy in which industry, society, and the environment benefit (Mármol et al. 2021).

The vegetable-and fruit-based losses and wastes generated during harvesting, handling, and postharvest processing, represent at least 40% of the total production (Wu et al. 2022), particularly those obtained from *Capsicum* spp. (Chel-Guerrero et al. 2021, Akhtar, Waqas and Nauman, 2021). The *Capsicum* genus belongs to the Solanaceae family. It contains more than 30 species, where *Capsicum annum, C. Chinense, C. baccatum, C. pubescens,* and *C. frutescens* are the most commercially cultivated worldwide. The *Capsicum genus* includes pungent (chili or hot pepper) and non-pungent (sweet peppers), depending on their capsaicinoid content (Chel-Guerrero et al. 2021). On the other hand, it has been reported that waste materials generated during the harvesting of these crops are higher than 18%, which include leaves, stems, seeds, and defective fruits (Cortés-Ferré et al. 2021), also during industrial processing, peduncles, seeds, placenta, and peels (bagasse) are generated (Chel-Guerrero et al. 2021, Jasna et al. 2018). These wastes and byproducts can be used for extracting valuable and innovative components for food (human and animal), pharmaceutical, and bioenergetics applications (Akhtar, Waqas and Nauman, 2021).

This chapter aims to describe and discuss the current scientific advances on the phytochemical profile of chilies and peppers byproducts, their related biological activities, and technological uses as an alternative to revalorize losses and wastes from chilies and peppers harvesting and processing.

Chilies and Peppers

Capsicum genus belongs to the Solanaceae family and holds more than 31 species and some varieties, highlighting *C. annum* (paprika, jalapeño, cayenne, or chiltepin), *C. chinense* (habanero, naga, and Scoth bonnet), *C. pubescens* (Peruvian rocotto and Mexican manzano), *C. baccatum* (ají pineapple, sugar rush peach, and aji fantasy), and *C. frutescens* (tabasco chili and malagueta pepper) as the most domesticated and economically important species worldwide (Sanatombi and Rajkumari. 2020, Özcan et al. 2021). Although each *Capsicum* species differs from the others (Table 1), they are an annual or perennial herbaceous plant cultivated in tropical and subtropical climates worldwide with agricultural and economic importance (Baenas et al. 2019).

Table 1. Distinctive characteristics between the most cultivated *Capsicum* species

Capsicum spp.	Flower color	N° flowers	Seed color	Fruit shape	Fruit weight (g)	Reference
C. annum[1]	White	1	Tan	Quadrangular	100 to 500	a, b
C. chinense[2]	White-green	2-5	Tan	Undulating surface	5 to 11	b, c
C. frutescens[3]	Green	2-5	Tan	Oblong-conical	0.5 to 5	b, d
C. baccatum[4]	White with a yellow spot	1-2	Tan	Elongate-campanulate	4.40 to 6	b, c
C. pubscens[5]	Purple	1-2	Black	Triangular and bell-shaped	40	b, f

[1]Bell pepper; [2]Habanero pepper; [3]Tabasco pepper; [4]Locoto pepper; [5]Rocoto pepper

[a]Anaya-Esparza et al. 2021, [b]Echave et al. 2012, [c]Kumar et al. 2012, [d]Jarret et al. 2007, [e]Albrecht et al. 2018, [f]Escalera-Ordaz et al. 2019

Table 2. Nutrimental composition and energy value (per 100 g of edible portion) of different *Capsicum* species

Parameter	*C. annum*[1]	*C. Chinense*[2]	*C. baccatum*[3]	*C. pubenecens*[4]	*C. frutescens*[5]
Moisture (g)	89.31	83.26	89.60	91.40	10.78
Total protein (g)	18.30	1.86	0.95	2.5	26.02
Carbohydrates (g)		8.19	9.10		55.09
Total fiber (g)	60.96	1.94		1.3	
Total fat (g)	11.04	0.12	0.14	1.8	4.01
Ash (g)	4.42	1.18	0.14	2.4	4.10
Vitamin C (mg)		110.19	158	~30.0	
Reference	a	b	c	d	e

[1]Red pepper; [2]King chili pepper; [3]Cera pepper; [4]Kitaiskii fonarik pepper; [5]Cambuci pepper

[a]Azabou et al. 2017, [b]Malakar et al. 2018, [c]Nascimento et al. 2012, [d]Pérez-Vázquez et al. 2022, [e]Firatligil-Durmus and Evranuz et al. 2010

Capsicum species are characterized by their edible fruits with nutritional importance (Table 2) (Özcan et al. 2021); they vary in color, size, shape, pungency degree, aroma, and taste (Ramandeep and Kaur 2020). Anatomically, *Capsicum* fruits are divided into exocarp (skin), mesocarp, endocarp, and seeds adhered to the placenta. The pericarp contains up to 20% (dry weight) cellulose and other fibrous components, whereas the skin is a rich source of dietary fiber (Sandoval-Castro et al. 2017). In this context, *Capsicum* spp. fruits are an important source of dietary fiber, vitamins (B, C, E, and provitamin A), minerals (Ca, Mg, P, K, Fe, Na, Se, and Zn), and natural antioxidants (vanillic, chlorogenic, and ferulic acids, catechin, rutin, kaempferol, and quercetin). Moreover, orange, red, and yellow-colored peppers can contain violaxanthin, neoxanthin, zeaxanthin, lutein, β-carotene, and lycopene (Baenas et al. 2019, Razola-Díaz et al. 2022), while green peppers contain chlorophylls (Vazquez-Flores et al. 2020); the concentration of these compounds depends on the ripening state of the fruit (Kavuncuoglu, Capar and Yalçin 2021). *Capsicum* fruits can be classified as hot or pungent (typically those with small fruits) or sweet or non-pungent (typically those with large fruits) peppers (Baenas et al. 2019), according to their Scoville heat units that depend on their capsaicinoids content such as capsaicin, dihydrocapsaicin, nordihydrocapsaicin, homocapsaicin, and homodihydrocapsaicin (Cortés-Ferré et al. 2021, Alam et al. 2020, Alam et al. 2018).

Traditional Uses of Chilies and Peppers

Traditionally, chilies and peppers are used for food and medicinal applications, as shown in Figure 1. Chilies and peppers are typically eaten raw in salads, cooked or baked, fermented, dried, or as a spice (Sanatombi and Rajkumari. 2020, Özcan et al. 2021, Jiménez et al. 2021). They exhibited great gastronomic-cultural relevance worldwide (Vazquez-Flores et al. 2020). For example, in Mexico, chilies and peppers are essential for making various sauce types (Cárdenas-Castro et al. 2019); moreover, *Capsicum* fruits are often used to add aroma and color to soups and pizzas (Askin, Buket and Sercan 2021). Furthermore, due to their typical taste, color, and pungency, various processed products (dried, canned, pasted, or powdered) are prepared with chilies and peppers, including puree (Ramandeep and Kaur. 2020), hot sauces (Cárdenas-Castro et al. 2019), ketchup, pickles, condiments, paste, soups, and dehydrated products (Alam et al. 2020, Cvetkovi 2022). They are also used as a natural colorant and antioxidant compounds for food applications (Jiménez et al. 2021, Aşkin 2018).

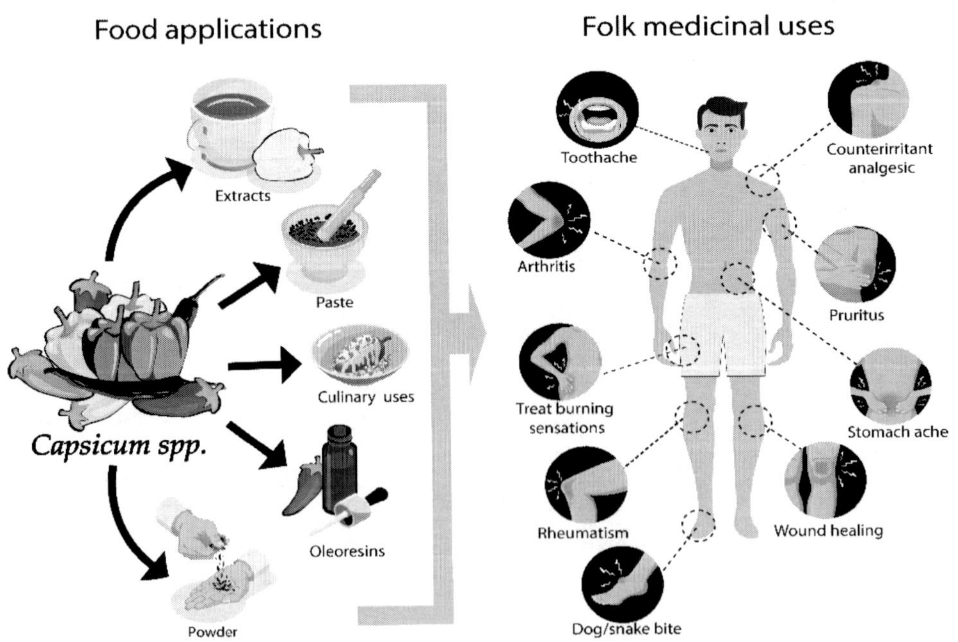

Figure 1. Traditional food and medicinal uses of chilies and peppers.

Regarding medicinal uses, the decoction, infusion, or extracts from *Capsicum* species and their byproducts (fruit, leaves, and stem) have been used as a rubefacient, local stimulant, and healing agent, as well as to treat arthritis, coughs, asthma, sore throats, fatigue, stomachaches, toothaches, and parasitic infections (Sanatombi and Rajkumari. 2020, Özcan et al. 2021, Zimmer et al. 2012). Moreover, it exhibited wound healing, anti-inflammatory, and analgesic effects (Zimmer et al. 20012). Additionally, in Africa, chilies and peppers are given to sick animals such as goats, chickens, and sheep to increase their resistance to diseases and parasites (Afolabi et al. 2017). In general, ethnopharmacological uses are associated with bioactive compounds and their therapeutic properties (Akhtar, Waqas and Nauman, 2021). Furthermore, *Capsicum* fruits exhibited nutraceutical and cosmetic usages (Balaky et al. 2021).

Chilies and Peppers Byproducts: Nutrimental and Phytochemical Characterization

According to Cotés-Ferré et al. (2021) wastes that arose during chili and sweet pepper harvesting (leaves, seeds, and defective fruits) correspond to 18% of the total production; in contrast, after processing in commercial products (*i.e.*, juice, hot sauce, puree, and paste), placenta, seeds, and peduncles (Figure 2) are generated (Cortés-Ferré et al. 2021, Zhao, Zhang and Zhang 2010) these correspond to approximately 65% of the total weight of raw material (Ksibi et al. 2015). Moreover, it is estimated that around 450 to 500 g of waste per kilogram of the total pepper weight is produced during pepper coloring and powder processing (Balaky et al. 2021).

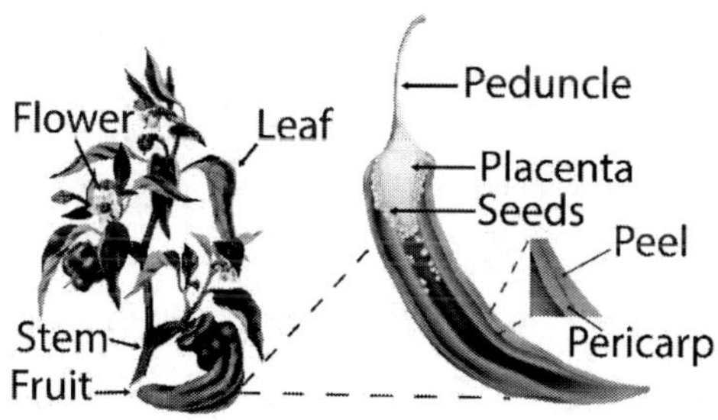

Figure 2. Chilies and peppers byproducts generated during harvesting and processing.

During capsanthin production, capsicum waste in powder form is generated (98.6% of inputted capsicum powder) (Zhao, Zhang and Zhang 2010). For example, Chenguang Natural Pigment Limited Company is the biggest capsanthin producer worldwide in China, generating 30000 tons of capsicum powder waste per year (Zhao, Zhang and Zhang 2010). Furthermore, in Mexico, Jalapeño pepper (*C. annum*) is highly consumed, mainly as an industrialized product, which is sliced, pickled (but not fermented), and canned; nonetheless, during the preliminary operation of pickled Jalapeño pepper, more than 500 tons of seed and placenta byproducts per year are produced and discarded as waste (only in the Northwest of Mexico) (Sandoval-Castro et al. 2017). These waste and byproducts contain protein (from 13 to 24%), dietary fiber (from 26 to 31%), vitamins (B, C, and E), and minerals (Ca, Mg, P, K, Fe, Na, Se, and Zn) (Akhtar et al. 2021, Cvetkovi 2022), but also bioactive compounds, as listed in Table 3.

Table 3. Some reported phytochemicals of chilies and peppers byproducts

Capsicum species	Byproduct	Bioactive compounds	Reference
C. annum	Seeds	Capsaicin, dihydrocapsaicin, β-carotene, zeaxanthin, lutein, epicatechin, rutin, quercetin, myricetin, gallic, vanillic, protocateuchuic, sinapinic, caffeic, rosmaniric, and chlorogenic acids	Šeregelj et al. (2019)
	Placenta	Capsaicin, dihydrocapsaicin, protocatechuic, caffeic, vanillic, chologenic, gallic, coumaric, and sinapic acids, epicatechin, catechin, quercetin, kaempferol, rutin, β-carotene	Sandoval-Castro et al. (2017)
	Peduncle	Caffeic acid, diosmetin, apigenin, kaempferol, rutin, α-carotene, β-carotene	Jiménez et al. (2021), Simonovska et al. (2016)
	Leaf	Feruloyl putrescine, 5 caffeoylquinci acid, luteolin, apigenin, luteolin-7-O-(2" apiosyl)glucoside,	Assefa et al. (2021)
C. chinense	Seeds	*alkaloids, saponins, flavonoids, phenols, tannins, and terpenoids	Ukwuani and Hassan (2018)
	Peduncle	Protocatechuic, chlorogenic, gallic, cinnamic, coumaric, acids, myricetin, catechin, luteolin, rutin, naringenin, hesperetin, vanillin, diosmetin, diosmin, apegnin	Chel-Guerrero et al. (2021), Chel-Guerrero et al. (2022)
	Leaf	Protocatechuic, chlorogenic, gallic, cinnamic, coumaric, acids, myricetin, catechin, quercetin, kaempferol, luteolin, rutin, naringenin, hesperetin, vanillin, diosmetin, diosmin, apegnin, chrysoeriol, vanillic acid, 5-caffeolquinic acid	Chel-Guerrero et al. (2021), Chel-Guerrero et al. (2022), Herrera-Pool et al. (2021)
	Stem	Protocatechuic, chlorogenic, gallic, cinnamic, coumaric acids, myricetin, quercetin, kaempferol, luteolin, rutin, catechin, naringenin, hesperetin, vanillin, diosmetin, diosmin, apegnin	Chel-Guerrero et al. (2021, Chel-Guerrero et al. (2022)
C. baccatum	Seeds	Total phenols and flavonoids	Zimmer et al. (2012)
C. pubescens	Fruit flour	Quercetin, apigenin, irilone, kaempferol, naringenin, hesperidin, jaceosidin, 5, caffeoylquinic, cinnamic, vanillic, caffeic, isoferullic, ferullic, and benzoic acids	de Sá Mendes et al. (2020)
C. frutescens	Seeds	Total phenols, capsaicin, dihydrocapsaicin, and chrysoeriol	Nascimento et al. (2014)
	Peel	Total phenols, capsaicin, dihydrocapsaicin, and chrysoeriol	Nascimento et al. (2014)
	Placenta	Capsaicin, dihydrocapsaicin, homodihydrocapsaicin, β-Carotina	Cervantes-Hernández et al. (2019)
	Leaf	3-tetradecene, diethyl phthalate, 1,2-benzene dicarboxylic acid, and diethyl ester	Velsankar et al. (2021)

*Qualitative analysis.

Razola-Díaz et al. (2022) informed that yellow pepper waste (*C. annum*) contained vanillic, chlorogenic, and ferulic acids, catechin, rutin, quercetin, violaxanthin, neoxanthin, meso-zeaxanthin, zeaxanthin, lutein, β-carotene, and lycopene. Kavuncuoğlu et al. (2021) reported that bell pepper seeds (*C. annum*) exhibited a total phenolic content of 151 mg gallic acid equivalent (GAE) per gram of dry extract. Similarly, Šeregelj et al. (2019) informed that *C. annum* seed contained capsaicin, dihydrocapsaicin, β-carotene, zeaxanthin, lutein, and epicatechin, rutin, quercetin, myricetin, gallic, vanillic, protocatechuic, sinapinic, caffeic, rosmaniric, and chlorogenic acids. Moreover, red hot pepper seeds have been reported to contain tannins and anthocyanins (Balaky et al. 2021). Additionally, it has been reported that

C. annum seeds contained fatty acids (C:16:0, C:16:1, C:18:0, C:18:1, C:18:2, and C:20:0) (Lee et al. 2017), α-tocopherol, γ-tocopherol, and α-tocotrienol (Cvetković et al. 2020).

Sandoval-Castro et al. (2017) informed that a mixture of scalded byproducts (seeds and placenta) from Jalapeño pepper (*C. annum*) contains bioactive compounds, where the major compounds identified were rutin (1144 mg/100 g), epicatechin (884 mg/100 g), and catechin (790 mg/100 g); moreover, other compounds such as gallic, caffeic, ferulic, 3,4-dihydroxybenzoic, chlorogenic, and coumaric acids, kaempferol, capsaicin, and dihydrocapsaicin were identified; these compounds exhibited strong antioxidant capacity.

Additionally, Valdez-Morales et al. (2021) recovered oil from Jalapeño peppers (*C. annum*) byproducts (seeds and placenta) after industrial scalding during canned pepper processing. The Jalapeño pepper byproducts contained 22% of oil (it exhibited a greenish color) with the presence of fatty acids, highlighting the linoleic acid (70% of the total fatty acids); moreover, oil is composed of β-carotene (1.85 mg/100 g of oil), α-tocopherol (46.12 mg/100 g of oil), capsaicin (83.2 mg/100 g of oil), and dihydrocapsaicin (79.78 mg/100 of oil). Furthermore, Zhang et al. (2019) reported that oils obtained from roasted *C. annum* seeds exhibited improved oxidative stability and antioxidant capacity during storage (30 days at 63°C).

Chel-Guerrero et al. (2021) characterized the phytochemical profile of habanero pepper (*C. chinenense*) byproducts (leaves, stem, and peduncles) generated during harvesting. They reported that peduncles predominantly contain chlorogenic acid, catechin, and rutin; likewise, the leaves of pepper plants are rich in chlorogenic acid, myricetin, β-carotene, and vitamin E. Stems contain chlorogenic and protocatechuic acids. These compounds have demonstrated antioxidant capacity and can be used as functional additives in food applications (Chel-Guerrero et al. 2021, Akhtar, Waqas and Nauman 2021). On the other hand, capsaicinoids such as capsaicin and dihydrocapsaicin were not detected in habanero chili byproducts (Chel-Guerrero et al. 2021). In contrast, Cervantes-Hernández et al. (2019) informed that the placenta of habanero chili contained capsaicin, dihydrocapsaicin, homodihydrocapsaicin, β-carotina. Recently, there has been a growing trend for extracting capsaicinoids from habanero chili using eco-friendly and emerging technologies such as ultrasound, microwave, pressurized hot water, and supercritical fluid extraction (Fabela-Morón et al. 2020). Moreover, in *C. chinense* seeds, the qualitative presence of alkaloids, saponins, flavonoids, phenols, tannins, and terpenoids has been reported (Ukwuani and Hassan 2018).

Velsakar et al. (2021) identified more than 20 phytochemical compounds during the characterization of the leaf extract from *C. frutescens* by gas chromatography-mass spectrometry, where diethyl phthalate, 1.2-benzene-dicarboxylic acid, and diethyl ester were the most abundant. In contrast, placenta of *C. frutescens* contained capsaicin, dihydrocapsaicin, homodihydrocapsaicin, and β-carotina (Cervantes-Hernández et al. 2019). Nascimento et al. (2014) informed that seeds and peels of *C. frutescens* contain total phenols, capsaicin, dihydrocapsaicin, and chrysoeriol.

de Sá Mendes et al. (2020) characterized the phytochemical profile of *C. pubescens* pepper flour. They reported the presence of 57 phenolic compounds and 4 lignans, including quercetin, apigenin, irilone, kaempferol, naringenin, hesperidin, jaceosidin, 5-caffeoylquinic, cinnamic, vanillic, caffeic, isoferullic, ferullic, and benzoic acids, lariciresinol, isolariciresinol, schisandrol B, and 7-hydroxyenterolactone, among others. In *C. baccatum* seeds, total phenols and total flavonoids have been reported (Zimmer et al. 2012). On the other hand, information about the phytochemical characterization of *C. pubescens* byproducts is scarce (de Sá Mendes et al. 2020).

According to these data, *Capsicum* byproducts are promising sources for obtaining nutritional and functional compounds for diverse applications (Anaya-Esparza et al. 2021).

Biological Activities of Chilies and Peppers Byproducts

Capsicum spp. wastes and byproducts contain bioactive compounds with interesting *in vitro* and *in vivo* biological activities (Akhtar, Waqas and Nauman 2021), as shown in Figure 3.

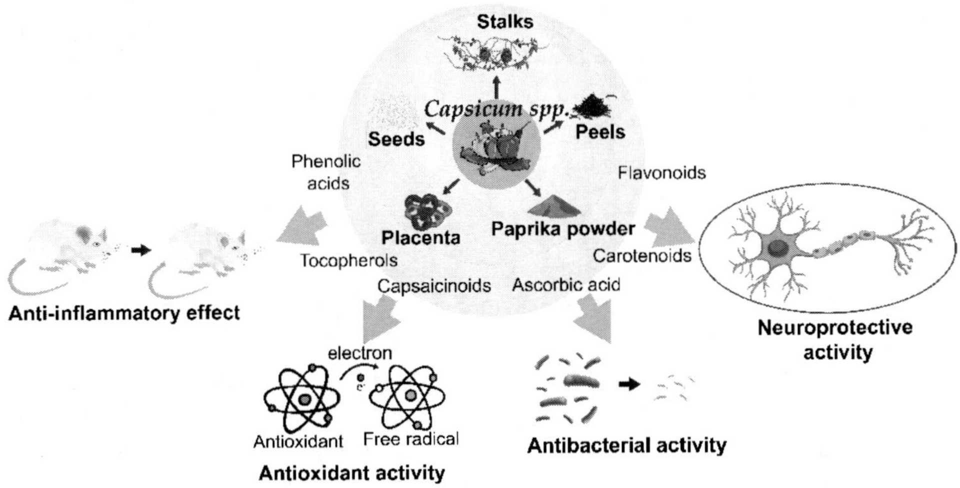

Figure 3. Biological activities of *Capsicum*-based byproducts.

Table 4 lists some reported *in vitro* and *in vivo* biological activities of chili and pepper byproducts, including antioxidant, anti-inflammatory, antibacterial, and α-glucosidase inhibitory, among others.

Regarding *Capsicum annum* byproducts, it has been reported that bell pepper seeds exhibited antioxidant properties by DPPH radical scavenging test, associated with their phenolic content (Kavuncuoglu, Capar and Yalçin 2021). Furthermore, the oil obtained from jalapeño pepper byproducts (seeds and placenta) exhibited strong antioxidant properties (by DPPH, ABTS, and ORAC methods) with higher values than olive oil; these properties were attributed to the fatty acids and phenolic compounds contents (Valdez-Morales et al. 2021). Additionally, red hot pepper seed extracts exhibited α-glucosidase inhibitory activity (a key enzyme responsible for carbohydrates digestion), attributed to the hypoglycemic properties of capsaicin (Liu et al. 2020). Hazekawa et al. (2017) informed that aqueous extract of leaves from bell peppers exerted anti-inflammatory activity (*in vitro*), inhibiting inflammatory cytokine production and suppressing the expression of inflammatory proteins; these effects were associated with the bioactivity of capsaicin. Moreover, the aqueous extract from chili pepper seeds exhibited antimicrobial properties against *Staphylococcus aureus, Pseudomonas aeruginosa, Enterococcus faecalis,* and *Enterobacter aerogenes*, mainly associated with phenols, flavonoids, and tannin contents (Balaky et al. 2021).

Table 4. Some reported biological activities of chilies and peppers byproducts

Capsicum sp.	Common name	*Capsicum* byproduct	Biological activity	Reference
C. annum	Bell pepper	Seeds	Antioxidant	Kavuncuoglu, Capar and Yalçin (2021)
	Jalapeño pepper	Seeds and placenta (oil)	Antioxidant	Valdez-Morales et al. (2021)
	Red hot pepper	Seeds	α-glucosidase inhibitory	Liu et al. (2020)
	Chili pepper	Seeds	Antibacterial	Balaky et al. (2021)
	Bell pepper	Leaves	Anti-inflammatory	Hazekawa et al. (2017)
C. chinense	Habanero chili	Leaves, stem, and peduncles	Antioxidant	Chel-Guerrero et al. (2021)
	Habanero chili	Leaves	Antioxidant	Herrera-Pool et al. (2021)
	Habanero chili	Leaves, stem, and peduncles	Anti-inflammatory	Chel-Guerrero et al. (2022)
	Habanero chili	Seeds, fruits, and leaves	Anti-inflammatory	Ukwuani and Hassan (2018)
	Habanero chili	Seeds	Antifungal	Dias et al. (2013)
C. baccatum	Ají	Seeds	Antioxidant and anti-inflammatory	Zimmer et al. (2012)
	Red pepper	Fruit extract	Anti-inflammatory	Allemand et al. (2016)
	Red pepper	Fruit juice	Anti-inflammatory	Spiller et al. (2010)
C. frutescens	NI	Leaves	Antifungal	Soumya and Nair (2012)
	Long pepper	Fuit extract	Antimicrobial	Bello, Boboye and Akinyosoye (2015)
	Pimenta malgueta	Seeds and peel	Antioxidant	Nascimento et al. (2014)
	Hot-short pepper	Pericarp and seeds	Neuroprotective	Oboh, Omodesola and Ogunruku (2010)
C. pubescens	Pepper	Pepper flour	Antioxidant	de Sá Mendes et al. (2020)

NI: no information.

Concerning *C. chinense* byproducts, Che-Guerrero et al. (2021) informed that habanero pepper byproducts (leaves, stem, and peduncles) contain antioxidant compounds for food applications (Chel-Guerrero et al. 2021, Akhtar, Waqas and Nauman 2021). Herrera-Pool et al. (2021) reported that the leaf extract from habanero chili exhibited strong antioxidant activity by $ABTS^{+}$ and DPPH radical inhibition methods, attributed to N-caffeoyl putrescine, apigenin, luteolin, and diosmetin derivatives. Moreover, leaf, stem, and peduncle extracts from habanero chili byproducts exhibited anti-inflammatory activity in a TPA-induced ear inflammation rat model at a dose of 1 mg/ear; however, the effect was in a byproduct-dependent response, where peduncle (28.91%) and leaf (24.60%) extracts were the most active, probably associated with the profile and concentration of phytochemicals in each byproduct (Chel-Guerrero et al. 2022). Furthermore, the hydromethanolic extracts from seeds, fruits, and leaves from *C. chinense* exhibited *in vitro* anti-inflammatory activities using fresh human blood samples (Ukwuani and Hassan 2018). Moreover, the protein extract from *C. chinense* seeds exhibited antifungal activity against *Candida tropicalis* in a dose-dependent response, mainly associated with the amino acid sequence of bioactive peptides (Dias et al. 2013).

For *C. baccatum* byproducts, Zimmer et al. (2012) reported that the extract from Ají seeds showed antioxidant (DPPH scavenging assay) and anti-inflammatory (carrageenan-induced mice) activities. These effects were associated with the phenolic, flavonoids, and capsaicin content and their ability to reduce oxidative stress and inhibit pro-inflammatory mediators (IL-6, TNF-α, PEG2, and nitric oxide) production. Similar trends were reported by Allemand et al. (Allemand et al. 2016), who informed that extracts from *C. baccatum* fruit powder significantly inhibited the TNF-α and nitric oxide production in carrageenan-induced peritonitis and paw edema-induced rat models associated with the capsaicin and phenolic compounds. These results agree with Spiller et al. (2010), who informed that *C. baccatum* fruit juice showed anti-inflammatory properties in a carrageenan-induced peritonitis rat model, inhibiting neutrophil migration and reducing vascular permeability and pro-inflammatory cytokines production.

Regarding *C. frutescens* byproducts, Soumya and Nair (2012) reported that the aqueous leaf extract from *C. frutescens* exhibited potent antifungal activity against *Aspergillus flavus* and *Aspergillus niger*; this effect was attributed to the leaf extract phytochemicals. Likewise, it has been reported that fruit extracts from *C. frutescens* have antibacterial activity against *Pseudomonas aeruginosa* and *Salmonella typhi* (Bello, Boboye and Akinyosoye 2015). Furthermore, seeds and peel extracts from *C. frutescens* exhibited antioxidant activity by DPPH radical scavenging assay. Oboh, Omodesola and Ogunruku (2010) informed that the hot short pepper powder and seeds showed neuroprotective effects, reducing cyclophosphamide-induced oxidative stress in the brain, attributed to the antioxidant properties of short pepper and seeds.

Related to *C. pubescens* byproducts, de Sá Mendes et al. (2020) informed that pepper flour showed antioxidant properties by ABTS and FRAP methods due to their phytochemicals.

According to these data, *Capsicum* byproducts have great potential for industrial uses in the food and non-food applications, reducing food loss and waste (Valdez-Morales et al. 2021, Sandoval-Castro et al. 2017).

Technological Uses of Chilies and Peppers Byproducts

Valorization of food wastes for industrial applications represents a challenge (Šeregelj et al. 2019). During harvesting or industrial processing of chilies and peppers, significant amounts of defective fruits, leaves, stems, pepper placenta, powder, and seeds are generated and usually discarded as organic waste (Lee et al. 2017). Therefore, it is possible to find a use for these byproducts (Figure 4), making their processing more efficient and productive (Zhang et al. 2019). On the other hand, the main limiting factor of *Capsicum* spp.-based wastes and byproducts is their stabilization due to their high moisture content that promotes microbial deterioration (Jasna et al. 2018). For this purpose, these products are commonly dehydrated before further use (Zhao, Zhang and Zhang 2010, Raak et al. 2017). In this context, *Capsicum* spp.-based byproducts could be valuable, technological, and functional products for diverse applications, as discussed below.

Food Applications

In the last few decades, chili and pepper byproducts as functional ingredients for food applications have been investigated (Ziegenfuss, Friedhelm, Greaves 2000, Cocan et al. 2021).

They have been used as an antimicrobial agent, food coloring additive, antioxidant agent, and dietary fiber source (Table 5).

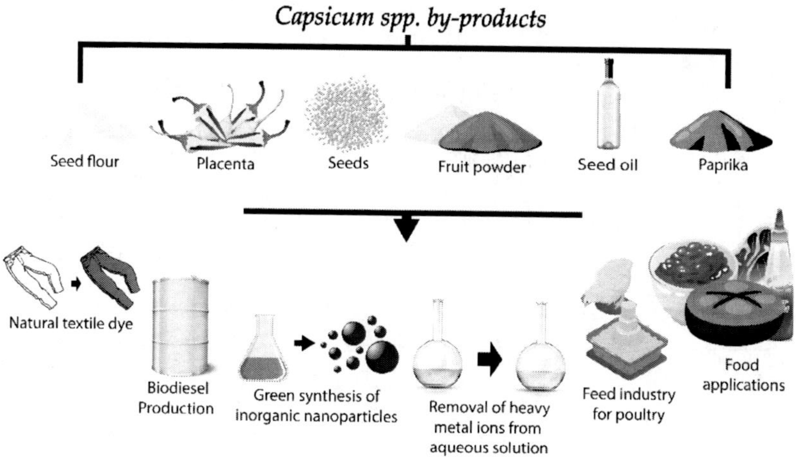

Figure 4. Technological uses of chili and pepper byproducts.

In 1998, Ziegenfuss et al. (2000) patented the process for obtaining a proteinic extract with antimicrobial properties against *Salmonella enterica* Typhimurium and *Escherichia coli* from pepper (*C. annum*) seeds obtained after industrial processing. According to the authors, this extract could be an antimicrobial agent for food applications. In 2008, Tae-I patented a methodology to produce a natural spice product from red pepper seeds byproduct.

C. annum seeds are considered a sustainable source to produce edible oil with potent antioxidant properties (Valdez-Morales et al. 2021, Azabou et al. 2017). Cocan et al. (2021) used chili and sweet pepper (*C. annum*) seed oil (100, 200, or 300 ppm) as a natural antioxidant agent to improve the thermo-oxidative stability of sunflower oil. The authors informed that the sunflower oil increased its stability when supplemented at 300 ppm of chili and sweet pepper seed oil, preventing the lipid oxidation of sunflower oil during heat treatment at 180°C. These results were associated with fatty acids (C16:0, C18:0, C18:1, C18:2, and C22:1) of *Capsicum* seeds oil.

It has been reported that fiber consumption exerts human health benefits, including intestinal microbiota modulation, blood glucose, and lipid level regulation (Padalino et al. 2013). The use of dietary fiber-rich supplements from pepper (*C. annum* cultivar Anita) processing waste has been investigated for making bakery products. Jasna et al. (2018) informed that adding pepper placenta-based supplement (1, 2, or 3%) increased the farinograph water absorption, positively improved the bread crust's color, and reduced the pore size of the breadcrumb compared to the control group. It also decreased the firming of breadcrumbs during the shelf life of bread in a dose-dependent manner. On the other hand, the use of pepper placenta-based powder promoted an increase in dough resistance and decreased breadcrumb cohesiveness but remained in an acceptable range. Furthermore, Danza et al. (2014) reported that adding yellow pepper (*C. annum*) flour (25%) significantly improved bread sensory attributes (color, taste, appearance, crust firmness, large bubbles, and overall quality) and glycemic index in a formulation-dependent response; moreover, the bread contained carotenes and capsaicin, exhibiting antioxidant properties. Similar trends were reported by Padalino et al. (2013), who informed that adding yellow pepper powder (15% wt.) improved nutritional quality (increase in dietary fiber, zeaxanthin, lutein, and β-carotene) and sensory attributes of gluten-free-spaghetti. Likewise, de Sá Mendes et al. (2020) fortified fruit-vegetable-based flour with pepper (*C. baccatum*) powder, improving its antioxidant properties, stability, and appearance (reddish color).

Table 5. Applications of chili and pepper byproducts in the food industry

Source	Application	Treatment	Results	Ref.
Bell pepper seeds (*C. annum*)	Antimicrobial agent	NI	Protein extract from pepper seeds exhibited antimicrobial activity against *Salmonella typhimurium* and *Escherichia coli*	Ziegenfuss, Brinkhaus and Greaves (2000)
Pepper seeds (*C. annum*)	Condiment	NI	Condiment for culinary uses	Tae-I (2008)
Chili and sweet pepper (*C. annum*) seed oil	Antioxidant agent	100, 200, and 300 ppm of seed oil were added to sunflower oil free from synthetic antioxidants	Seed oil can stabilize sunflower oil at low concentrations similar to those of synthetic antioxidants	Cocan et al. (2021)
C. annum	Edible oil	Solvent extraction	Oil is rich in oleic and palmitic acids	Chávez-Mendoza et al. (2015)
Pepper placenta (*C. annum* cultivar Anita)	Bakery products	Dried pepper placenta-based dietary fiber at 1, 2, or 3% in weight	Dietary fiber-rich supplement improved water absorption, color, and pore size of bread	Jasna et al. (2018)
Red pepper waste	Functional ingredient for yogurt production	The encapsulated extract was added at 10% w/v	Yogurt fortified exhibit improved nutritional quality, color, and shelf life	Šeregelj et al. (2019)
Pepper powder	Functional ingredient for flour fortification	Encapsulated pepper powder was added 1:1 ratio	Pepper powder improved the stability, appearance, and antioxidant properties of fruit-vegetable flour	de Sá Mendes et al. (2020)
Red pepper powder (*C. annum*)	Fresh pork sausage production	Pepper powder was added at 0.1, 05, 2% of the total weight	Pepper powder improved the shelf life of sausage without affecting sensory attributes	Martínez et al. (2006)
Paprika (*C. annum*)	Spanish-type sausage	Paprika was added at 30 g/kg	Paprika improved the color of sausage	Fernández-López et al. (2002)
Paprika (*C. annum*)	Meat	Pimiento paprika was added at 1.5 and 2 g/kg	Paprika improved the color of meat and prevented lipid oxidation in a dose-dependent manner	Bázan-Lugo et al. (2012)
Red pepper (*C. annum*) powder	Mutton nuggets	Pepper powder was added at 9% of the total weight	Help to improve sensory attributes of mutton nuggets	Mendiratta, Shinde and Mane (2012)
C. annum	Wheat bread	Yellow pepper flour was added at 25% of the total weight	Pepper flour improved the sensory attributes and glycemic index of bread	Danza et al. (2014)
Yellow pepper (*C. annum*) powder	Gluten-free-spaghetti (GFS)	Pepper powder was added at 15% of the total weight	Pepper flour improved the nutritional quality of GFS	Padalino et al. (2013)
Paprika oleoresin	Prepared poultry meat coloring	Oleoresin combined in milk (3g/100 mL)	Mixture enhanced marinating poultry meat performance	Yusop et al. (2012)
Pepper seed flour (*C. annum*)	Food additive for frying products	Combination of wheat flour and pepper seed flour (1:2 ratio)	Pepper seed flour improved the sensory attributes of fried chicken meatballs	Kilincceker and Kurt (2016)
Capia pepper seed flour	Spreadable paste	NI	The paste is rich in dietary fiber and phytochemicals	Yilmaz, Bostanci and Ok (2019)
Capia pepper seed flour	Breakfast spicy sauce	Vegetable type sauce contains 20% weight of pepper seed flour	The sauce contains dietary fiber, protein, and antioxidant compounds	Yilmaz (2020)

NI: no information.

Chili and pepper byproducts have been used as functional ingredients in dairy products. Šeregelj et al. (2019) used encapsulated red pepper waste extracts (obtained from the pepper processing industry in Serbia) to improve shelf-life, nutritional quality, and yogurt color. Fortified yogurt retained carotenoids (71.4%) and phenolic compounds (123%) with good sensory attributes (color, appearance, and overall acceptance). Moreover, the addition of encapsulated red pepper seeds waste extract increased the antioxidant properties of yogurt and positively influenced the number of lactic acid bacteria during 21 days of storage.

Regarding meat products, Martínez et al. (2006) informed that adding red sweet pepper (*C. annum*) powder (at 0.1, 0.5, or 2% of total weight) improved the color and prevented lipid oxidation and microbial growth of sausages packaged in a modified atmosphere (80% oxygen and 20% carbon dioxide) and stored for 16 days at 2°C without affecting their sensory attributes, associated with the antioxidant and antimicrobial properties of pepper powder phytochemicals. However, these results were in a dose-dependent response, where the best results were observed at 2% of pepper powder. Similar trends were informed in Spanish-type sausage after adding paprika at 30 g/kg (Fernández-López et al. 2002). Additionally, it has been reported that adding *C. annum* powder to mutton nuggets significantly improved sensory attributes (appearance and flavor) (Mendiratta, Shinde and Mane 2013). Bázan-Lugo et al. (2012) reported that paprika powder effectively enhanced the color and prevented lipid oxidation of nitrite-reduced meat batters. Yusop et al. (2012) used paprika oleoresin (3 g/100 mL of milk) to marinate poultry meat. They mentioned that the carrier systems improved the color quality and sensory attributes of chicken meat due to oleoresin phytochemicals.

Moreover, paprika is widely used to improve color in several dehydrated foods (Koncsek et al. 2019). Kilincceker and Kurt (2016) investigated the potential use of pepper seed flour as a food additive in fried products. They informed that pepper seed flour improved the quality of deep-fat fried chicken meatballs, decreasing frying time and fat absorption. Moreover, pepper seed flour improved fried products´ appearance, color, odor, flavor, and texture.

Additionally, Yilmaz, Bostanci and Ok (2019) developed a new spreadable paste (chocolate and molasses type) using capia pepper (*C. annum*) seed flour. This product is rich in soluble and insoluble dietary fiber, tocopherols, linoleic acid, and sterols compared to commercially available similar products. Yilmaz (2020)· using pepper seed flour (20% in weight), developed a spicy sauce with a significant contribution of dietary fiber, protein, and antioxidant compounds. Moreover, spicy sauce contained essential amino acids (methionine, glycine, and alanine) and minerals (Ca, P, Na, and P).

According to these data, valorization of food processing waste and byproducts is an essential part of the food industry; these *Capsicum*-based byproducts can be used to develop new functional food products.

Feed Applications

The feed industry for poultry production has been rapidly growing to meet egg and meat supplies (Zhao, Zhang and Zhang 2010). Chili and pepper byproducts have been investigated as a natural supplement for poultry nutrition (Table 6).

Afolabi et al. (2017) informed that the broilers supplemented with hot red pepper waste powder (from 0.1 to 0.3% of feed) exhibited a weight gain compared to birds fed the control diet. Furthermore, an increase has been reported in the carcass of chicken broiler supplemented with hot red pepper powder waste at 1g/kg of feed (Goran, Abo and Diakalia 2016). Moreover, the intestinal villi length of broilers supplemented with hot red pepper powder significantly improved due to the reduction of the crypt depth (Soliman and Shalafifi 2020). These effects suggested that the hot red pepper powder can modify the morphology of the small intestine of birds (Abd et al. 2022).

Table 6. Uses of chili and pepper byproducts in the poultry industry

Source	Application	Treatment	Results	Ref.
C. annum	Broiler skin pigmentation	Pigments at different levels of 4.5, 9, and 13.5 ppm of *Capsicum* xanthophylls were added to the basal diet	Natural pigments exhibited skin coloration in a dose-dependent response	Calafat, Vilà and Fontgibell (2005)
Hot red pepper (*C. annum*)	Poultry production	Pepper powder at 0.1, 0.2, or 0.3% in total diet	Pepper powder enhanced broilers weight gain	Afolabi et al. (2017)
Hot red pepper (*C. annum*)	Poultry production	Diets were supplemented with 1g of pepper powder per kg of fed	Supplementation improved the intestinal health of birds	Goran, Abo and Diakalia (2016)
Hot red pepper (*C. annum*)	Poultry production	Pepper powder at 0.5, 1, or 2% in total diet	Supplementation increased villi length and reduced crypth size of birds	Soliman and Shalafifi (2020)
C. frutescens powder	Poultry nutrition	Capsicum powders (5% v/v) were fermented with yeast strains for 24 h	The protein produced contains essential amino acids in good quantities	Zhao, Zhang and Zhang (2010)
Hot pepper (*C. annum*)	Japanese quail egg quality	HPCAWP at 1, 2, or 4 g/kg diet was added to the basal diet	HPCAWP improved egg quality parameters	Filik (2020)

HPCAWP: hot *C. annum* waste powder.

Additionally, poultry product pigmentation is widely obtained by adding carotenoids (natural or synthetic) to the diet. Thus, the efficiency of natural red pigments of *Capsicum annum* waste was investigated for broiler skin pigmentation by Calafat et al. (2005), who informed that broilers supplemented with *C. annum* pigments exhibited yellow-orange skin coloration in a dose-dependent response (4.5, 9, and 1.5 ppm of *Capsicum* xanthophylls), similar to those of synthetic canthaxanthin.

Zhao, Zhang, and Zhang (2010) produced a single-cell protein for poultry-fed purposes by a non-conventional fermentation method (starter culture growth in YM broth and supplemented with 5% v/v of capsicum powder) using *C. frutescens* wastes generated during capsanthin processing. The authors informed that the single-cell protein contained essential amino acids such as methionine, valine, isoleucine, leucine, phenylalanine, histidine, and arginine with higher lysine contents and threonine, improving the protein quality of poultry feed.

For its part, Filik (2020) investigated the effect of dietary hot pepper (*C. annum*) waste powder supplementation (HPCAWP at 1, 2, or 4 g/kg diet) on egg production of Japanese quail layers. They informed that the dietary supplementation of HPCAWP had significant effects on laying egg weight, total egg yield, albumen weight, and egg shape and color in a dose-dependent manner, where best results were observed at doses of 2 g/kg diet.

In general, chili and pepper byproducts can be used for poultry nutrition, improving egg and skin quality, associated with the bioactive compounds in *Capsicum* byproducts. Moreover, these byproducts improved broiler performance due to their antimicrobial activities that may prevent intestinal diseases in birds (Anaya-Esparza et al. 2021, Filik 2020).

Environmental Applications: Bioremediation

Water pollution is a global concern that affects human health and the environment. Therefore, the potential use of chili seeds of *C. annum* as a low-cost bioadsorbent to remove heavy metals from aqueous solutions has been investigated (Medellin-Castillo et al. 2017, Alves et al. 2021). Medellin-Castillo et al. (2017) informed that chili seeds (dried at 80°C/24 h, crushed, and sieved in particle size of 1.1 mm) exhibited good adsorption capacity towards Cd(II) and Pb(II) ions; moreover, these effects were dependent on the pH and temperature conditions. The authors explain that the negatively charged chili seeds' surface and the positively charged surface of heavy metals in the aqueous solution favored the chemical reaction, possibly by ion-exchange mechanism and the electrostatic attraction, and π-cation interactions; nonetheless, chili seeds showed up to five-fold time affinity for Pb(II) removal than Cd(II). Similarly, Alves et al. (2021) reported that *C. baccatum* wastes exhibited bioabsorption properties and could be used to remove Cr^{3+} from aqueous solutions; however, the effect was pH-dependent, with better results at pH 6.

Textile Applications as a Natural Dye

Ksibi et al. (2015) investigated the potential use of pepper byproducts extracts as a dye for colored textile and clothing production. Byproducts were generated in the production of canned foods from Tunisia, and they contain a mixture of seeds (15.9%), leftovers (34.7%), and stems (49.4%). The authors mentioned that fabrics dyed (at acid pH and 60°C for 60 min) exhibited a red-yellow color, associated with the interaction between pigments with the fibers of wool fabrics; moreover, the dye impregned in wool substrate showed antimicrobial activity against *Pseudomona aeruginosa* and *Staphylococcus aureus*, attributed to the functional groups of phenolic compounds of *Capsicum* byproducts. These findings indicate that using pepper waste/byproducts can develop functional textile materials.

Bioenergetics

Industrialized biofuels currently use food waste and byproducts as an initial feedstock to produce biodiesel as an alternative to their revalorization. In the Republic of Korea, ≥17000 tons of pepper seeds are generated each year during the production of red pepper powder (a traditional Korean ingredient cooking). Lee et al. (2017) reported that it is possible to synthesize biodiesel using waste pepper seeds without previous lipid isolation steps by thermal-induced transmethylation on a mesoporous silica-based material. The peppers seeds contained 27% lipids (C16:0, C16:1, C18:0, C18:1, C18:2, and C20:0), and these lipids could be converted into biodiesel (91.7% of yield) by *in-situ* transmethylation at 390°C for 1 min. According to the authors, pepper seed waste are a desirable feedstock for diesel production, and this proposed method exhibited high tolerance against impurities presented in the pepper seeds waste (Lee et al. 2017). Ponnam et al. (2021) explained that *C. annum* waste exhibited adequate kinetic parameters for designing and developing pyrolyzers, a key factor for converting pepper waste biomass to biodiesel.

Capping and Stabilizing Agent for Green Synthesis of Nanoparticles

In recent years, the green synthesis of nanoparticles exhibited a growing trend because it is an easy, rapid, eco-friendly, and low-cost alternative. Velsankar et al. (2021) used leaf extract of *C. frutescens* as a capping and stabilizing agent during the synthesis of CuO nanoparticles. The nanoparticles exhibited spherical and rectangular shapes ranging from 20 to 40 nm; moreover, they demonstrated antimicrobial (against *Bacillus anthracis*, *Listeria monocytogenes*, and *Klebsiella pneumoniae*) and antioxidant activities in a dose-dependent response. According to the authors, capping and stabilizing properties of leaf *C. frutescens* during CuO nanoparticles synthesis are attributed to the reaction of phytochemicals and copper nitrate precursor (forming copper hydroxide by a bioreduction stage), facilitating the formation of CuO during the calcined process. In this context, natural extracts, mainly from *Capsicum*-based byproducts, could be a suitable and non-toxic alternative to synthesizing inorganic nanoparticles.

Conclusion and Perspectives

The revalorization of chili and pepper byproducts is an area of opportunity for research through technological applications, adding value to productive processes (harvest, handling, postharvest, and processing), reducing food losses and waste. Significant evidence indicates that different *Capsicum* spp. byproducts (defective fruits, stem, seeds, placenta, peduncles, leaves, and oleoresin) are a good source of bioactive compounds. The main phytochemicals reported include phenolic compounds, flavonoids, carotenoids, capsaicinoids, fatty acids, and essential oils; also, bioactive peptides are reported. Moreover, *in vitro* and *in vivo* studies (antioxidant, anti-inflammatory, neuroprotective, antimicrobial, and α-inhibitory properties) supported the traditional medicinal uses of chilies and peppers byproducts. On the other hand, *Capsicum* byproducts can be used as an antimicrobial agent for food purposes, as an ingredient for developing functional foods for humans and animals, as a bioadsorbent for removal of heavy metal ions from aqueous media, as a natural dye for textile and clothing applications, for biodiesel production, and as a stabilizing agent for nanoparticle development by green synthesis. The increasingly trendy use of *Capsicum* byproducts is related to their phytochemical profile and biological activities. However, further studies are still needed to promote the technological benefits of these valuable food byproducts.

References

Aşkin, R. 2018. "Preservation of sweet red pepper paste quality: effect of packing paterial, ozone gas and protective agent use." *Food Science and Technology* 38 (4):698-703. doi:10.1590/1678-457x.13917.

Abd, E. H., E. Mohamed, T. E. S. Mohamed, R. E. Ahmed, R. G. Ahmed, M. N. Maha, M. S. Ahmed, M. E. T. Amira, E. T. Ayman, M. S. Heba, and A. E. T. Khaled. 2022. "hot red pepper powder as a safe alternative to antibiotics in organic poultry feed: An updated review." *Poultry Science* 101 (4):101684. doi:10.1016/j.psj.2021.101684.

Afolabi, K. D., K. N. Eugenes, M. A. Olufemi, and O. Rotimi. 2017. "Hot red pepper (*Capsicum annum* L.) meal enhanced the immunity, performance and economy of broilers fed in phases." *Journal of Biology, Agriculture and Healthcare*, 7 (8):1-7.

Akhtar, A, A. Waqas, and k. Nauman. 2021. "Phytochemical constituents and biological properties of domesticated *Capsicum* species: A review." *Bioactive Compounds in Health and Disease* 4 (9):201-225. doi:10.31989/bchd.v4i9.837

Alam, Md A., F. S. Nur, H. M. Nor, A. B. Noor, A. M. Kamarul, A. Nadiawati, A. Farzad, and P. Asaduzzaman. 2018. "Evaluation of antioxidant compounds, antioxidant activities and capsaicinoid compounds of chili (*Capsicum* sp.) germplasms available in Malaysia." *Journal of Applied Research on Medicinal and Aromatic Plants* 9:46-54. doi:10.1016/j.jarmap.2018.02.001.

Alam, Md. A., M. Saleh, Md. M. Gazi, T. A. Nadirah, A. Farzad, M. M. Rahman, K. R. Siddhartha, S. J. Abdul, and M. Z. Alam. 2020. "Evaluation of phenolics, capsaicinoids, antioxidant properties, and major macro-micro minerals of some hot and sweet peppers and ginger land-races of Malaysia." *Journal of Food Processing and Preservation* 44 (6):1-11. https://doi.org/10.1111/jfpp.14483.

Albrecht, E., D. Zhang, A. D. Mays, R. A. Saftner, and J. R. Stommel. 2018. "Genetic diversity in *Capsicum baccatum* is significantly influenced by its ecogeographical distribution." *Nature* 16 (68):1-15.

Allemand, A., B. Franco, A. Rigon, S. Moreno, P. R. Torres, and G. Gosmann. 2016. "Red pepper (*Capsicum baccatum*) extracts present anti-inflammatory effects *in vivo* and inhibit the production of TNF-α and NO *in vitro*." *Journal of Medicinal Food* 19 (8):759-767. doi:10.1089/jmf.2015.0101.

Alves, A., D. Maia, A. Henrique, C. Hitomi, S. Tonello, and L. Cardoso. 2021. "*Capsicum baccatum* Var. pendulum waste used to capture Cr3+ from aqueous solutions," *Chemistry Research Journal* 6 (1):159-171.

Anaya-Esparza, L. M, Z. Villagrán, O. Vázquez-Paulino, F. Ascencio, and A. Villarruel-López. 2021. "Bell peppers (*Capsicum annum* L.) losses and wastes: Source for food and pharmaceutical applications." *Molecules* 26 (17):1-23. doi:10.3390/molecules26175341.

Askin, B., and O. Y. Sercan. 2021. "Bioactive compounds, antioxidant potential and color properties of dried red pepper (*Capsicum annuum* L.)." *International Journal of Innovative Approaches in Agricultural Research* 5 (2):158-171. doi:10.29329/ijiaar.2021.358.1.

Assefa, S. T., E. Y. Yang, G. Asamenew, H. W. Kim, M. C. Cho, and J. Lee. 2021. "Identification of α-glucosidase inhibitors from leaf extract of pepper (*Capsicum* spp.) through metabolomic analysis." *Metabolites* 11 (10). doi:10.3390/metabo11100649.

Azabou, S., F. B. Taheur, M. Jridi, and M. Bouaziz. 2017. "Discarded seeds from red pepper (*Capsicum annum*) processing industry as a sustainable source of high added-value compounds and edible oil." *Environmental Science and Pollution Research*, 22196-22203. doi:10.1007/s11356-017-9857-9.

Baenas, N., M. Belović, N. Ilic, D. A. Moreno, and C. García-Viguera. 2019. "Industrial use of pepper (*Capsicum annum* L.) derived products: Technological benefits and biological advantages." *Food Chemistry* 274:872-885. doi:10.1016/j.foodchem.2018.09.047.

Balaky, H. H., Y. Galali, E. Karaoğul, E. Altuntaş, N. H. Rasol, and A. A. Mustafa. 2021. "Study of microwave technique in hot pepper seed by different solvents with an assessing antioxidant and antibacterial properties." *Polytechnic Journal* 11 (1):80-86. doi:10.25156/ptj.v11n1y2021.

Bázan-Lugo, E., I. García-Martínez, R. H. Alfaro-Rodríguez, and A. Totosaus. 2012. "Color compensation in nitrite-reduced meat batters incorporating paprika or tomato paste." *Journal of the Science of Food and Agriculture* 92 (8):1627-1632. doi:10.1002/jsfa.4748.

Bello, I., B. E. Boboye, and F. A. Akinyosoye. 2015. "Phytochemical screening and antibacterial properties of selected Nigerian long pepper (*Capsicum frutescens*) fruits." *African Journal of Microbiology Research* 9 (38):2067-2078. doi:10.5897/ajmr2014.7286.

Calafat, F. A, B. Vilà, and A. Fontgibell. 2005. "Efficiency of natural red pigments of *Capsicum annum* in broiler pigmentation," *In XVIIth European Symposium on the Quality of Poultry Meat*, 2005, pp. 118-123.

Cárdenas-Castro, A. P., G. C. Perales-Vázquez, L. A. de la Rosa, V. M. Zamora-Gasga, V. M. Ruiz-Valdiviezo, E. Álvarez-Parrilla, and S. G. Sáyago-Ayerdi. 2019. "Sauces: An undiscovered healthy complement in Mexican cuisine." *International Journal of Gastronomy and Food Science*, 17:100154 doi:10.1016/j.ijgfs.2019.100154.

Cervantes-Hernández, F., P. Alcalá-González, O. Martínez, and J. J. Ordaz-Ortiz. 2019. "Placenta, pericarp, and seeds of Tabasco chili pepper fruits show a contrasting diversity of bioactive metabolites." *Metabolites* 9 (10):206. doi:10.3390/metabo9100206.

Chávez-Mendoza, C., E. Sánchez, E. Muñoz-Márquez, J. P. Sida-Arreola, and M. A. Flores-Córdova. 2015. "Bioactive compounds and antioxidant activity in different grafted varieties of bell pepper." *Antioxidants* 4 (2):427-46. doi:10.3390/antiox4020427.

Chel-Guerrero, L. D., J. E. Oney-Montalvo, and I. M. Rodríguez-Buenfil. 2021. "Phytochemical characterization of byproducts of Habanero pepper grown in two different types of soils from Yucatán, Mexico." *Plants* 10 (4):779 doi:10.3390/plants10040779.

Chel-Guerrero, L. D., G. Castañeda-Corral, M. López-Castillo, M. Scampicchio, K. Morozova, J. E. Oney-Montalvo, G. Ferrentino, J. J. Acevedo-Fernández, and I. M. Rodríguez-Buenfil. 2022. "*In vivo* anti-Inflammatory effect, antioxidant activity, and polyphenolic content of extracts from *Capsicum chinense* byproducts." *Molecules* 27 (4):1323. doi:10.3390/molecules27041323.

Cocan, I., M. Negrea, A. Cozma, E. Alexa, M. A. Poiana, D. Raba, C. Danciu, I. Popescu, A. I. Cadariu, D. Obistioiu, and I. Radulov. 2021. "Chili and sweet pepper seed oil used as a natural antioxidant to improve the thermo-oxidative stability of sunflower oil." *Agronomy* 11 (12):1-13. doi:10.3390/agronomy11122 579.

Cortés-Ferré, H. E., D. Guajardo-Flores, G. Romero-de la Vega, and J. A. Gutiérrez-Uribe. 2021. "Recovery of capsaicinoids and other phytochemicals involved with TRPV-1 receptor to re-valorize chili pepper waste and produce nutraceuticals." *Frontiers in Sustainable Food Systems* 4:1-17. doi:10.3389/fsufs. 2020.588534.

Cvetković, T. 2022. "Quality of pepper seed byproducts: A review." *Foods*, 11 (5):1-16. https://doi.org/ 10.3390/foods11050748.

Cvetković, T., J. Ranilović, D. Gajari, H. Tomić-Obrdalj, D. Šubarić, T. Moslavac, A. M. Cikoš, and S. Jokić.. 2020. "Podravka and Slavonka varieties of pepper seeds (*Capsicum annuum* L.) as a new source of highly nutritional edible oil." *Foods* 9 (9):1262. doi:10.3390/foods9091262.

Danza, A., M. Mastromatteo, F. Cozzolino, L. Lecce, V. Lampignano, J. Laverse, and M. A. del Nobile. 2014. "Processing and characterization of durum wheat bread enriched with antioxidant from yellow pepper flour." *LWT - Food Science and Technology* 59 (1):479-485. doi:10.1016/j.lwt.2014.06.001.

de Sá Mendes, N., L. C. Favre, G. Rolandelli, C. dos Santos., É. C. B. A. Gonçalves, and M. P. Buera. 2020. "Flour from 'fruits and vegetables' waste with addition of a South-American pepper (*Capsicum baccatum*) proposed as food ingredient." *International Journal of Food Science and Technology* 55 (3):1230-1237. doi:10.1111/ijfs.14358.

de Sá Mendes, N., P. P. S. Coimbra, M. C. B. Santos, L. C. Cameron, M. S. L. Ferreira, M. P. Buera, and É. C. B. A. Gonçalves. 2020. "*Capsicum pubescens* as a functional ingredient: Microencapsulation and phenolic profiling by UPLC-MSE." *Food Research International* 135:109292. doi:10.1016/j.foodres.2020.109292.

Dias, G. B., V. Moreira, U. Zottich, S. Ferreira, A. O. Carvalho, R. Rodrigues, O. L. Tavares-Machado, K. T. Sales, A. T. S. Ferreira, J. Perales, and M. Da Cunha. 2013. "Isolation, characterization and antifungal activity of proteinase inhibitors from *Capsicum chinense* Jacq. seeds." *Protein Journal* 32 (1):15-26. doi:10.1007/s10930-012-9456-z.

Escalera-Ordaz, A. K., H. Guillén-Andrade, M. B. N. Lara-Chávez, C. Lemus-Flores, J. G. Rodríguez-Carpena, and R. Valdivia-Bernal. 2019. "Characterization of cultivated varieties of *Capsicum pubescens* in Michoacán, Mexico." *Mexican Journal of Agricultural Sciences* 23:239-251.

European Commission, and Technical Report. 2010. *Preparatory study on food waste across Eu 27*. October. Vol. 33.

Fabela-Morón, M. F., J. C. Cuevas-Bernardino, T. Ayora-Talavera, and N. Pacheco. 2020. "Trends in capsaicinoids extraction from Habanero chili pepper (*Capsicum chinense* Jacq.): Recent advanced techniques." *Food Reviews International* 36 (2):105-134. doi:10.1080/87559129.2019.1630635.

FAO The State of Food and Agriculture 2019. *Moving forward on food loss and waste reduction*; Rome, Italy, 2019.

Fernández-López, J., J. A. Pérez-Álvarez, E. Sayas-Barberá, and F. López-Santoveña. 2002. "Effect of paprika (*Capsicum annum*) on color of Spanish-type sausages during the resting stage." *Journal of Food Science* 67 (6):2410-2414. doi:10.1111/j.1365-2621.2002.tb09562.x.

Filik, G. 2020. "The effects of dietary hot pepper *Capsicum annuum* waste powder supplementation on egg production traits of Japanese quail layers." *Ciencia Rural* 50 (9):1-8. https://doi.org/10.1590/0103-8478cr20190945

Firatligil, E., and O. Evranuz. 2010. "Response Surface Methodology for Protein Extraction Optimization of Red Pepper Seed (*Capsicum frutescens*)." *LWT - Food Science and Technology* 43 (2): 226-231. doi:10.1016/j.lwt.2009.08.017.

Goran, E. N., P. Abo, and B. Djakalia. 2016. "Effect of *Capsicum annum*, *Allium sativum*, and *Thymus vulgaris* on the zootechnic performance of broiler in growth phase." *Global Journal of Advanced Research* 3 (12):1070-1077.

Echave, J., A. G. Pereira, M. Carpena, M. A. Prieto, and J. Sinal-Gandara. 2012. "*Capsicum* seeds as a source of bioactive compounds: Biological properties, extraction systems, and industrial application." *IntechOpen* 13. http://dx.doi.org/10.5772/55358.

Hazekawa, M., Y. Hideshima, K. Ono, T. Nishinakagawa, T. Kawakubo-Yasukochi, T. Takatani-Nakase, and M. Nakashima. 2017. "Anti-inflammatory effects of water extract from bell pepper (*Capsicum annuum* L. Var. grossum) leaves *in vitro*." *Experimental and Therapeutic Medicine* 14 (5):4349-55. doi:10.3892/etm.2017.5106.

Herrera-Pool, E., A. L. Ramos-Díaz, M. A. Lizardi-Jiménez, S. Pech-Cohuo, T. Ayora-Talavera, J. C. Cuevas-Bernardino, U. García-Cruz, and N. Pacheco. 2021. "Effect of solvent polarity on the ultrasound assisted extraction and antioxidant activity of phenolic compounds from Habanero pepper peaves (*Capsicum chinense*) and its identification by UPLC-PDA-ESI-MS/MS." *Ultrasonics Sonochemistry* 76:105658. doi:10.1016/j.ultsonch.2021.105658.

Jarret, R. L., E. Baldwin, B. Perkins, R. Bushway, and K. Guthrie. 2007. "Diversity of fruit quality characteristics in *Capsicum frutescens*." *HortScience* 42 (1):16-19. doi:10.21273/hortsci.42.1.16.

Jasna, M., K. Žarko, V. Milan, I. Marko, and R. Jelena. 2018. "Possibilities for utilization of dietary fiber-rich supplement from pepper (*Capsicum annum* L.) processing waste in bakery products." *Journal of Engineering & Processing Managment* 10 (1):28-33. doi:10.7251/JEPM1810028M.

Jiménez, D., R. Vardanega, F. Salinas, C. Espinosa-Álvarez, W. Bugueño-Muñoz, J. Palma, A. A. Meireles, M. C. Ruíz-Domínguez, and P. Cerezal-Mezquita. 2021. "Effect of drying methods on biorefinery process to obtain capsanthin and phenolic compounds from *Capsicum annuum* L." *Journal of Supercritical Fluids* 174:105241. doi:10.1016/j.supflu.2021.105241.

Kavuncuoglu, H., T. D. Capar, and H. Yalçin. 2021. "Total phenolic contents, antioxidant activities and antioxidant capacities of some selected pepper (*Capsicum annuum* L.) pulps and seeds." *European Journal of Science and Technology* 21:581-586. doi:10.31590/ejosat.793148.

Sanatombi K., and Rajkumari, S. 2020. "Effect of processing on quality of pepper: A review." *Food Reviews International* 36 (6). Taylor & Francis: 626-43. doi:10.1080/87559129.2019.1669161.

Kilincceker, O., and S. Kurt. 2016. "The effects of red pepper seed flour on the some quality characteristics of deep fat fried chicken meatballs." *Advances in Food Science* 38(2):76-81.

Koncsek, A., H. G. Daood, Z. H. Horváth, M. Fekete, A. Véha, and L. Helyes. 2019. "Improvement of antioxidant content and color stability in spice paprika powder by rosemary extract supplementation." *Journal of Food Processing and Preservation* 43 (8):1-10. doi:10.1111/jfpp.14000.

Ksibi, I. E., R. B. Slama, K. Faidi, M. B. Ticha, and M. F. M'henni. 2015. "Mixture approach for optimizing the recovery of colored phenolics from red pepper (*Capsicum annum* L.) byproducts as potential source of natural dye and assessment of its antimicrobial activity." *Industrial Crops and Products* 70:34-40. doi:10.1016/j.indcrop.2015.03.017.

Kumar, S., K. J. Kiran, J. Sahu, M. Rudrapal, and M. Debnath. 2012. "A short review on *Capsicum chinense* Jacq." *Journal of Herbal Medicine and Toxicology* 6 (2):7-10.

Lee, J., J. Kim, Y. S. Ok, and E. E. Kwon. 2017. "Rapid biodiesel synthesis from waste pepper seeds without lipid isolation step." *Bioresource Technology* 239:17-20. doi:10.1016/j.biortech.2017.05.011.

Liu, Y., Y. Chen, Y. Wang, J. Chen, Y. Huang, Y. Yan, Y. L. Li, Z. Li, Y. Ren, and Y. Xiao. 2020. "Total phenolics, capsaicinoids, antioxidant activity, and α-glucosidase inhibitory activity of three varieties of pepper seeds." *International Journal of Food Properties* 23 (1):1016-1035. doi:10.1080/10942912.2020.1775646.

Malakar, S., S. Sarkar, N. Kumar, S. Malakar, and R. Jaganmohan. 2018. "Studies of biochemical characteristics and identification of active phyto-compounds of King chili (*Capsicum chinense* Jacq.) using GC-MS." *Journal of Pharmacognosy and Phytochemistry* 7 (3):3100-3104.

Mármol, I., J. Quero, R. Ibarz, P. Ferreira-Santos, J. A. Teixeira, C. M. R. Rocha, M. Pérez-Fernández, S. García-Juiz, J. Osada, O. Martín-Belloso, and M. J. Rodríguez-Yoldi. 2021. "Valorization of agro-food byproducts and their potential therapeutic applications." *Food and Bioproducts Processing* 128: 247-258. doi:10.1016/j.fbp.2021.06.003.

Martínez, L., I. Cilla., J. A. Beltrán, and P. Roncalés. 2006. "Effect of *Capsicum annuum* (red sweet and Cayenne) and *Piper nigrum* (black and white) pepper powders on the shelf life of fresh pork sausages packaged in modified atmosphere." *Journal of Food Science* 71 (1):1-6. doi:10.1111/j.1365-2621.2006.tb12405.x.

Medellín-Castillo, N. A., E. Padilla-Ortega, M. C. Regules-Martínez, and C. Carranza-Álvarez. 2017. *"Single and competitive adsorption of Cd (II) and Pb (II) ions from aqueous solutions onto industrial chili seeds (Capsicum annuum) waste"* 27(2):61-69. doi:10.1016/j.serj.2017.01.004.

Mendiratta, S. K., A. T. Shinde, and B. G. Mane. 2013. "Effect of added vegetable (carrot, radish and capsicum) as functional ingredients in mutton nuggets." *Journal of Meat Science and Technology* 1 (2):71-76.

Nascimento, D. O. K., V. Juarez, T. Saldanha, J. Lucena, J. L. Barbosa, and M. I. Martins-Jacintho. 2012. "Chemical characterization and nutritional information of organic cambuci pepper jelly (*Capsicum baccatum* L.)" *Revista Verde de Agroecologia e Desenvolvimento Sustentável* 7:283-288.

Nascimento, P. L. A., C. E. S. T. Nascimento, N. S. M. Ramos, G. R. Silva, J. E. Galindo-Gomes, R. E. A. Falcão, K. A. Moreira, A. L. F. Porto, and T. M. S. Silva. 2014. "Quantification, antioxidant and antimicrobial activity of phenolics isolated from different extracts of *Capsicum frutescens* (Pimenta malagueta)." *Molecules* 19 (4):5434-5447. doi:10.3390/molecules19045434.

Oboh, G., and O. O. Ogunruku. 2010. "Cyclophosphamide-induced oxidative stress in brain: Protective effect of hot short pepper (*Capsicum frutescens* L. Var. abbreviatum)." *Experimental and Toxicologic Pathology* 62 (3):227-233. doi:10.1016/j.etp.2009.03.011.

Özcan, M. M., N. Uslu, N. S. Efe, A. N. Erdem, Z. Değerli, D. A. Kulluk, and N. C. Sağlık. 2021. "Effect of thermal processing on the bioactive compounds and color parameters of types of three sweet pepper." *Journal of Food Processing and Preservation* 45 (7): e15661. doi:10.1111/jfpp.15661.

Padalino, L., M. Mastromatteo, L. Lecce, F. Cozzolino, and M. A. Del Nobile. 2013. "Manufacture and characterization of gluten-free spaghetti enriched with vegetable flour." *Journal of Cereal Science* 57 (3):333-342. doi:10.1016/j.jcs.2012.12.010.

Pérez-Vázquez, M. A. K., Y. Pacheco-Hernández, E. Lozoya-Gloria, C. Mosso-González, S. A. Ramírez-García, O. Romero-Arenas, and N. Villa-Ruano. 2022. "Peppermint essential oil and its major volatiles as protective agents against soft rot caused by *Fusarium sambucinum* in cera pepper (*Capsicum pubescens*)." *Chemistry and Biodiversity* 19 (1):e202100991. doi:10.1002/cbdv.202100835.

Ponnam, V., P. Ghodke, S. Tondepu, and R. N. Mandapati. 2021. "Thermal behaviour kinetic modeling of *Capsicum annuum* waste biomass using an iso-conversion method." *Journal of Thermal Engineering* 7 (2):18-29. doi:10.18186/thermal.865555.

Raak, N., C. Symmank, S. Zahn, J. Aschemann-Witzel, and H. Rohm. 2017. "Processing - and product - related causes for food waste and implications for the food supply chain." *Waste Management* 61:461-72. doi:10.1016/j.wasman.2016.12.027.

Ramandeep K., and K. Kamaljit. 2020. "Effect of processing on color, rheology and bioactive compounds of different sweet pepper purees." *Plant Foods for Human Nutrition* 75 (3):369-375. doi:10.1007/s11130-020-00824-0.

Razola-Díaz, M. C., A. M. Gómez-Caravaca, J. López-de Andrés, A. Voltes-Martínez, A. Zamora, G. M. Pérez-Molina, D. J. Castro, J. A. Marchal, and V. Verardo. 2022. "Evaluation of phenolic compounds and pigments content in yellow bell pepper wastes." *Antioxidants* 11 (3):557. doi:10.3390/antiox11030557.

Reguengo, L. M., M. K. Salgaço, K. Sivieri, and M. R. Maróstica-Júnior. 2022. "Agro-industrial byproducts: Valuable sources of bioactive compounds." *Food Research International* 152:110871. doi:10.1016/j.foodres.2021.110871.

Sandoval-Castro, C. J., M. Valdez-Morales, D. Oomah, R. Gutiérrez-Dorado, S. Medina-Godoy, and G. Espinosa-Alonso. 2017. "Bioactive compounds and antioxidant activity in scalded jalapeño pepper industrial byproduct (*Capsicum annuum*)." *Journal of Food Science and Technology* 54 (7):1999-2010. doi:10.1007/s13197-017-2636-2.

Šeregelj, V., V. T. Šaponjac, S. Lević, A. Kalušević, G. Ćetković, J. Čanadanović-Brunet, V. Nedović, S. Stajčić, J. Vulić, and A. Vidaković. 2019. "Application of encapsulated natural bioactive compounds from red pepper waste in yogurt." *Journal of Microencapsulation* 36 (8):704-714. doi:10.1080/02652048.2019.1668488.

Simonovska, J., M. Škerget, Ž. Knez, M. Srbinoska, Z. Kavrakovski, A. Grozdanov, and V. Rafajlovsk. 2016. "Physicochemical characterization and bioactive compounds of stalk from hot fruits of *Capsicum annuum* L." *Macedonian Journal of Chemistry and Chemical Engineering* 35 (2):199-208. doi:10.20450/mjcce.2016.944.

Soliman, N., and S. AlAfifi. 2020. "The productive performance, intestinal bacteria and histomorphology of broiler chicks feed diets containing hot red pepper." *Egyptian Poultry Science* 40 (I): 345-357.

Soumya, S. L., and B. R. Nair. 2012. "Antifungal efficacy of *Capsicum frutescens* L. extracts against some prevalent fungal strains associated with groundnut storage." *Journal of Agricultural Technology* 8 (2):739-750.

Spiller, F., M. K. Alves, S. M. Vieira, T. A. Carvalho, C. E. Leite, A. Lunardelli, J. A. Poloni, F. Q. Cunha, and J. R. de Oliveira. 2010. "Anti-Inflammatory effects of red pepper (*Capsicum baccatum*) on carrageenan- and antigen-induced inflammation." *Journal of Pharmacy and Pharmacology* 60 (4):473-478. doi:10.1211/jpp.60.4.0010.

Tae-I., "Patents preparation method of natural condiments prepared by using geum kang pine tree, South Korea Patent KR20080010772A, June 19, 2008.

Ukwuani, A. N., and I. B. Hassan. 2018. "In *vitro* anti-Inflammatory activity of hydromethanolic seed, fruit and leave extracts of *Capsicum chinense* (red pepper)." *European Journal of Biomedical and Pharmaceutical Sciences* 2 (2):57-65.

Valdez-Morales, M., C. J. Sandoval-Castro, R. Gutiérrez-Dorado, J. A. Mendoza-Espinoza, and S. Medina-Godoy. 2021. "Oil recovery from Jalapeño pepper by products and analysis of the industrial scalding process on its nutraceutical potential." *Waste and Biomass Valorization* 12 (8):4475-87. doi:10.1007/s12649-020-01323-8.

Vázquez-Flores, A. A., O. Góngora-Pérez, I. Olivas-Orduña, O. A. Muñoz-Bernal, P. Osuna-Ávila, J. Rodrigo-García, L. A. de la Rosa, and E. Álvarez-Parilla. 2020. "Pytochemical profile and antioxidant activity of Chiltepin chili (*Capsicum annuum* Var. Glabriusculum), Sonora, Mexico." *Journal of Food Bioactives* 11:57-65. doi:10.31665/jfb.2020.11237.

Velsankar, K., S. Suganya, P. Muthumari, S. Mohandoss, and S. Sudhahar, 2021. "Ecofriendly green synthesis, characterization and biomedical applications of CuO nanoparticles synthesized using leaf extract of *Capsicum frutescens*." *Journal of Environmental Chemical Engineering* 9 (5):106299. doi:10.1016/j.jece.2021.106299.

Wu, J., K. Blackshaw, J. Cho, N. Koolaji, J. Yun, A. Schindeler, P. Valtchev, and F. Dehghani. 2022. "Recovery of high-value compounds from food byproducts." *In Food Engineering Innovations Across the Food Supply Chain*, edited by Pablo Juliano, Roman Buckow, Minh H. Nguyen, Kai Knoerzer, and Jay Sellahewa, 61-88, Academic Press.

Yilmaz, E., H. Bostanci, and S. Ok. 2019. "Valorization of Capia pepperseed flour-II: Sensory properties and storage stability of the new spreadable pastes." *Waste and Biomass Valorization* 10 (10):3163-3171. doi:10.1007/s12649-018-0337-3.

Yilmaz, E. 2020. "Valorization of Capia pepperseed flour in breakfast sauce production." *Waste and Biomass Valorization* 11 (12):6803-6813. doi:10.1007/s12649-020-00939-0.

Yusop, S. M., M. G. O'Sullivan, M. Preuß, H. Weber, J. F. Kerry, and J. P. Kerry. 2012. "Assessment of nanoparticle paprika oleoresin on marinating performance and sensory acceptance of poultry meat." *LWT - Food Science and Technology* 46 (1):349-355. doi:10.1016/j.lwt.2011.08.014.

Zhang, R. Y., H. M. Liu, Y. X. Ma, and X. D. Wang. 2019. "Effects of roasting on composition of chili seed and storage stability of chili seed oil." *Food Science and Biotechnology* 28 (5):1475-1486. doi:10.1007/s10068-019-00578-9.

Zhao, G., W. Zhang, and G. Zhang. 2010. "Production of single cell protein using waste capsicum powder produced during capsanthin extraction." *Letters in Applied Microbiology* 50 (2):187-91. doi:10.1111/j.1472-765X.2009.02773.x.

Ziegenfuss, S., F. Brinkhaus, J. Greaves. 2000. "Anti-bacterial protein extracts from seeds of marigold and paprika," U. S. Patent 6,086,885, July 11, 2000.

Zimmer, A. R., B. Leonardi, D. Miron, E. Schapoval, J. Rodrigues, and G. Gosmann. 2012. "Antioxidant and anti-Inflammatory properties of *Capsicum baccatum*: From traditional use to scientific approach." *Journal of Ethnopharmacology* 139 (1):228-33. doi:10.1016/j.jep.2011.11.005.

Chapter 5

Coffee Industry Byproducts

César Ozuna[1,2,*]
Rey Castañeda-Rodríguez[1]
and Esaú Oseguera-Sánchez[1]
[1]Posgrado en Biociencias, Universidad de Guanajuato, Irapuato, Guanajuato Mexico
[2]Departamento de Alimentos, Universidad de Guanajuato, Irapuato, Guanajuato Mexico

Abstract

The coffee industry generates a large number of byproducts that, due to their high content in bioactive compounds with antioxidant and antimicrobial capacity, require long times and large amounts of oxygen for their degradation. Thus, in recent decades, various scientific research works have focused on the generation of solutions that would allow for their recovery. The main objective of this chapter is to review the main byproducts generated by the coffee industry (coffee silverskin, pulp, spent coffee grounds, among others), their bioactive compounds, and the main non-conventional techniques used for their extraction. Additionally, this chapter reviews the social and environmental impacts of coffee byproducts. According to the research carried out, the main bioactive compounds present in these byproducts are hydroxycinnamic acids, carotenoids, chlorogenic acids, anthocyanins, polyphenolic compounds, among others. Various investigations have demonstrated the ability of non-conventional extraction techniques (power ultrasound, microwaves, supercritical fluids, among others) to make the extraction processes of bioactive compounds present in coffee byproducts more efficient, profitable, and environmentally friendly. However, the great challenge for these technologies is that they are scalable and affordable for the industry. As for the economic, environmental, and social impact of coffee byproducts, it is important to carry out a complete restructuring in the search for sustainable production processes. Likewise, there is a need to create multidisciplinary strategies that would involve post-production stages that facilitate the bioconversion of coffee byproducts to biomaterials, functional ingredients, and new products in the chemical, pharmaceutical, and food industries.

Keywords: bioactive compounds, coffee silverskin, coffee pulp, coffee spent grounds, extraction, environmental impact

---

* Corresponding Author's Email: cesar.ozuna@ugto.mx.

In: Food Byproducts
Editors: S. Teresita Martín-del-Campo, A. Cardador-Martínez and J. del Pilar Ramírez-Anaya
ISBN: 979-8-88697-612-0

Introduction

Coffee is one of the most popular beverages worldwide, and its consumption has been increasing steadily, presenting high commercial importance in the last 150 years (Castañeda-Rodríguez, Mulík, and Ozuna 2020). During 2021-2022, global coffee production was 167.2 million bags of 60 kg (International Coffee Organization 2022). From an economic point of view, the most important species of coffee in the world are *Coffea arabica* and *Coffea canephora* (Murthy and Madhava Naidu 2012). The coffee beans of both species are very similar once they are toasted, but they differ in taste, aroma, and texture of the final product (Ky et al. 2001; Giacalone et al. 2019). The coffee bean consists of a smooth surface (outer hard skin), pericarp (pulp), mucilage (pectin layer), endocarp (parchment coffee) and silver skin (straw) (Esquivel and Jiménez 2012; Arya et al. 2021), most of which represent a potential coffee processing byproduct.

The coffee industry generates a large amount of waste in its processing (Lachenmeier et al. 2021). Coffee industry byproducts arise from coffee grain processing and brewing (Bondam et al. 2022). These solid and liquid wastes may also contain the same amount of bioactive compounds as the beverage itself (Ozuna et al. 2020). However, the byproducts derived from the production of coffee grains are generally considered agro-industrial residues and may have a negative impact when discarded directly into the environment, representing an important environmental problem for coffee-producing countries (Murthy and Naidu, 2012c; Pedraza-Beltrán et al., 2012). It is estimated that from the annual amount of ground coffee consumed, more than 6 million tons of waste spent on coffee grounds will be generated (Mussatto, Ballesteros, et al. 2011). Therefore, the exploitation of processing byproducts is currently a key issue in several food industries, thanks to their potential (Gemechu 2020; Arya et al. 2021). Once the bioactive compounds are extracted from the byproducts, they can be further used for different purposes, such as functional food ingredients, drugs, nutraceuticals or cosmetic products (Iriondo-Dehond, Iriondo-Dehond, and Del Castillo 2020; Bondam et al. 2022).

The main objective of this chapter is to review the main byproducts generated by the coffee industry (coffee silverskin, pulp, spent coffee grounds, among others), paying special attention to their content in bioactive compounds with antioxidant capacity and the use of non-conventional techniques for their extraction. In addition, this chapter reviews the economic, environmental, and social impacts of the generation of the byproducts of the coffee industry.

Byproducts Generated during Coffee Processing

There are two basic methods of coffee cherry processing, the wet method and the dry method, which vary in complexity and quality of the final product (Mussatto, Machado, et al. 2011; Murthy and Madhava Naidu 2012; Bondam et al. 2022; Arya et al. 2021). Figure 1 shows the byproducts generated by these two methods.

In the dry method, fresh coffee berries are dried directly in the sun (Esquivel and Jiménez 2012; Bondam et al. 2022), constantly moving and furrowed. Coffee berries are considered dry when they produce a sound of the internal coffee grain while being stirred. Usually, the drying process lasts between 12-15 days. Once the coffee husk (dry coffee pulp) dries the remaining green grain is separated from the husk with the aid of a mechanical crusher (Murthy and

Madhava Naidu 2012; Bondam et al. 2022). On the other hand, the wet method uses large amounts of water and a microbial fermentation to remove the mucilage still present on the grains (Mussatto, Machado, et al. 2011; Arya et al. 2021). Fresh coffee berries are mixed with water and fermentation occurs, usually between 12-48 h. Next, a mechanical crusher is used to detach the pulp from the grain. The fermentation process allows the hydrolysis of the mucilage by enzymatic and microbial action (Esquivel and Jiménez 2012). In both methods, the obtained coffee grain undergoes roasting. This processing step varies according to the region but is always time-temperature dependent. After this, the coffee grains are cooled to avoid excessive roasting. Roasted coffee is ground and packaged for distribution (Mussatto, Machado, et al. 2011). The type of processing byproduct is determined by the method used to obtain the coffee grain. If the processing is dry, the main waste generated is the coffee silverskin, the pulp, and the husk. On the other hand, wet processing allows recovery of the skin and the pulp as the first single fraction, the mucilage and soluble sugars in a second fraction, and the husk in a third fraction (Esquivel and Jiménez 2012).

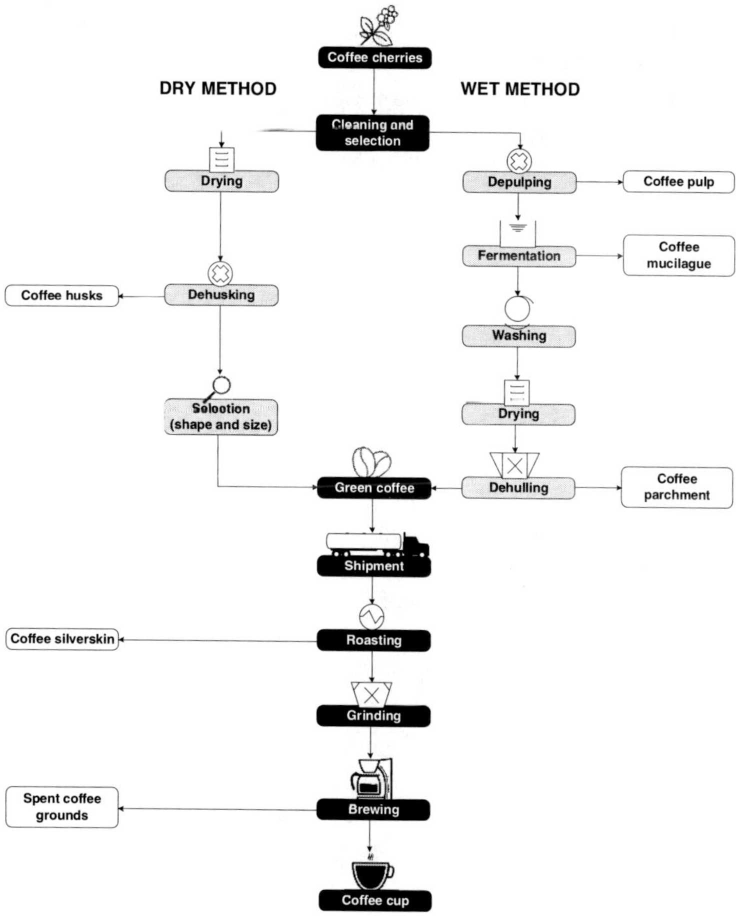

Figure 1. Diagram of the byproducts generated during coffee processing. Adapted from Bondam et al. (2022), Gemechu (2020), Esquivel and Jiménez (2012), and Arya et al. (2021).

Depending on the method of the coffee grain processing or the beverage preparation, various types of solid waste are generated, including coffee silverskin, pulp, parchment, and spent coffee grounds; while liquid waste generated includes water used in pulping and coffee honey, that is, the mucilage released during the fermentation process (Mussatto, Ballesteros, et al. 2011; Bondam et al. 2022; Gemechu 2020; Arya et al. 2021).

The byproducts generated during the processing of the coffee grain represent more than 45% of the coffee cherry weight (Esquivel and Jiménez 2012). The coffee pulp is the first byproduct obtained during the processing of the berries and represents 43.58% of the total wet weight of the fresh fruit. Every year, more than 225 million tons of pulp are generated, and 162,900 tons of fresh pulp are generated per million bags (Cubero-Abarca, Moya, R., Valaret, and Tomazello Filho 2014; Navia P., Velasco M., and Hoyos C. 2011). In contrast, the coffee silverskin, which represents 4.2% of the fresh fruit weight with a caloric capacity of 17.9 MJ/kg, is considered another of the byproducts belonging to the coffee grain process and can be used as fuel (Lee et al. 2021; Navia P., Velasco M., and Hoyos C. 2011).

As for the byproducts generated during the coffee brewing process, spent coffee grounds are a residue that has a very fine particle size, high humidity, acidity, and organic load. They are obtained by treating the roasted coffee powder with hot water or steam used to prepare instant coffee (Mussatto, Machado, et al. 2011). Spent coffee grounds show sequestering free radicals and antioxidant capacity, rich in saccharides (mannose and galactose) and proteins (Campos-Vega et al. 2015). Their chemical composition varies from plant to plant, and between different parts of the plant, there are variations according to geographic location, the age of the plant, climate, and soil conditions (Murthy and Madhava Naidu 2012).

Economic, Environmental, and Social Implications of Coffee Industry Byproducts

According to the International Coffee Organization, in 2021, 10.2 million tons of coffee were consumed worldwide (International Coffee Organization 2022). Obtaining a cup of good coffee represents a process that is meticulously cared for, from the harvest of the coffee berry to the brewing process (Castañeda-Rodríguez, Mulík, and Ozuna 2020). However, the main added value of this product is generated after export: during the commercialization of green coffee, its roasting, and the preparation of the beverage itself. Therefore, this economic and social impact of the coffee industry does not benefit the coffee-producing countries but the coffee processing countries instead. Furthermore, the current techniques used in coffee processing do not come from a sustainable economic model. In this sense, the producers, whose capacity for action is exceeded by the demand for coffee consumption and the imminent climate change, are the ones who are the most adversely affected in their economy and well-being. Therefore, the generation of added value in coffee byproducts could contribute to the economic improvement of coffee producers. On the other hand, sustainable control of industrial waste management contributes significantly to the development of a country's economy (Murthy and Madhava Naidu 2012). In this sense, the application of chemical or biotechnological methods for the recovery or production of bioactive compounds with added value from byproducts, residues, and effluents represents a viable strategy for developing countries (Mussatto, Machado, et al. 2011).

As for environmental implications, the byproducts generated during coffee production represent around 90% of the total weight of the cherry and are typically discarded into the environment (Iriondo-DeHond et al. 2019). The biodegradation processes of coffee residues require extensive periods and quantities of oxygen (Esquivel and Jiménez 2012). Unsustainable practices in the production of coffee beans cause serious environmental problems since their byproducts contain large amounts of organic matter that makes them highly polluting. Also, the anti-nutrients present in coffee byproducts (polyphenols, caffeine, and tannins) make them highly toxic and reduce their possible use as livestock feed (Wen, Zhang, Rai, et al. 2019). In addition, during their decomposition, compounds are produced that can leach into aquifers and generate unpleasant odors that can affect the quality of life of those people who live in communities near the producing areas (Murthy and Madhava Naidu 2012; Haddis and Devi 2008). In this regard, it has been reported that these people may present serious health problems, feelings of dizziness and nausea, eye and skin irritation, stomach and respiratory problems, among others (Haddis and Devi 2008). Coffee residues distributed in the environment also represent a high risk for flora and fauna present in the ecosystems (Murthy and Naidu, 2012c).

During the generation of the coffee beverage itself, spent coffee grounds are another byproduct that is generated by tons (Ozuna et al. 2020), and it is estimated that one ton of green coffee beans will produce about 650 kg of spent coffee grounds as waste from their consumption (Mata, Martins, and Caetano 2018). In addition, to produce 1 kg of soluble coffee, 2 kg of spent coffee grounds will be generated (Stylianou et al. 2018). In this sense, 50% of spent coffee grounds are produced by the soluble coffee industry and by cafeteria chains, and the rest corresponds to domestic consumption (Miladi et al. 2021). Thus, coffee is an extremely valuable product whose production and subsequent sale represent an improvement to the economy of various countries. In some American regions, coffee production is of the utmost importance to achieve development goals at the economic and social levels. However, the current environmental problem requires a complete restructuring in the search for sustainable production processes. Likewise, there is a need to create multidisciplinary strategies that involve post-production stages that facilitate the bioconversion of coffee byproducts to biomaterials, functional ingredients, and new products in the chemical, pharmaceutical, and food industries.

Bioactive Compounds in Coffee Industry Byproducts

The recovery of bioactive compounds from agro-food waste is now one of the most important issues in the area of food science and technology (Vilas-Boas et al. 2022; Tiwari and Khawas 2021). Bioactive compounds are often synthesized in small amounts in nature. They are present as conjugates or mixtures in extracts which require long and intensive purification procedures. In addition, these molecules' structural diversity and complexity make their chemical synthesis rather costly (Sasidharan et al. 2010).

Some of the main bioactive compounds present in coffee byproducts are hydroxycinnamic acids (Hoseini et al. 2021), carotenoids (Obruca et al. 2015), chlorogenic acids (Bondam et al. 2022; Santos da Silveira et al. 2019), anthocyanins (Prata and Oliveira 2007), polyphenolic compounds and antioxidants (Murthy and Madhava Naidu 2012; Gemechu 2020), among others. The byproducts can contain phenolic compounds such as chlorogenic acid, solids, and esters of trans cinnamic and quinic acids (Mazzafera and Silvarolla 2010). Phenolic compounds, including flavonoids are considered antioxidants with specific health benefits (Ferruzzi 2010). Their most interesting property is the antioxidant activity due to the effect of their chemical structure, namely several phenolic hydroxyl groups that react with free radicals (Panusa et al. 2013).

During coffee processing, the distribution of phenolic compounds changes in the grain and its byproducts. The main alkaloid in green coffee grains is caffeine, representing between 1 and 4%. Other phenolic compounds are chlorogenic acid, solids and esters of trans cinnamic and quinic acids (Mazzafera and Silvarolla 2010). In turn, in toasted coffee beans, we can find diterpenes in a concentration of 1-12 mg/cup and some free phenolic acids like caffeic, ferulic, and dimethoxycinnamic acids (Esquivel and Jiménez 2012).

The chlorogenic acid is the main phenolic compound present in coffee byproducts, with reported levels of 2.3-3% in coffee silverskin and pulp (Murthy and Naidu, 2012b) and 34.2% in waste spent coffee grounds (Ramalakshmi et al. 2009). The natural compounds in coffee byproducts contain multiple bioactive compounds, most of which are not exploited (Campos-Vega et al. 2015).

In the spent coffee grounds, chlorogenic, isochlorogenic and feruloilquinic acids have been (Campos-Vega et al. 2015), whereas the presence of hydroxycinnamic acids has been found in the coffee pulp (Murthy and Madhava Naidu 2012). The nature of the polyphenolic compounds can range from simple molecules (phenolic acids) to highly polymerized compounds (tannins) and it is known that these compounds have been attributed antioxidant and antimicrobial properties (Murthy and Manonmani 2009; Rufián-Henares and De La Cueva 2009).

Bioactive Compound Recovery from Coffee Industry Byproducts

Both coffee and the byproducts released during its processing represent an important and valuable source of bioactive compounds, including antioxidants and antimicrobials (Bravo et al. 2013; 2012). Different research areas strive to recover bioactive compounds in extracts obtained from different coffee byproducts, which have added value to said compounds (Magalhães et al. 2016; Arya et al. 2021). Some of the strategies for the use of these byproducts are the production of biofuel (Lee et al. 2021), growth of fungal strains (Machado et al. 2012), production of unicellular proteins and edible fungi (Janissen and Huynh 2018), extraction of bioactive compounds (Murthy and Madhava Naidu 2012), among others. In the case of the food industry, various applications have been investigated that would allow for the incorporation of coffee byproducts as an ingredient or additive in the development of new products (Table 1).

The extraction of bioactive compounds utilizing conventional extraction techniques usually requires large amounts of solvents and energy (Wen, Zhang, Sun, et al. 2019). Therefore, innovative technologies for the extraction of bioactive compounds seek to replace the traditional technologies (Ameer, Shahbaz, and Kwon 2017). Recently, green technologies for the extraction of bioactive compounds have been developed (Gil-Chávez et al. 2013; Wen, Zhang, Sun, et al. 2019). Among these, there is high power ultrasound (Bhargava et al. 2021), microwave-assisted extraction (Pettinato, Casazza, and Perego 2019; Pettinato et al. 2019), and the use of supercritical fluids (Vandeponseele et al. 2020).

Unlike conventional extraction techniques, green technologies avoid using high extraction temperatures and large amounts of organic solvents, which may adversely affect the quality of the bioactive extracts (Galanakis 2012). Moreover, green technologies use shorter extraction times, are more cost-effective and environment-friendly (Pettinato et al. 2019; Bhargava et al. 2021).

High-Power Ultrasound

High-power ultrasound is a green technology that has been used with great success in the bioactive compound extraction of plant origin (Bhargava et al. 2021). The main mechanism that contributes to improving extraction processes is cavitation. This phenomenon consists of reaching high pressures (>100 MPa) and temperatures (5000 K) at the molecular level, which facilitate the mass and heat transfer between the plant matrix and the liquid medium (Ozuna et al. 2015).

The effectiveness of high-power ultrasound on the extraction of bioactive compounds depends on certain variables during the process, such as ultrasonic power, medium temperature, type of solvent, product properties, and solid-liquid ratio (Bhargava et al. 2021). Thus, in order to design an optimal extraction process assisted by high-power ultrasound, it is important to optimize the main process variables to achieve better extraction yields of compounds (Bhargava et al. 2021).

High-power ultrasound technology has been applied in the extraction processes of bioactive compounds from coffee industry byproducts (mainly coffee silverskin and spent coffee grounds) with the main objective of accelerating the mass transfer process and reducing the treatment times (Al-Dhabi, Ponmurugan, and Maran Jeganathan 2017; Okur et al. 2021). According to the revised articles, high-power ultrasound has improved the conventional extraction methods of bioactive compounds from coffee byproducts by up to 35%. Regarding the generated extracts, due to the fact that high-intensity ultrasound allows controlling the process temperature, this technology avoids overheating of the sample and, therefore, there is no loss of thermosensitive compounds.

Microwaves

Microwaves are defined as electromagnetic waves with a frequency between 300 MHz and 300 GHz and a length of 0.001 to 1 m. Microwaves are located between X-rays and infrared rays on the electromagnetic spectrum (Letellier and Budzinski 1999). When microwaves pass through a medium, their energy can be absorbed and converted into heat, which can contribute to mass transport phenomena (Ameer, Shahbaz, and Kwon 2017).

Microwaves are a green technology that has been used as a fast, effective, and profitable extraction technique that allows for generating added value to the byproducts of the coffee industry. Extraction time and temperature are two of the most important variables in achieving a successful extraction. In general, when high temperatures are applied, high extraction yields are obtained due to the increase in energy applied to the system, which improves mass transfer phenomena. However, thermosensitive molecules could be degraded (Pettinato, Casazza, and Perego 2019; Pettinato et al. 2019).

Regarding these byproducts, microwaves have been used to extract compounds with antioxidant capacities such as phenols and flavonoids in spent coffee grounds (Pettinato, Casazza, and Perego 2019; Pettinato et al. 2019) and coffee silverskin (Guglielmetti et al. 2017) (Table 3). Microwave-assisted extraction has been shown to increase the process by around 35% compared to conventional extraction methods.

Table 1. Applications of coffee byproducts in the food industry

Coffee byproducts compounds	Applications	Technological and biological function	References
Carbohydrates (antioxidant dietary fiber)	Cakes, breads, biscuits, sugars (mannose, galactose, arabinose, and glucose homopolymer)	Fat replacement, wheat flour substitution, prebiotic and antioxidant activity	(Mussatto, Machado, et al. 2011; Ateş and Elmacı 2018; Garcia-Serna et al. 2014)
Non-protein nitrogenous compounds	Food ingredient	Antioxidant activity	(Iriondo-Dehond, Iriondo-Dehond, and Del Castillo 2020)
Minerals (Mg, P, Ca, Na, Fe, Mn, and Cu)	Food ingredient	Anti-inflammatory and antioxidant activity	(Iriondo-Dehond, Iriondo-Dehond, and Del Castillo 2020; Nzekoue et al. 2022)
Phenolic compounds (natural phenolic antioxidants, total polyphenols, chlorogenic acids)	Yogurt, fermented and distilled beverages, bread, and biscuits	Antioxidant activity	(Sampaio et al. 2013; E. Machado et al. 2018; Bertolino et al. 2019)
Caffeine	Beverages	Antioxidant activity and body weight control	(Martinez-Saez et al. 2014)
Lipids (linoleic, palmitic, and oleic acids)	Active packaging films	O_2 barrier properties, antioxidant activity	(Vidal et al. 2022; Al-Hamamre et al. 2012; Todaka et al. 2013)

Table 2. Bioactive compound extraction from coffee byproducts assisted by high-intensity ultrasound

Coffee byproduct	Device	Experimental conditions	Maximal extraction concentration				Reference
			Bioactive compound	Value	Antioxidant activity	Value	
Spent coffee grounds	Ultrasonic probe	S: 80% methanol; F: 25 kHz; R: 1:10 (w/v); t: 5-15 min; T: 25°C	Total phenolic content	9.51 mg GAE/g FW	TEAC (DPPH)	~90%	(Okur et al. 2021)
					FRAP	0.89 mmol $FeSO_4$/100 g FW	
Coffee silverskin	Ultrasonic bath	S: Water, methanol, 50% methanol, 70% ethanol; F: 40 kHz; R: 1:5 (w/v); t: 120 min; T: 20°C	Total phenolic content	73.4 mg GAE/g DW	Radical scavenging activity (DPPH)	101.7 IC_{50} (µg/mL)	(Kamgang et al., 2020)
			Total flavonoid content	7.6 mg RE/g DW			
Coffee silverskin	Ultrasonic bath	S: Water, methanol, 50% methanol, 70% ethanol; F: 40 kHz; R: 1:5 (w/v); t: 120 min; T: 20°C	Total phenolic content	35.68 mg GAE/g DW	TEAC (DPPH)	48.73 mg TE/g DW	(Zengin et al., 2020)
			Total flavonoid content	2.32 mg RE/g DW	TEAC (ABTS)	95.05 mg TE/g DW	
Spent coffee grounds	Ultrasonic bath	S: Water, methanol, 50% methanol, 70% ethanol; F: 40 kHz; R: 1:5 (w/v); t: 120 min; T: 20°C	Total phenolic content	93.55 mg GAE/g DW	TEAC (DPPH)	296.78 mg TE/g DW	(Zengin et al., 2020)

Coffee byproduct	Device	Experimental conditions	Maximal extraction concentration				Reference
			Bioactive compound	Value	Antioxidant activity	Value	
Coffee silverskin	Ultrasonic probe	S: Water, 80% methanol; F: 20 kHz; R: 1:50 (w/v); t: 10 min	Total flavonoid content	2.32 mg RE/g DW	TEAC (ABTS)	276.19 mg TE/g DW	(Wen, Zhang and Rai 2019)
			Total phenolic content	6.08 GAE/g extract			
Spent coffee grounds	Ultrasonic probe	S: Ethanol; F: 20 kHz; R: 1:5-1:30 (w/v); t: 5-45 min; T: 30-50°C	Total phenolic content	33.84 GAE/g DW	NR	NR	(Al-Dhabi, Ponmurugan, and Maran Jeganathan 2017)
			Total flavonoid content	5.04 QE/g DW			
			Chlorogenic acid content	1.43 mg/g DW			

Where ABTS: 2,2'-azino-di (3-ethylbenzthiazoline-6-sulfonic acid); DPPH: 2 2-diphenyl-1-picrylhydrazyl; DW: dry weight; F: frequency; GAE: galic acid equivalent; NR: not reported; QE: quercetin equivalent; R: ratio; RE: rutin equivalent; S: solvent; T: temperature; t: time; TE: Trolox equivalent; TEAC: Trolox equivalent antioxidant capacity.

Table 3. Bioactive compound extraction from coffee byproducts assisted by microwaves

Coffee byproduct	Device	Experimental conditions	Maximal extraction concentration				Reference
			Bioactive compound	Value	Antioxidant activity	Value	
Spent coffee grounds	Microwave laboratory oven	S: Water, 40% ethanol, 80% ethanol; P: 500 W; R: 1:10 (w/v); t: 30-90 min; T: 90-150°C	Total phenolic content	46 mg CAE/g DW	TEAC (ABTS)	55.8 mg TE/L	(Pettinato et al., 2019a)
			Total flavonoid content	16.4 mg QE/g DW			
Spent coffee grounds	Microwave laboratory oven	S: 54% ethanol; P: 500 W; R: 1:10 (w/v); t: 1-20 min; T: 119.85-149.85°C	Total flavonoid content	~43 mg CAE/g DW	TEAC (ABTS)	0.88 µg TE/g DW	(Pettinato et al., 2019b)
Coffee silverskin	Microwave digestion system	S: 60% ethanol; P: 280 W; R: 1:35 (w/v); t: 15-30 min: T: 37-55°C	Total phenolic content	10.58 mg GAE/g DW	TEAC (DPPH)	9.47 µmol TE/g DW	(Guglielmetti et al. 2017)
Spent coffee grounds	Microwave household oven	S: 20-80% ethanol; P: 80 W; R: 1:9 (w/v); t: 0.66-6 min.	Total phenolic content	398.95 mg GAE/g extract	FRAP	2.86 mmol TE/L extract	(Pavlović et al. 2013)
					Radical scavenging activity (DPPH)	90.15%	
Spent coffee grounds	Microwave household oven	S: 20% ethanol; P: 240-560 W; R: 1:4.76-1:13.24 (w/v); t: 0.18-3.48 min; T: 90-150°C	Total phenolic content	79.83% dry extract	Radical scavenging activity (DPPH)	98.24%	(Ranic et al. 2014)
					FRAP	6.66 mmol Fe^{2}/g DW	

Where DPPH: 2 2-diphenyl-1-picrylhydrazyl; DW: dry weight; FRAP: ferric reduction antioxidant power; GAE: galic acid equivalent; P: power; R: ratio; S: solvent; T: temperature; t: time; TE: Trolox equivalent; TEAC: Trolox equivalent antioxidant capacity.

Table 4. Bioactive compound extraction from coffee byproducts assisted by subcritical water and supercritical fluids

Coffee byproduct	Device	Experimental Conditions	Maximal extraction				Reference
			Bioactive compounds	Value	Antioxidant activity	Value	
Spent coffee grounds	Superfluid extractor	S: Water; P: 5 MPa; FR: 1:10–1:70 (w/v); t: 15-75 min; T: 110-190°C	Total phenolic content	86.23 mg GAE/g DW	TEAC (ABTS)	81.38 mmol TE/100 g DW	(Xu et al. 2015)
					TEAC (DPPH)	42.13 mmol TE/100 g DW	
Coffee silverskin	Superfluid extractor	S: Water; P: 1-5.3 MPa; FR: 1:50 (w/v); t:17-42 min; T: 180-270°C	Total phenolic content	130.0 mg GAE/g extract	Radical scavenging activity (DPPH)	379.0 µmol TE/g extract	(Narita and Inouye 2012)
					H-ORAC	2629.0 µmol TE/g extract	
Coffee husks	Superfluid extractor	S: CO_2 (4-8% ethanol); P: 10-30 MPa; t: 258 min; T:40-60°C	Total phenolic content	36.0 mg CAE/g extract	Radical scavenging activity (DPPH)	630.0 (IC_{50}) µg/mL extract	(Andrade et al. 2012)
					TEAC (ABTS)	141.0 µmol TE/g extract	
Spent coffee grounds	Superfluid extractor	S: CO_2 (8-15% ethanol); P: 10-30 MPa; t: 258 min; T: 40-60°C	Total phenolic content	57.0 mg CAE/g extract	Radical scavenging activity (DPPH)	478.2 (IC_{50}) µg/mL extract	(Andrade et al. 2012)
					TEAC (ABTS)	275.1 µmol TE/g extract	

Where ABTS: 2,2'-azino-di (3-ethylbenzthiazoline-6-sulfonic acid); CAE: caffeic acid equivalents; DPPH: 2 2-diphenyl-1-picrylhydrazyl; DW: dry weight; P: pressure; R: ratio; S: solvent; T: temperature; t: time; TE: Trolox equivalent; TEAC: Trolox equivalent antioxidant capacity.

Subcritical Water and Supercritical Fluids

Subcritical water and supercritical fluids are two operations that are based on subjecting fluids to particular conditions of pressure (~7.4 MPa) and temperature (300-650 K) to achieve a change in their dielectric properties without affecting their state of aggregation. Both technologies are environmentally friendly extraction technology that offers high selectivity, short processing times, and reduced organic solvent use. However, one of the main disadvantages of these technologies is the difficulty of water in reaching a supercritical temperature, leaving it at a subcritical point. On the other hand, CO_2 has the disadvantage of not having a polarity that would allow for the extraction of various bioactive compounds; therefore, it is necessary to use co-solvents to improve its effectiveness (Cardenas-Toro et al. 2014).

Subcritical water and supercritical fluids have been applied in extraction processes of bioactive compounds from coffee industry byproducts (coffee silverskin, coffee husks, and spent coffee grounds) (Xu et al. 2015; Narita and Inouye 2012; Andrade et al. 2012). Compared to high-power ultrasound and microwaves, subcritical water and supercritical fluids have proven to be less effective in improving the extraction processes (<10%). Therefore, future research should focus on a systematic study of bioactive compound extraction by means of green technology combinations, always seeking the most efficient and cost-effective applications and bearing in mind the extract quality.

Conclusion

Despite representing one of the main productive sectors worldwide, the coffee industry needs to move towards more sustainable processes where there is full use of the byproducts it generates and, thus, contribute to the Sustainable Development Goals promoted by the United Nations (United Nations 2022).

Although the biological, functional, and technological properties of coffee byproducts have already been demonstrated in various investigations, more research is needed to generate viable applications that can be commercialized and positioned in the market. These developments can generate greater added value in the production chain, especially in the first links, and thus contribute to the economic income of coffee producers, who are currently in a situation of poverty.

Regarding unconventional technologies for the extraction of bioactive compounds, it is important that these can be scaled to an industrial level and that they are affordable for the industry. It depends on this that these extraction technologies can be a viable option to be able to generate added value to coffee byproducts. Scientific research should be focused on the scaling of these technologies and the profitability that they can have compared to traditional extraction techniques.

In conclusion, the comprehensive use of coffee byproducts continues to be a topic of great interest for coffee producers, local governments, the scientific community, and various industries. However, this interest must materialize in joint actions that would make it possible to find viable and profitable solutions, thus contributing to generating the least environmental impact and to guaranteeing the well-being of the people from coffee-producing countries worldwide. Finally, continued study of the impact generated by coffee byproducts on the health of producers and residents living in nearby sites is also highly relevant.

References

Al-Dhabi, Naif Abdullah, Karuppiah Ponmurugan, and Prakash Maran Jeganathan. 2017. "Development and Validation of Ultrasound-Assisted Solid-Liquid Extraction of Phenolic Compounds from Waste Spent Coffee Grounds." *Ultrasonics Sonochemistry* 34 (January): 206–13. https://doi.org/10.1016/J.ULTSONCH.2016.05.005.

Al-Hamamre, Zayed, Sascha Foerster, Franziska Hartmann, Michael Kröger, and Martin Kaltschmitt. 2012. "Oil Extracted from Spent Coffee Grounds as a Renewable Source for Fatty Acid Methyl Ester Manufacturing." *Fuel* 96 (June): 70–76. https://doi.org/10.1016/J.FUEL.2012.01.023.

Ameer, Kashif, Hafiz Muhammad Shahbaz, and Joong Ho Kwon. 2017. "Green Extraction Methods for Polyphenols from Plant Matrices and Their Byproducts: A Review." *Comprehensive Reviews in Food Science and Food Safety* 16 (2): 295–315. https://doi.org/10.1111/1541-4337.12253.

Andrade, Kátia S., Ricardo T. Gonalvez, Marcelo Maraschin, Rosa Maria Ribeiro-Do-Valle, Julian Martínez, and Sandra R. S. Ferreira. 2012. "Supercritical Fluid Extraction from Spent Coffee Grounds and Coffee Husks: Antioxidant Activity and Effect of Operational Variables on Extract Composition." *Talanta* 88 (January): 544–52. https://doi.org/10.1016/J.TALANTA.2011.11.031.

Arya, Shalini S., Rahul Venkatram, Pavankumar R. More, and Poornima Vijayan. 2021. "The Wastes of Coffee Bean Processing for Utilization in Food: A Review." *Journal of Food Science and Technology 2021 59:2* 59 (2): 429–44. https://doi.org/10.1007/S13197-021-05032-5.

Ateş, Gizem, and Yeşim Elmacı. 2018. "Physical, Chemical and Sensory Characteristics of Fiber-Enriched Cakes Prepared with Coffee Silverskin as Wheat Flour Substitution." *Journal of Food Measurement and Characterization 2018 13:1* 13 (1): 755–63. https://doi.org/10.1007/S11694-018-9988-9.

Bertolino, Marta, Letricia Barbosa-Pereira, Daniela Ghirardello, Cristian Botta, Luca Rolle, Alessandro Guglielmetti, Stefania Borotto Dalla Vecchia, and Giuseppe Zeppa. 2019. "Coffee Silverskin as Nutraceutical Ingredient in Yogurt: Its Effect on Functional Properties and Its Bioaccessibility." *Journal of the Science of Food and Agriculture* 99 (9): 4267–75. https://doi.org/10.1002/JSFA.9659.

Bhargava, Nitya, Rahul S. Mor, Kshitiz Kumar, and Vijay Singh Sharanagat. 2021. "Advances in Application of Ultrasound in Food Processing: A Review." *Ultrasonics Sonochemistry*. Elsevier B.V. https://doi.org/10.1016/j.ultsonch.2020.105293.

Bondam, Aline Felten, Daiele Diolinda da Silveira, Jaqueline Pozzada dos Santos, and Jessica Fernanda Hoffmann. 2022. "Phenolic Compounds from Coffee By-Products: Extraction and Application in the Food and Pharmaceutical Industries." *Trends in Food Science & Technology* 123 (May): 172–86. https://doi.org/10.1016/J.TIFS.2022.03.013.

Bravo, Jimena, Isabel Juániz, Carmen Monente, Bettina Caemmerer, Lothar W. Kroh, M. Paz De Peña, and Concepción Cid. 2012. "Evaluation of Spent Coffee Obtained from the Most Common Coffeemakers as a Source of Hydrophilic Bioactive Compounds." *Journal of Agricultural and Food Chemistry* 60 (51): 12565–73. https://doi.org/10.1021/JF3040594/ASSET/IMAGES/MEDIUM/JF-2012-040594_0003.GIF.

Bravo, Jimena, Carmen Monente, Isabel Juániz, M. Paz De Peña, and Concepción Cid. 2013. "Influence of Extraction Process on Antioxidant Capacity of Spent Coffee." *Food Research International* 50 (2): 610–16. https://doi.org/10.1016/J.FOODRES.2011.04.026.

Campos-Vega, Rocio, Guadalupe Loarca-Piña, Haydé A. Vergara-Castañeda, and B. Dave Oomah. 2015. "Spent Coffee Grounds: A Review on Current Research and Future Prospects." *Trends in Food Science & Technology* 45 (1): 24–36. https://doi.org/10.1016/J.TIFS.2015.04.012.

Cardenas-Toro, Fiorella P., Tania Forster-Carneiro, Mauricio A. Rostagno, Ademir J. Petenate, Francisco Maugeri Filho, and M. Angela A. Meireles. 2014. "Integrated Supercritical Fluid Extraction and Subcritical Water Hydrolysis for the Recovery of Bioactive Compounds from Pressed Palm Fiber." *The Journal of Supercritical Fluids* 93 (September): 42–48. https://doi.org/10.1016/J.SUPFLU.2014.02.009.

Castañeda-Rodríguez, Rey, Stanislav Mulík, and César Ozuna. 2020. "*Brewing Temperature and Particle Size Affect Extraction Kinetics of Cold Brew Coffee in Terms of Its Physicochemical, Bioactive, and Antioxidant Properties.*" Https://Doi.Org/10.1080/15428052.2020.1848683. https://doi.org/10.1080/15428052.2020.1848683.

Cubero-Abarca, R., J. Moya, R., Valaret, and M. Tomazello Filho. 2014. "Use of Coffee (Coffea Arabica) Pulp for the Production of Briquettes and Pellets for Heat Generation." *Ciência e Agrotecnologia* 38 (5): 461–70.

Esquivel, Patricia, and Víctor M. Jiménez. 2012. "Functional Properties of Coffee and Coffee By-Products." *Food Research International* 46 (2): 488–95. https://doi.org/10.1016/J.FOODRES.2011.05.028.

Ferruzzi, Mario G. 2010. "The Influence of Beverage Composition on Delivery of Phenolic Compounds from Coffee and Tea." *Physiology & Behavior* 100 (1): 33–41. https://doi.org/10.1016/J.PHYSBEH.2010.01.035.

Galanakis, Charis M. 2012. "Recovery of High Added-Value Components from Food Wastes: Conventional, Emerging Technologies and Commercialized Applications." *Trends in Food Science & Technology* 26 (2): 68–87. https://doi.org/10.1016/J.TIFS.2012.03.003.

Garcia-Serna, Esther, Nuria Martinez-Saez, Marta Mesias, Francisco J. Morales, and M. Dolores Del Castillo. 2014. "Use of Coffee Silverskin and Stevia to Improve the Formulation of Biscuits." *Polish Journal of Food and Nutrition Sciences* 64 (4): 243–51. https://doi.org/10.2478/PJFNS-2013-0024.

Gemechu, Feyera Gobena. 2020. "Embracing Nutritional Qualities, Biological Activities and Technological Properties of Coffee Byproducts in Functional Food Formulation." *Trends in Food Science & Technology* 104 (October): 235–61. https://doi.org/10.1016/J.TIFS.2020.08.005.

Giacalone, Davide, Tina Kreuzfeldt Degn, Ni Yang, Chujiao Liu, Ian Fisk, and Morten Münchow. 2019. "Common Roasting Defects in Coffee: Aroma Composition, Sensory Characterization and Consumer Perception." *Food Quality and Preference* 71 (January): 463–74. https://doi.org/10.1016/J.FOODQUAL.2018.03.009.

Guglielmetti, A., V. D'Ignoti, D. Ghirardello, S. Belviso, and G. Zeppa. 2017. "Optimisation of Ultrasound and Microwave-Assisted Extraction of Caffeoylquinic Acids and Caffeine from Coffee Silverskin Using Response Surface Methodology." *Italian Journal of Food Science* 29 (3): 2017–2410. https://doi.org/10.14674/IJFS-727.

Haddis, Alemayehu, and Rani Devi. 2008. "Effect of Effluent Generated from Coffee Processing Plant on the Water Bodies and Human Health in Its Vicinity." *Journal of Hazardous Materials* 152 (1): 259–62. https://doi.org/10.1016/J.JHAZMAT.2007.06.094.

Hoseini, Marziyeh, Stefania Cocco, Cristiano Casucci, Valeria Cardelli, and Giuseppe Corti. 2021. "Coffee By-Products Derived Resources. A Review." *Biomass and Bioenergy* 148 (May): 106009. https://doi.org/10.1016/J.BIOMBIOE.2021.106009.

International Coffee Organization. 2022. "*Monthly Coffee Market Report (2021/22).*" 2022. https://www.ico.org/Market-Report-21-22-e.asp.

Iriondo-DeHond, Amaia, Natalia Aparicio García, Beatriz Fernandez-Gomez, Eduardo Guisantes-Batan, Francisco Velázquez Escobar, Gracia Patricia Blanch, Manuel Ignacio San Andres, Sebastián Sanchez-Fortun, and María Dolores del Castillo. 2019. "Validation of Coffee By-Products as Novel Food Ingredients." *Innovative Food Science & Emerging Technologies* 51 (January): 194–204. https://doi.org/10.1016/J.IFSET.2018.06.010.

Iriondo-Dehond, Amaia, Maite Iriondo-Dehond, and María Dolores Del Castillo. 2020. "Applications of Compounds from Coffee Processing By-Products." *Biomolecules 2020, Vol. 10, Page 1219* 10 (9): 1219. https://doi.org/10.3390/BIOM10091219.

Janissen, Brendan, and Tien Huynh. 2018. "Chemical Composition and Value-Adding Applications of Coffee Industry by-Products: A Review." *Resources, Conservation and Recycling* 128 (January): 110–17. https://doi.org/10.1016/J.RESCONREC.2017.10.001.

Joana Gil-Chávez, G., José A. Villa, J. Fernando Ayala-Zavala, J. Basilio Heredia, David Sepulveda, Elhadi M. Yahia, and Gustavo A. González-Aguilar. 2013. "Technologies for Extraction and Production of Bioactive Compounds to Be Used as Nutraceuticals and Food Ingredients: An Overview." *Comprehensive Reviews in Food Science and Food Safety* 12 (1): 5–23. https://doi.org/10.1111/1541-4337.12005.

Ky, C. L., J. Louarn, S. Dussert, B. Guyot, S. Hamon, and M. Noirot. 2001. "Caffeine, Trigonelline, Chlorogenic Acids and Sucrose Diversity in Wild *Coffea arabica* L. and *C. canephora P. accessions.*" *Food Chemistry* 75 (2): 223–30. https://doi.org/10.1016/S0308-8146(01)00204-7.

Lachenmeier, Dirk W., Steffen Schwarz, Jörg Rieke-Zapp, Ennio Cantergiani, Harshadrai Rawel, María Angeles Martín-Cabrejas, Maria Martuscelli, Vera Gottstein, and Simone Angeloni. 2021. "Coffee By-Products as Sustainable Novel Foods: Report of the 2nd International Electronic Conference on Foods; Future Foods and Food Technologies for a Sustainable World" *Foods 2022, Vol. 11, Page 3* 11 (1): 3. https://doi.org/10.3390/FOODS11010003.

Lee, Xin Jiat, Hwai Chyuan Ong, Wei Gao, Yong Sik Ok, Wei Hsin Chen, Brandon Han Hoe Goh, and Cheng Tung Chong. 2021. "Solid Biofuel Production from Spent Coffee Ground Wastes: Process Optimisation, Characterisation and Kinetic Studies." *Fuel* 292 (May): 120309. https://doi.org/10.1016/J.FUEL.2021.120309.

Letellier, M., and H. Budzinski. 1999. "Microwave Assisted Extraction of Organic Compounds." *Analusis* 27 (3): 259–70. https://doi.org/10.1051/ANALUSIS:1999116.

Machado, Ercília M. S., Rosa M. Rodriguez-Jasso, José A. Teixeira, and Solange I. Mussatto. 2012. "Growth of Fungal Strains on Coffee Industry Residues with Removal of Polyphenolic Compounds." *Biochemical Engineering Journal* 60 (January): 87–90. https://doi.org/10.1016/J.BEJ.2011.10.007.

Machado, Ercília, Solange I. Mussatto, José Teixeira, Mar Vilanova, and José Oliveira. 2018. "Increasing the Sustainability of the Coffee Agro-Industry: Spent Coffee Grounds as a Source of New Beverages." *Beverages 2018, Vol. 4, Page 105* 4 (4): 105. https://doi.org/10.3390/BEVERAGES4040105.

Magalhães, Luís M., Sandia Machado, Marcela A. Segundo, João A. Lopes, and Ricardo N. M. J. Páscoa. 2016. "Rapid Assessment of Bioactive Phenolics and Methylxanthines in Spent Coffee Grounds by FT-NIR Spectroscopy." *Talanta* 147 (January): 460–67. https://doi.org/10.1016/J.TALANTA.2015.10.022.

Martinez-Saez, Nuria, Mónica Ullate, María A. Martin-Cabrejas, Patricia Martorell, Salvador Genovés, Daniel Ramon, and María Dolores Del Castillo. 2014. "A Novel Antioxidant Beverage for Body Weight Control Based on Coffee Silverskin." *Food Chemistry* 150 (May): 227–34. https://doi.org/10.1016/J.FOODCHEM.2013.10.100.

Mata, Teresa M., António A. Martins, and Nídia S. Caetano. 2018. "Bio-Refinery Approach for Spent Coffee Grounds Valorization." *Bioresource Technology* 247 (January): 1077–84. https://doi.org/10.1016/J.BIORTECH.2017.09.106.

Mazzafera, Paulo, and Maria Bernadete Silvarolla. 2010. "Caffeine Content Variation in Single Green Arabica Coffee Seeds." *Seed Science Research* 20 (3): 163–67. https://doi.org/10.1017/S0960258510000140.

Miladi, Malek, António A. Martins, Teresa M. Mata, Miguel Vegara, María Pérez-Infantes, Rania Remmani, Antonio Ruiz-Canales, and Dámaris Núñez-Gómez. 2021. "Optimization of Ultrasound-Assisted Extraction of Spent Coffee Grounds Oil Using Response Surface Methodology." *Processes 2021, Vol. 9, Page 2085* 9 (11): 2085. https://doi.org/10.3390/PR9112085.

Murthy, Pushpa S., and M. Madhava Naidu. 2012. "Sustainable Management of Coffee Industry By-Products and Value Addition—A Review." *Resources, Conservation and Recycling* 66 (September): 45–58. https://doi.org/10.1016/J.RESCONREC.2012.06.005.

Murthy, Pushpa S., and H. K. Manonmani. 2009. "Physico-Chemical, Antioxidant and Antimicrobial Properties of Indian Monsooned Coffee." *European Food Research and Technology 2009 229:4* 229 (4): 645–50. https://doi.org/10.1007/S00217-009-1098-9.

Mussatto, Solange I., Lina F. Ballesteros, Silvia Martins, and José A. Teixeira. 2011. "Extraction of Antioxidant Phenolic Compounds from Spent Coffee Grounds." *Separation and Purification Technology* 83 (1): 173–79. https://doi.org/10.1016/J.SEPPUR.2011.09.036.

Mussatto, Solange I., Ercília M.S. Machado, Silvia Martins, and J. A. Teixeira. 2011. "Production, Composition, and Application of Coffee and Its Industrial Residues." *Food and Bioprocess Technology 2011 4:5* 4 (5): 661–72. https://doi.org/10.1007/S11947-011-0565-Z.

Narita, Yusaku, and Kuniyo Inouye. 2012. "High Antioxidant Activity of Coffee Silverskin Extracts Obtained by the Treatment of Coffee Silverskin with Subcritical Water." *Food Chemistry* 135 (3): 943–49. https://doi.org/10.1016/J.FOODCHEM.2012.05.078.

Navia P., D.P., R. J. Velasco M., and José L Hoyos C. 2011. "Production and Evaluation of Ethanol from Coffee Processing By-Products." *Vitae* 18 (3): 287–94.

Nzekoue, Franks Kamgang, Germana Borsetta, Luciano Navarini, Doaa Abouelenein, Jianbo Xiao, Gianni Sagratini, Sauro Vittori, Giovanni Caprioli, and Simone Angeloni. 2022. "Coffee Silverskin: Characterization of B-Vitamins, Macronutrients, Minerals and Phytosterols." *Food Chemistry* 372 (March): 131188. https://doi.org/10.1016/J.FOODCHEM.2021.131188.

Obruca, Stanislav, Pavla Benesova, Dan Kucera, Sinisa Petrik, and Ivana Marova. 2015. "Biotechnological Conversion of Spent Coffee Grounds into Polyhydroxyalkanoates and Carotenoids." *New Biotechnology* 32 (6): 569–74. https://doi.org/10.1016/J.NBT.2015.02.008.

Okur, Ilhami, Betul Soyler, Purlen Sezer, Mecit Halil Oztop, and Hami Alpas. 2021. "Improving the Recovery of Phenolic Compounds from Spent Coffee Grounds (SCG) by Environmentally Friendly Extraction Techniques." *Molecules 2021, Vol. 26, Page 613* 26 (3): 613. https://doi.org/10.3390/MOLECULES 26030613.

Ozuna, C., S. Mulík, B. Valdez-Rodríguez, M. D. R. Abraham-Juárez, and C. L. Fernández-López. 2020. "The Effect of Organic Farming on Total Phenols, Total Flavonoids, Brown Compounds and Antioxidant Activity of Spent Coffee Grounds from Mexico." *Biological Agriculture and Horticulture* 36 (2). https://doi.org/10.1080/01448765.2019.1704876.

Ozuna, C., I. Paniagua-Martínez, E. Castaño-Tostado, L. Ozimek, and S. L. Amaya-Llano. 2015. "Innovative Applications of High-Intensity Ultrasound in the Development of Functional Food Ingredients: Production of Protein Hydrolysates and Bioactive Peptides." *Food Research International* 77. https://doi.org/10.1016/j.foodres.2015.10.015.

Panusa, Alessia, Antonio Zuorro, Roberto Lavecchia, Giancarlo Marrosu, and Rita Petrucci. 2013. "Recovery of Natural Antioxidants from Spent Coffee Grounds." *Journal of Agricultural and Food Chemistry* 61 (17): 4162–68. https://doi.org/10.1021/JF4005719/ASSET/IMAGES/MEDIUM/JF-2013-005719_0002. GIF.

Pavlović, Marija D., Aneta V. Buntić, Slavica S. Šiler-Marinković, and Suzana I. Dimitrijević-Branković. 2013. "Ethanol Influenced Fast Microwave-Assisted Extraction for Natural Antioxidants Obtaining from Spent Filter Coffee." *Separation and Purification Technology* 118 (October): 503–10. https://doi.org/10.1016/J.SEPPUR.2013.07.035.

Pettinato, Margherita, Alessandro Alberto Casazza, Pier Francesco Ferrari, Domenico Palombo, and Patrizia Perego. 2019. "Eco-Sustainable Recovery of Antioxidants from Spent Coffee Grounds by Microwave-Assisted Extraction: Process Optimization, Kinetic Modeling and Biological Validation." *Food and Bioproducts Processing* 114 (March): 31–42. https://doi.org/10.1016/J.FBP.2018.11.004.

Pettinato, Margherita, Alessandro Alberto Casazza, and Patrizia Perego. 2019. "The Role of Heating Step in Microwave-Assisted Extraction of Polyphenols from Spent Coffee Grounds." *Food and Bioproducts Processing* 114 (March): 227–34. https://doi.org/10.1016/J.FBP.2019.01.006.

Prata, Emille R. B. A., and Leandro S. Oliveira. 2007. "Fresh Coffee Husks as Potential Sources of Anthocyanins." *LWT - Food Science and Technology* 40 (9): 1555–60. https://doi.org/10.1016/J.LWT. 2006.10.003.

Ramalakshmi, K., L. Jagan Mohan Rao, Yuko Takano-Ishikawa, and Masao Goto. 2009. "Bioactivities of Low-Grade Green Coffee and Spent Coffee in Different in Vitro Model Systems." *Food Chemistry* 115 (1): 79–85. https://doi.org/10.1016/J.FOODCHEM.2008.11.063.

Ranic, Marija, Marina Nikolic, Marija Pavlovic, Aneta Buntic, Slavica Siler-Marinkovic, and Suzana Dimitrijevic-Brankovic. 2014. "Optimization of Microwave-Assisted Extraction of Natural Antioxidants from Spent Espresso Coffee Grounds by Response Surface Methodology." *Journal of Cleaner Production* 80 (October): 69–79. https://doi.org/10.1016/J.JCLEPRO.2014.05.060.

Rufián-Henares, José A., and Silvia P. De La Cueva. 2009. "Antimicrobial Activity of Coffee Melanoidins - A Study of Their Metal-Chelating Properties." *Journal of Agricultural and Food Chemistry* 57 (2): 432–38. https://doi.org/10.1021/JF8027842/ASSET/IMAGES/MEDIUM/JF-2008-027842_0008.GIF.

Sampaio, Armando, Giuliano Dragone, Mar Vilanova, José M. Oliveira, José A. Teixeira, and Solange I. Mussatto. 2013. "Production, Chemical Characterization, and Sensory Profile of a Novel Spirit Elaborated from Spent Coffee Ground." *LWT - Food Science and Technology* 54 (2): 557–63. https://doi.org/10.1016/J.LWT.2013.05.042.

Santos da Silveira, Jessica, Noël Durand, Stella Lacour, Marie Pierre Belleville, Ana Perez, Gérard Loiseau, and Manuel Dornier. 2019. "Solid-State Fermentation as a Sustainable Method for Coffee Pulp Treatment and Production of an Extract Rich in Chlorogenic Acids." *Food and Bioproducts Processing* 115 (May): 175–84. https://doi.org/10.1016/J.FBP.2019.04.001.

Sasidharan, S., Y. Chen, D. Saravanan, K. M. Sundram, and L. Yoga Latha. 2010. "Extraction, Isolation And Characterization Of Bioactive Compounds From Plants' Extracts." *African Journal of Traditional, Complementary and Alternative Medicines* 8 (1): 1–10. https://doi.org/10.4314/ajtcam.v8i1.60483.

Stylianou, Marinos, Agapios Agapiou, Michalis Omirou, Ioannis Vyrides, Ioannis M. Ioannides, Grivas Maratheftis, and Dionysia Fasoula. 2018. "Converting Environmental Risks to Benefits by Using Spent Coffee Grounds (SCG) as a Valuable Resource." *Environmental Science and Pollution Research 2018 25:36* 25 (36): 35776–90. https://doi.org/10.1007/S11356-018-2359-6.

Tiwari, Ajita, and Roshna Khawas. 2021. "Food Waste and Agro By-Products: A Step towards Food Sustainability." *Innovation in the Food Sector Through the Valorization of Food and Agro-Food By-Products*, July. https://doi.org/10.5772/INTECHOPEN.96177.

Todaka, Masatoshi, Wasana Kowhakul, Hiroshi Masamoto, Mikiji Shigematsu, and Siaw Onwona-Agyeman. 2013. "Thermal Decomposition of Biodiesel Fuels Produced from Rapeseed, Jatropha, and Coffee Oils with Different Alcohols." *Journal of Thermal Analysis and Calorimetry 2013 113:3* 113 (3): 1355–61. https://doi.org/10.1007/S10973-013-3056-9.

United Nations. 2022. "*Sustainable Development Goals*." 2022. https://www.un.org/sustainabledevelopment/.

Vandeponseele, Alexandre, Micheline Draye, Christine Piot, and Gregory Chatel. 2020. "Subcritical Water and Supercritical Carbon Dioxide: Efficient and Selective Eco-Compatible Solvents for Coffee and Coffee by-Products Valorization." *Green Chemistry* 22 (24): 8544–71. https://doi.org/10.1039/D0GC03146A.

Vidal, O. L., Barros Santos, M. C., Batista, A. P., Andrigo, F. F., Baréa, B., Lecomte, J., Figueroa-Espinoza, M. C., Gontard, N., Villeneuve, P., Guillard, V., Rezende, C. M., Bourlieu-Lacanal, C., & Larraz Ferreira, M. S. 2022. "Active Packaging Films Containing Antioxidant Extracts from Green Coffee Oil By-Products to Prevent Lipid Oxidation." *Journal of Food Engineering* 312 (January): 110744. https://doi.org/10.1016/J.JFOODENG.2021.110744.

Vilas-Boas, Ana A., Débora A. Campos, Ana Silva Oliveira, and Manuela Pintado. 2022. "Agro-Food Wastes: New Sources of Antioxidants." *Membrane Engineering in the Circular Economy*, January, 197–227. https://doi.org/10.1016/B978-0-323-85253-1.00020-4.

Wen, Le, Zhihang Zhang, Dilip Rai, Da Wen Sun, and Brijesh K. Tiwari. 2019. "Ultrasound-Assisted Extraction (UAE) of Bioactive Compounds from Coffee Silverskin: Impact on Phenolic Content, Antioxidant Activity, and Morphological Characteristics." *Journal of Food Process Engineering* 42 (6): e13191. https://doi.org/10.1111/JFPE.13191.

Wen, Le, Zhihang Zhang, Da Wen Sun, Saravana Periaswamy Sivagnanam, and Brijesh K. Tiwari. 2019. "Combination of Emerging Technologies for the Extraction of Bioactive Compounds." *Https://Doi.Org/10.1080/10408398.2019.1602823* 60 (11): 1826–41. https://doi.org/10.1080/10408398.2019.1602823.

Xu, Honggao, Weiyou Wang, Xin Liu, Fang Yuan, and Yanxiang Gao. 2015. "Antioxidative Phenolics Obtained from Spent Coffee Grounds (*Coffea arabica* L.) by Subcritical Water Extraction." *Industrial Crops and Products* 76 (December): 946–54. https://doi.org/10.1016/J.INDCROP.2015.07.054.

Chapter 6

Pomegranate Byproducts

Ma. Claudia Castañeda-Saucedo[1]
Ernesto Tapia-Campos[2,*]
Jessica del Pilar Ramírez-Anaya[3,†]
Adriana Cavazos-Garduño[4]
Julio C. Serrano-Niño[4]
Daniel Guajardo-Flores[5]
Sandra Teresita Martín del Campo[6]
María Luisa Pita-López[7]
Octavio Nuñez-Maciel[8]
and Martha Leticia Rujano-Silva[8]

[1]Department of Natural Sciences, University of Guadalajara, México
[2]Plant Biotechnology Unit, Center for Research and Assistance in Technology and Design, State of Jalisco (CIATEJ), Guadalajara, Jalisco, Mexico
[3]Science, Computational and Technological Innovation, University of Guadalajara, Mexico
[4]Department of Pharmacobiology, CUCEI-University of Guadalajara, Guadalajara, Jalisco, Mexico
[5]School of Engineering and Sciences, Tecnológico de Monterrey
[6]Tecnologico de Monterrey, School of Engineering and Sciences, Querétaro, Qro. Mexico
[7]Department of Basic Health Sciences, University of Guadalajara, México
[8]Economic and Administrative Sciences, CUSUR-University of Guadalajara, Ciudad Guzmán, Jalisco, Mexico

Abstract

Pomegranate (*Punica granatum* L.) is an important source of bioactive compounds and has been used in traditional medicine for centuries. The edible part of the pomegranate is 50% of the fruit; the rest is inedible (waste, seeds, membranes, as well as cracked fruits). Industrial waste is an environmental pollution problem since the material is prone to microbial deterioration. However, these wastes contain large amounts of bioactive

* Corresponding Author's Email: etapia@ciatej.mx.
† Corresponding Author's Email: jessicar@cusur.udg.mx.

In: Food Byproducts
Editors: S. Teresita Martín-del-Campo, A. Cardador-Martínez and J. del Pilar Ramírez-Anaya
ISBN: 979-8-88697-612-0

compounds such as polyphenols, ellagitannins, vitamins, minerals and polyunsaturated fatty acids, etc., that have nutritional characteristics and biofunctional compounds that promote health through their antimicrobial, antifungal, anticancer, and antidiabetic properties. They also aid in the prevention of osteoporosis; they have cosmetic benefits, slow progression of Alzheimer's disease, lower cholesterol and systolic pressure, maintain dental health, and increase the nutritional quality and shelf life of food products by slowing the oxidation process and limiting growth of microflora. Pomegranate industrial wastes and by-products offer an opportunity to generate products of high functional value, such as powerful natural antioxidants, nutraceuticals, and natural preservatives that can be added to dairy products, meat, fish, cereals and nuts or made into films and coatings for food packaging. In conclusion, pomegranate by-products contain valuable compounds that can be used to improve the nutritional and functional properties of food products for animal or human consumption, and by converting them into nutraceutical products, pollution caused by wastes from industrial processing of pomegranate can be reduced.

Keywords: bioactive compounds, industrial wastes, antioxidants, health benefits, *Punica granatum*

Introduction

Pomegranate (*Punica granatum*) is a deciduous shrub native to Iran. It is highly adaptable to varying climate conditions and is currently cultivated worldwide. The largest producers are the Mediterranean countries (Costa et al. 2019). Since ancient times, its consumption has been associated with numerous health benefits (Kandylis and Kokkinomagoulos, 2020), improving the overall quality of food products (Parafati et al. 2021) and contributing to skin health (Ko et al. 2021). The consumption of pomegranate has increased due to its sensorial attributes and remarkable contents of bioactive compounds (Gullón et al. 2020). There are also non-commercial pomegranate fruits with some physical defects and fruit disorders but with good quality juice and seeds. However, they generate at the same time huge amounts of waste (Dhineshkumar and Ramasamy, 2016).

The agroindustry of pomegranate generates huge amounts of waste, such as peels and seeds, leading to disposal issues and environmental pollution (Kaderides et al. 2021). For this reason, sustainability in the food and agriculture industry is compromised. However, recycling and converting these wastes into value-added products for reuse in the food chain would revitalize sustainability (Ko et al. 2021) and contribute to a circular economy model based on sustainable processes (Gullón et al. 2020)

Pomegranate is mostly consumed fresh, although it is also consumed as minimally processed pomegranate seeds, jams, marmalades, natural juices, jellies, juice concentrates, frozen seeds, refrigerated seeds, seeds in syrup, candied arils, carbonated beverages, arils in brandy and in vinegar (Gumienna et al. 2016, Dhineshkumar and Ramasamy, 2016). Pomegranate is one of the world's healthiest fruits (Kaderides et al. 2021), displaying a wide range of phytochemicals including phenolic acids, anthocyanins, flavonoids (Fourati et al. 2020), tannins (El moujahed et al. 2021) ellagic and gallic acid (Meselhy et al. 2018). Pomegranate peel has higher antioxidant activity and higher amounts of bioactive compounds than the edible fleshy parts (Table 1) (Kaderides et al. 2021, Yan et al. 2017).

Both non-commercial pomegranate fruits with some physical defects and fruit disorders and industrial residues are promising sources of nutrients and bioactive components (Bostami et al. 2015) with potential functional properties (Gullón et al. 2020). Pomegranate waste can be reutilized to benefit human health (Dhineshkumar and Ramasamy, 2016) in food (Kandylis and Kokkinomagoulos, 2020) and in skin and health applications (Ko et al. 2021, Houston et al. 2017, Leong-Perng, et al. 2021).

Table 1. Total contents of phenols (TPC, mg GAE g^{-1}), flavonoids (TFC, mg RE g^{-1}), and proanthocyanidins TPR (mg CE g^{-1}), and antioxidant activity by DPPH, ABTS and FRAP in pomegranate juice, peel and seed

	Juice	Peel	Seed
TPC (mg GAE g^{-1})	0.39-1.29	57.7-155.9	2.96-3.78
TFC (mg RE g^{-1})	0.19-0.68	45.5-84.0	3.9-5.24
TPR (mg CE g^{-1})	2.75-10.43	84.9-97.3	2.25-4.71
DPPH (mg TEAC g^{-1})	1.4-8.0	428.5-1207.7	11.9-19.7
ABTS (mg TEAC g^{-1})	1.3-5.2	422.8-1105.1	11.4-16.1
FRAP (mg TEAC g^{-1})	2.3-9.4	357.0-1663.8	22.1-31.1

GAE (gallic acid equivalent), RE (rutin equivalent), TEAC (Trolox equivalent antioxidant capacity). Constructed by authors with data from Yan et al. 2017.

Both *in vitro* and *in vivo* studies have demonstrated that pomegranate seeds possess antioxidants, anti-cardiovascular diseases, anti-osteoporosis, antidiabetic, anti-inflammatory and anticancer activities, among others (Fourati et al. 2020). Pomegranate by-products possess antioxidant and antimicrobial properties that inhibit the process of lipid peroxidation in fatty foods, scavenging free radicals and preventing growth of microorganisms (Kaderides et al. 2021). Moreover, pomegranate by-products have great potential as animal feed, replacing some of the ingredients in diets and contributing to better quality meat.

Composition of Pomegranate

The pomegranate fruit consists of peel and membranes, arils, and seeds (Figure 1) (Tapia-Campos et al. 2016). Pomegranate is a source of vitamins, minerals, lipids (Table 2), and bioactive components (Table 3) that make it highly nutritive (Khadivi et al. 2020). Pomegranate juice contains 13.7-18.3 of total soluble solids, 124-145 mg g^{-1} fresh weight of total reducing sugar, 19.8-60.99 of titratable acid, and pH 3.23-3.72 (Yan et al. 2017)

Biofunctional compounds have been reported in practically all parts of the pomegranate (Table 3): in peel and pericarp (anthocyanidins, phenolic, catechin, flavones, flavanones), juice (quercetin, anthocyanins, glucose, ascorbic acid, ellagic), seed oil (sterols, ellagic acid), flowers (gallic acid, ursolic acid, triterpenoids including maslinic and Asiatic acid), root and bark (piperidine alkaloids, ellagitannins, punicalin and punicalagin acid, gallic acid, catechin, minerals, amino acids) and leaves (apigenin, tannins, flavone glycosides, luteolin) (Jauhar et al. 2018).

Generation of agro-industrial by-products is an economic and environmental problem. However, these raw materials can be a suitable source of bioactive compounds for technological or nutritional purposes (Figure 2) (Jeria et al. 2022). The pomegranate peel is rich in polyphenols, and punicalagin (505.89 mg g^{-1} DW) is the most abundant compound (Liu et al. 2022).

Figure 1. Pomegranate fruits, arils, peel, and seeds.

Table 2. Food value of pomegranate fruit var. "Wonderful" (raw and juice)

	Raw	Juice		Raw	Juice
(g)					
Water,	77.9	86	Carbohydrates	18.7	13.1
Energy kcal-kJ	83-346	54-228	Dietary fiber, total	4	0.1
Protein	1.67	0.15	Sugars, total	13.7	12.6
Total lipid	1.17	0.29	Glucose	--	6.28
Ash	0.53	0.49	Fructose	--	6.37
Minerals (mg)					
Calcium	10	11	Sodium	3	9
Iron	0.3	0.1	Zinc	0.36	0.09
Magnesium	12	7	Copper	0.158	0.021
Phosphorus	36	11	Manganese	0.119	0.095
Potassium	236	214	Selenium	0.0005	0.0003
Vitamin (mg)					
Vitamin C	10.2	0.1	Vitamin B-6	0.075	0.04
Thiamin	0.067	0.015	Folate, total	0.038	0.024
Riboflavin	0.053	0.015	Choline, total	7.6	4.8
Niacin	0.293	0.233	Vitamin E	0.6	0.38
Pantothenic acid	0.377	0.285	Vitamin K	0.0164	0.0104
Fatty acids (g)					
Total monounsaturated	0.093	0.0.059	Total saturated	0.12	0.077
MUFA 16:1	0.012	0.008	SFA 12:0	0.006	0.004
MUFA 18:1	0.077	0.049	SFA 14:0	0.006	0.004
MUFA 20:1	0.004	0.003	SFA 16:0	0.07	0.044
Total polyunsaturated	0.079	--	SFA 18:0	0.038	0.024
PUFA 18:2	0.079	--			
Campesterol	0.001	--			
Beta-sitosterol	0.004	--			

Vitamin C (total ascorbic acid), Vitamin E (alpha-tocopherol), Vitamin K (phylloquinone), USDA, 2019.

Table 3. Biofunctional compounds of pomegranate pulp, peel, juice, seed and leaves

	Compounds	References
Pulp	[bg]Gallic acid, [b]p-coumaric, [b]Ferulic acid, [b]Epicatechin, [bg]Rutin, [b]Punicalagin α, [bg]Punicalagin β, [b]Cyanidin-3,5-diglucoside, [b]Delphinidin 3-glucoside, [b]Pelargonidin 3,5-diglucoside, [b]Cyanidin 3-glucoside, [b]Pelargonidin 3-glucoside.	[b]Caldeira, et al. 2010, [g]Jauhar et al. 2018
Peel	[abcf]Gallic acid, [a]Chlorogenic acid, [acf]Ellagic acid, [ab]Ferulic acid, [b]p-coumaric, [a]Syringic acid, [b]Epicatechin, [ab]Rutin, [bcf]Punicalagin α, [bcf]Punicalagin β, [a]Methyl gallate, [a]catechin, [a]Pyrocatechol, [a]Vanillin, [a]Naringenin, [a]Taxifolin, [b]Cyanidin-3,5-diglucoside, [b]Delphinidin 3-glucoside, [b]Pelargonidin 3,5-diglucoside, [b]Cyanidin 3-glucoside, [b]Pelargonidin 3-glucoside.	[a]Farag et al. 2020, [b]Caldeira, et al. 2010 [c]Yan et al.2017, [f]Sabraoui et al. 2020
Seeds	Punicalagin β, Punicalagin α, Ellagic acid, Gallic acid.	[f]Sabraoui et al. 2020
Leaves	[a]Gallic acid, [a]Methyl gallate, [a]Catechin, [a]Syringic acid, [a]Pyrocatechol, [a]Rutin, [a]Ellagic acid, [a]Ferulic acid, [a]vanillin, [a]Naringenin, [a]Taxifolin.	[a]Farag et al. 2020
Oil (seeds)	[dg]Punicid acid, [d]α-Eleostearic acid, [d]catalpic, [d]β-Eleostearic, [d]3,4-Dihidroxy-phenylacetic, [d]p-Hidroxy-benzoic, [d]canillic, [d]syringic, [d]Ferulic, [d]m-Coumaric, [d]p-Coumaric, [d]Rosmarinic, [d]Quercitin, [d]Naringenin, [de]Tocoferoles (α, γ, δ, β), [e]Tocotrienol (α, γ, δ, β), [de]β-Caroteno, [e]Lutein, [e]Zeaxanthin, Campesterol, Stigmasterol, β-sitosterol, Brassicasterol, Campestanol, Delta-5 Avenasterol, Delta-7 Stigmasterol.	[d]Costa et al. 2019 [e]Badr et al. 2020

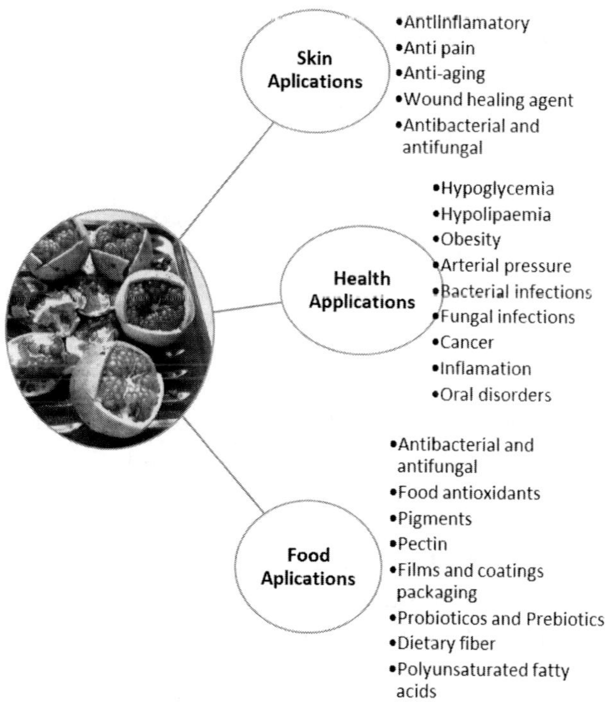

Figure 2. Health, skin, and food applications of pomegranate wastes.

Polyphenol Extraction Methods

Several extraction techniques have been employed to obtain valuable components from pomegranate residues. The most used methods are conventional, related to solid: liquid extraction techniques, which commonly use an organic solvent to obtain a variety of compounds. Recently, novel approaches have focused on the use of green technologies to obtain oils and phenolic compounds from pomegranate peel and seeds to improve the recovery of these compounds and their cost-effective extraction. Among the most used extraction techniques are the supercritical fluid (SFE), enzyme-assisted (EAE), ultrasound-assisted (UAE), pressurized liquid (PLE) and microwave-assisted (MAE) techniques (Jauhar et al. 2018). Each of these techniques considers several factors related to the optimization process in which the scale up process requires a good deal of statistical and modeling analysis for its implementation. Many scientific articles are related to the recovery optimization of oils and phenolic compounds from pomegranate and its residues. Here, we summarize the most recent studies on the recovery of phenolic compounds from pomegranate peel and seed residues.

One drawback in the use of natural antioxidants, both individual compounds and extracts, is their stability. Many antioxidant compounds start to degrade during extraction from the source and immediately after, which translates into loss of biological benefits. Structural changes in the bioactive compounds can occur at elevated water, air and oil temperatures, which are responsible for that loss. Furthermore, when it is processed at ambient temperatures, enzymes, chemicals, and the passage of time may trigger different chemical reactions within the materials, affecting the stability of the bioactive compounds. Treatments that reduce antioxidant activity are pasteurization, sterilization, blanching, evaporation, extrusion, drying, roasting, baking, frying, microwaving, infrared light applications, cutting, smoking and storage (even when frozen) (Carocho et al. 2017).

Pomegranate by-products contain oils and polyphenols that are of interest to many scientists as they may serve as raw materials for developing functional ingredients. Recently, different green technology approaches have been studied to maximize recovery of flavonoids and phenolic acids. For instance, Rashid et al. (2022) studied the UAE of polyphenols from pomegranate peel, using encapsulating agents for water-oil-water nanoemulsions to prevent a decrease in oxidative stability of the compounds and increase their antioxidant activity. Rifna and Dwivedi (2022) studied the use of PLE to optimize the extraction of tannins from pomegranate peel; they found that the use of PLE significantly increased antioxidant capacity by 67% compared with conventional extraction. Also, PLE extracts significantly increased the extracts' gallic acid and tannic acid derivatives. Furthermore, the PLE and UAE increased the extraction yields of phenolic compounds and peptides from pomegranate seed waste (Guzmán-Lorite et al. 2022). The combination of both bioactive molecules increased the antioxidant, antihypertensive and hypocholesterolemic activities relative to conventional solvent extraction. Vallarino et al. (2022) also used PLE and UAE in combination to obtain an extract rich in gallic acid and punicalagin. This extract was tested on mice with autoimmune encephalomyelitis, and efficiency was comparable to control spinal pathological issues. Liu et al. (2022) reported that UAE resulted in extraction of the highest yield of polyphenols (505.89 mg g dry weight), using 53% aqueous ethanol at a ratio of 1:25 w/v and ultrasonic power of 757 W over a period of 25 min. Moreover, Kupnik et al. (2022) compared conventional cold maceration and soxhlet extraction versus the UAE and SFE of polyphenols from pomegranate peels. This study showed that the total contents of phenols and proanthocyanidins, composed mainly of egallic acid, were higher using SFE at 20 MPa. In a similar study, the application of SFE at 20 MPa, 35°C, and 5.0 mL of ethanol using pomegranate peel, exhibited in a higher antioxidant capacity of the extract, which may be used as an ingredient to prevent oxidation of commercial biodiesel B100. (Witt et al. 2022). By this means, pomegranate seeds and peel may be a valuable resource to obtain bioactive compounds through sustainable green technologies.

Food Applications

About one-third of the world's food production is lost or wasted every year. Food waste is food that is still good for human consumption but is thrown away by retailers or consumers; according to FAO, this accounts for 40% of food losses in developed countries (Andrade et al. 2019). Pomegranate peel and mesocarp, considered fruit processing waste, are rich sources of beneficial phytochemicals, which can be employed to improve the overall quality of food products (Parafati et al. 2021).

Table 4. Effects of adding pomegranate by-products to food products

By-product	Product	Effect	References
Seed 5%	Cupcakes	↑Protein, fat and fiber contents	Ayoubi et al. 2022
Juice, 13%	Yogurt	↑Total phenolic contents by 4.3- to 6.1-fold, anthocyanin contents 2650–2870-fold, total acidity, syneresis, and redness value. ↓Streptococcus thermophilus count	Golmakani et al. 2021.
Peel and mesocarp	Cheese	Inhibition of S. aureus, S. hemolyticus, P. fluorescents, B. subtilis, and B. cereus	Parafati et al. 2021
Seed powder 7.5%	gluten-free bread	↑Springiness, total phenolics content, antioxidant activity. ↓Hardness and chewiness	Bourekoua et al. 2018
Peel 200 ppm	Sardine fish oil	Slowdown of lipid oxidation	Sarojini et al. 2019
Peel powder 1.5%	Fish Burger	↑Lipid, ash, and carbohydrate ↓Moisture, protein, and pH values	Abou-Taled, 2022
Peel 1-1.5%	Ground buffalo (Bubalus bubalis) meat	↑Content polyphenols and antioxidant activity Improved preservation of ground meat under 20 days of refrigeration	Ghimire et al. 2022
Peel 5 mg mL^{-1}	Carrot juice	↓Microbial counts at 28d. ↑Phenolic and flavonoid contents, as well as ABTS and FRAP antioxidant capacities, not affect color, total soluble solids and pH	Trigo et al. 2021

The application of pomegranate and its extracts, mainly as antioxidants and antimicrobials, has been extensively studied in different types of food products and has shown very promising results (Ko et al. 2021). In addition, many studies have shown that these additives can positively affect overall sensory quality and hence, the shelf life of food products (Kandylis and Kokkinomagoulos, 2020). Recent studies have demonstrated the potent antioxidant and antimicrobial effects of using pomegranate peel and seed as natural food additives (Ko et al. 2021). These are molecules introduced into foodstuffs to carry out specific technological functions, varying from conserving, imparting color and taste, sweetening, and texturizing (Table 4). They are added during the manufacture of the food and all steps of production and introduction in the foodstuff are strictly controlled (Carocho et al. 2017). They have also been used in bioplastics and edible coatings for food packaging; these components have shown great plasticizing effects on packaging materials while extending the shelf life of food through active packaging (Ko et al. 2021).

Food Antioxydants

Food antioxidants are used to prevent, among others, autoxidation and oxidation of foodstuffs, while stabilizing lipids and other food compounds to prevent the cascade of oxidative reactions. They also halt these reactions when they cannot be avoided. Food antioxidants added to foodstuffs have the same mission as the endogenous antioxidants of the human body, protecting food from oxidative attack and conserving its organoleptic, texture, and consumption safety properties (Carocho et al. 2017).

The use of preservatives in food has been widely used and plays a very important role in the food industry. The food industry's demand for antioxidants from natural sources has continuously grown, especially now with increasing numbers of adverse toxicological reports on many synthetic compounds (Kandylis and Kokkinomagoulos, 2020, 14). All food antioxidants have a common objective: to conserve food for the longest time without altering or conferring taste or color (Carocho et al. 2017). The use of pomegranate by-products in the fresh food industry could be a sustainable way to reduce environmental impact and costs associated with the disposal of by-products while making use of the beneficial effects on product quality and shelf life (Panza et al. 2021).

Lyophilized pomegranate peel nanoparticles at concentrations of 1 and 1.5% are powerful natural antioxidants. Their antimicrobial effect can retard oxidation in meatballs stored at $4 \pm 1^{o}C$ for 15 d. In addition, they increase the contents of phenolic compounds and antioxidant capacity and lower contents of peroxide, thiobarbituric acid reactive substances, and total volatile base nitrogen. They also reduce microbial load and generate acceptable sensory parameters in terms of odor and color (Morsy et al. 2018). Similarly, Panza et al. (2021) reported that final cooked cod sticks breaded with pomegranate peel powder contained three to five times more phenolic compounds, and two to seven times more flavonoids, and four times more antioxidant activity than the cooked control fish. Furthermore, microbial growth during refrigerated storage also slowed without negatively altering sensory characteristics.

Pigments

Synthetic pigments from petrochemicals have been extensively used in a wide range of food products (Sharma et al. 2021). However, these pigments have adverse effects on human health, making it obligatory for the scientific community to explore safer, natural, and eco-friendly pigments (Sharma et al. 2021). Valorization of plant wastes and plant by-products (e.g., peels, seeds or pomace) can meet the demands of natural pigment production at industrial levels for potential food, pharmaceutical, and cosmeceutical applications. These wastes/by-products are a rich source of natural pigments such as anthocyanins, betalains, carotenoids, and chlorophylls (Sharma et al. 2021). Anthocyanins are the most prevalent flavonoids in pomegranates and are responsible for the fruit's orange, red, and purple colors (Ko et al. 2021). In the natural pigment of pomegranate, Micó-Vicent et al. (2021) found the anthocyanins delphinidin-3-glucoside, cyanidin-3-glucoside, pelargonidin-3-glucoside, delphinidin-3,5-diglucoside, cyanidin3,5-diglucoside and pelargonidin-3,5-diglucoside, which can act as effective additives in films due to their inherent properties as flavonoids by changing the color of the packaging material to reduce food oxidation (Ko et al. 2021). Also, Hanani et al. (2019), with a fish gelatin film containing pomegranate peel powder (1-5%), reported a decrease in lightness and an increase in redness of the film, attributed to the presence of anthocyanins in pomegranate peel powder. Thus, the pomegranate's anthocyanins show great potential as an active ingredient for intelligent packaging that can slow oxidation and respiration rate and respond to changes in system pH through color changes in packaging films. Several other components in pomegranate rind and seed have plasticizing, strengthening, and elongating effects on bioplastics, which further improve the abilities of pomegranate biowaste as an effective active ingredient in food packaging system applications (Ko et al. 2021). It is envisaged that these natural pigments can contribute significantly to the development of functional foods with rich biotherapeutic potential (Sharma et al. 2021).

Pectin

Pectin is widely approved worldwide for use in foods as a safe additive (Ciriminna et al. 2015). Pectin is the methylated ester of polygalacturonic acid and has a wide range of applications in the pharmaceutical, cosmetic, and food industries: as a gelling agent in fruit-based products, stabilizer in fruit and milk beverages, and fruit filling for bakery and confectionary products (Marić et al. 2018), and for thickening and encapsulation of cosmetics (Ciriminna et al. 2015). Traditionally, pectin was mostly used in producing fruit jams and jellies, products with high sugar content, to secure the desired texture (Ciriminna et al. 2015). In detail, high-density pectin is used by the confectionery industry, while medium and low-density pectin is used to produce yogurt and fruit juice (Ciriminna et al. 2015). Large yields of pectin have been reported in pomegranate peel: 2.81-8.92% (Pereira et al. 2016) and 19.4% (Talekar et al. 2019). Therefore, commercial pectin could be replaced by that extracted from pomegranate peel to prepare pomegranate jam without altering its organoleptic properties. A jam with 0.2% pectin extracted from pomegranate peel exhibited overall acceptability like that of the jam prepared with commercial pectin. It was also redder and less firm (Abid et al. 2018).

Films and Coatings for Food Packaging

Valorization of industrial by-products has resulted in significant advances in the development of new food protectant formulations, which benefit from the synergistic effect between biocontrol agents and natural bioactive compounds (Kharchoufi et al. 2018). The application of edible coatings has recently been under considerable attention. An active packaging film can help maintain quality and prolong the shelf life of food products (Hanani et al. 2019, Ghorbani et al. 2021), preserving the natural pigments, reducing lipid oxidation in food products, and enhancing cell adhesion and durability of edible packaging (Kumar et al. 2020). An active packaging film prepared from fish gelatin with incorporated pomegranate peel powder (1-5%) exhibited higher tensile strength, which protects the physical integrity of the food. Also, the antioxidant properties of the films improved (Hanani et al. 2019). Similarly, Valdés et al. (2019) found high total phenolic content and antioxidant activity in energy bars with edible pectin-based coating with pomegranate peel extracts; ethanolic and aqueous extracts showed inhibitory effects on *S. aureus* and *Escherichia coli* (Ghorbani et al. 2021). Also, pomegranate peel powder (5%) in fish gelatin film increased antimicrobial activities against *Staphylococcus aureus*, *Listeria monocytogenes* and *E. coli* (Hanani et al. 2019). Similarly, Valdés et al. (2019)

found that an active packaging film of fish gelatin prepared with 10% pomegranate juice by-product inhibited *S. aureus*, and at 30% it enhanced the film's stiffness and water resistance and provided good ultraviolet and visible light carrier properties, resulting in improved opacity, which could be favorable for food with high lipid content. In another study, it was reported that the addition of 0.361 g pomegranate peel extract mL^{-1} to both chitosan and locust bean gum coatings reduced the incidence of *Penicillium digitatum* by 49 and 28%, respectively, and the addition of *Wickerhamomyces anomalus* cells to coatings strengthened the antifungal effect and reduced *P. digitatum* incidence (up to 95 and 75%, respectively) Kharchoufi et al. 2019).

Applications in Meat Production

The inclusion of whole pomegranate by-product (200g kg^{-1}) into a cereal-based concentrate diet can be a strategy to increase total polyunsaturated acids. This inclusion increased linolenic, rumenic, and vaccenic acids in lamb liver and muscle without affecting final body and carcass weight, dry matter intake, or average daily gain (Natalello et al. 2019). Similarly, Tabasum et al. (2015) found that supplementation of diets with 2% pomegranate by-products improved the meat composition of broiler meat, fatty acid profile, and reduced lipid oxidation, due to the increased crude protein and moisture contents, whereas ether extract and cholesterol decreased. Another study found that spent her breast meat samples dipped in 0.02% pomegranate fruit juice phenolics reduced protein oxidation, inhibited microbial growth, and were sensorily acceptable up to 12 days of refrigerated storage at 4 °C (Vaithiyanathan et al. 2011).

Likewise, the use of pomegranate-based marinades could lead to convenient upgraded ready-to-cook products with extended shelf life, while maintaining high sensory quality due the inhibition of *Pseudomonas* spp. and *Brochothrix thermosphacta*, which lead to the absence or reduced production of volatile compounds and organic acids that can cause off-odors typical of undesirable and degraded products (Lytou et al. 2018). In another study, Hayrapetyan et al. (2012) reported that pomegranate extract has the potential to be used as a natural preservative in meat products at 24.7 mg dry pomegranate extract mL^{-1} in meat paté stored at 4 °C for 46 days, as it inhibited the growth of *L. monocytogenes*. It thus has the potential to be used as a natural preservative in meat products.

Nutraceuticals

Nutraceutical is a broad term derived from the combination of nutrition and pharmaceutical (Ghani et al. 2019). Other authors define nutraceuticals as a group of products that are more than food but less than pharmaceuticals (Télessy, 2019). To date, Nutraceutical do not have a specific definition different from those of other food-derived categories, e.g., food supplements, herbal products, pre and probiotics, and functional foods (Daliu et al. 2018). Nutraceuticals are divided into three main categories: herbs, nutrients, and dietary supplements (Ghani et al. 2019). In most countries, nutraceuticals are included in the group of dietary supplements. Use frequency of nutraceuticals is 50-70% in the population of developed countries, and this percentage increases with age (Télessy, 2019).

Food products used as nutraceuticals are dietary fibers, prebiotics, probiotics, polyunsaturated fatty acids, antioxidants, spices, and other herbs and natural foods (Ghani et al. 2019). Clinical studies have been carried out on many nutraceuticals, and their results support their effectiveness and general safety (Télessy, 2019). The combined and concerted action of nutrient and biologically active compounds is flagged as an indicator of possible health benefits (Durazzo et al. 2020). Nutraceuticals are used with food to prevent and treat many acute chronic diseases (Ghani et al. 2019). Nutraceuticals are a challenge for the future of prevention and therapy and a triggering tool in medicine. The possibility of preventing disease and/or supporting pharmacological therapy, which is nowadays mainly based on pharmaceuticals, may be a powerful tool to combat pathological, chronic, long-term diseases in subjects who do not qualify for pharmacological therapy (Durazzo et al. 2020). Nutraceuticals cannot replace pharmaceuticals but can be a strong high-value tool for prevention and an aid in treatment of some pathological conditions (Daliu et al. 2018).

Probiotics

Probiotics are food compounds composed of live microbes. They have many beneficial effects on the human body. When these probiotic food supplements are taken in suitable concentrations, they can be highly beneficial for the host by maintaining microbial stability in the intestine (Ghani et al. 2019). Fermentation is a unique biotechnological process with great potential for recycling some industrial or agro-industrial process by-products into useful animal feeds in developing countries. Fermentation does not require the use of chemicals and is easy to manage under on-farm conditions or on an industrial scale. It can also improve nutritional quality and reduce the anti-nutritional effect of tannins contained in pomegranate (Bostami et al. 2015). A study tested a diet based on corn, wheat and soy flour with added fermented whole pomegranate by-product (2%) containing probiotics (*Lactobacillus acidophilus, Bacillus subtilis, Saccharomyces cerevisiae*) on broilers. It was found increase daily weight gain, lower fecal pH, and reduce emission of ammonia, hydrogen sulfide, and sulfur dioxide, minimizing air pollution (Bostami et al. 2015).

Dietary Fibers

Dietary supplements are substances that are synthesized by adding some essential constituents to our diet. Dietary fiber is the most important element of a healthy diet. With the help of dietary fiber, food can be digested efficiently, and waste can pass easily through the digestive system. This fiber is usually made up of carbohydrate compounds. They are classified into two types: insoluble and soluble dietary fibers. Many natural food sources, such as pomegranate peel and juice (USDA, 2019) contain two kinds of fibers in different concentrations (Ghani et al. 2019).

Fatty Acids

Fresh pomegranate and juice contain monounsaturated, polyunsaturated and saturated fatty acids (USDA, 2019), while seeds contain oil typically in the range of 12-20% (Ahmed et al.

2019b). Pomegranate seed oil is a non-traditional oil, rich in bioactive components with several functionalities that benefit health and achieve nutritional, antioxidant, and anti-mycotoxigenic activities in food (Badr et al. 2020). Conjugated fatty acids (>70%) are present in pomegranate seed oil, which contains higher a concentration in the form of punicic acid (9cis, 11trans, 13cis-conjugated linolenic acid), these fatty acids received remarkable interest due to valuable physiological effects on various diseases. Also, Momeni et al. (2021, 31) reported 73, 9.37, 7.74 and 0.95% linolenic, linoleic, oleic and eicosenoic acids, respectively, in Iranian pomegranates. Application of pomegranate seed oil on fungal media reduced aflatoxin secretion up to 63%, and zearalenone up to 78.5%. These results confirm the bio-functionality of the oil to regulate fungal secondary metabolite processes.

Health Applications

Fruits, in general, play a major role in maintaining a balanced diet. They provide abundant macro and micronutrients and bioactive compounds that promote health. Over the past few decades, there have been many studies indicating the importance of fruit consumption in the prevention of health-associated risks (Kandylis and Kokkinomagoulos, 2020). Recently, encouraging findings have increased interest in pomegranate. Numerous studies concerning the potential health benefits (Table 5) of pomegranate and its constituents, in addition to its nutritional value, have been conducted (Table 2 and 3) (Kandylis and Kokkinomagoulos, 2020).

Metabolic Syndrome

The pomegranate and especially its extracts have attracted the attention of the research community due to their numerous health benefits, some of them associated with obesity (Kandylis and Kokkinomagoulos, 2020), arterial pressure (Sahebkar et al. 2017, Grabež et al. 2020, Moazzen and Alizadeh, 2017), hypolipidemia (Grabež et al. 2020, Hou et al. 2019), and hypoglycemia (Faddladdeen and Ojaimi, 2019, Grabež et al. 2020).

Obesity and Lipid Metabolism

Obesity and metabolic syndrome are in continuous growth. Imbalances in lipid metabolism are profound features of metabolic disorders (Hou et al. 2019). The classical approach of hypocaloric diet and exercise has shown little long-term adherence; thus, there is an unending search for new strategies to prevent and treat obesity and related metabolic alterations (Jeria et al. 2022). Pomegranate peel extract has been found to be rich in a variety of phenolic substances, among which are ellagic acid, punicalagin (Xiang et al. 2022), and anthocyanin (Fabroni et al. 2016), which has recently been studied for its potential benefits to human health (Xiang et al. 2022).

In vivo and *in vitro* studies have shown that extracts of different pomegranate fractions (peels, flowers, juice, and seeds) regulate lipid metabolism in metabolic-disorder-associated diseases such as atherosclerosis, nonalcoholic fatty liver disease, and type 2 diabetes, helping to alleviate the development of diseases. Amelioration of oxidative stress and inflammatory response is believed to underlie the regulation of lipid metabolism by pomegranate extracts (Hou et al. 2019).

Table 5. Effects of pomegranate by-products on health

Disease	By-product	Method	Effect	References
Alzheimer's	Peel		↑Phenolic contents and antioxidant properties, inhibition of acetylcholinesterase, a key enzyme involved in the progress of Alzheimer's disease	Caldeira, et al. 2010
Lung cancer	Leaves	*In-vitro*	Suppressed microsomal prostaglandin E synthase-1 expression, apoptosis	Toda et al. 2020
Prostate cancer	Peel	*In-vitro*	Apoptosis and impairs metastasis	Deng et al. 2017
Breast and colon cancer	Peel	*In-vitro*	Decreased viability of cancer cells	Pinnamaneni 2020
Obesity	Seed oil	*In vitro*, human	Reduced the differentiation of preadipocytes to adipocytes, anti-obesity gene expression.	Khabeer et al. 2019
Liver dysfunction	Leaf extract plus gallic acid	*In vivo*, rats	Liver alterations restored more effectively using pomegranate	Prabhakar and Krishnamurthy 2019
Metabolic syndrome	Rind	*In-vitro*	Inhibition of metabolism-related enzymes: α-amylase, tyrosinase, and hyaluronidase	Ahmed et al. 2019a
Gastrointestinal nematodes	Whole fruit	*In vivo* small ruminants	Reduction by 50% of gastrointestinal nematode egg output	Castagna et al. 2021
Prevention of dental caries	Punicalagin	*In vitro*	Inhibition of Streptococcus mutans, S. Sanguinis at 500 mg mL-1 punicalagin gel equivalent	Millo et al. 2017
Anti-HIV-1	Ellagic acid, punicalagins	*In vitro*	Inhibition of RT-associated HIV-1 RNase H and HIV-1 IN LEDGF-dependent activities	Sanna et al. 2021

The phytotherapeutic treatment of excessive weight and/or moderate obesity is growing widely. For this reason, there is great interest in the phenolic compounds in fruits and vegetables, which may inhibit the pancreatic enzyme lipase, the main enzyme responsible for digestion of dietary triglycerides. Pomegranate juice extracts strongly inhibit pancreatic lipase *in vitro*, and inhibition of pancreatic lipase activity is positively correlated with high anthocyanin content (Fabroni et al. 2016).

In recent decades, pomegranate has been used in the prevention and treatment of metabolic syndrome (Hou et al. 2019). Purified pomegranate peel polyphenols, punicalagin and ellagic acid in different concentrations possess a lipid-lowering effect in human hepatic cells. These compounds decrease total cholesterol content and increase total bile acid content (Lv et al. 2016). In another study, Zhao et al. (2016) reported that macrophage cholesterol accumulation and foam cell formation are the hallmarks of early atherogenesis. Pomegranate peel polyphenols have been shown to inhibit macrophage foam cell formation and promote cholesterol efflux. Additionally, Grabež et al. (2020) reported that pomegranate peel extract has shown beneficial effects on blood levels of triglycerides, high-density lipoprotein cholesterol, thiobarbituric acid reactive species, and fatty acid profile in total plasma lipids of patients with diabetes mellitus type 2, suggesting antioxidative potential of pomegranate peel extract on hypolipidemia, hypoglycemia.

Diabetes

Diabetes mellitus can result from disorders in insulin secretion or receptors and is characterized by hyperglycemia. Natural antioxidants contained in pomegranate have a hypoglycemic effect (Faddladdeen and Ojaimi, 2019, Grabež et al. 2020). Pomegranate peel extract decreased glucose level in blood and showed effects against histological and functional alteration of diabetic-induced rat liver. Its use by diabetic patients has been advised. Its protective effect is most probably linked to the antioxidant activity of its polyphenol content (Faddladdeen and Ojaimi, 2019). Liu et al. (2022) suggest that punicalagin and the punicalagin-rich extract from pomegranate peel have the potential for development into functional foods for the prevention and management of diabetes because punicalagin has shown α-glucosidase inhibitory activity.

Blood Pressure Regulation

Pomegranate has potential benefits in dietary supplements or as therapeutic adjuvants against vascular diseases, such as hypertension, coronary artery disease, and peripheral artery disease (Wang et al. 2018); pomegranate juice or peel extract showed significant reductions in both systolic and diastolic readings (Sahebkar et al. 2017) after pomegranate juice or peel extract consumption in patients with type 2 diabetes (Grabež et al. 2020) or in patients with metabolic syndrome (Moazzen and Alizadeh, 2017). Pomegranate extract reduces oxidative stress and platelet aggregation, diminishes lipid uptake by macrophages, positively influences endothelial cell function, and is involved in blood pressure regulation. These effects are attributed to the presence of hydrolyzable tannins, ellagitannins and ellagic acid, anthocyanins, and flavonoids (Wang et al. 2018). Also, clinical studies have demonstrated that daily intake of pomegranate juice lessens hypertension and attenuates atherosclerosis in humans (Wang et al. 2018). This evidence suggests that it may be prudent to include this fruit juice in a heart-healthy diet (Sahebkar et al. 2017).

Anticancer

Flavonoids are naturally present in vegetables, fruits, and various plants. Flavonoids are a major class of polyphenols related to plant secondary metabolites and have many beneficial effects, inhibiting various types of cancers in humans (Ghani et al. 2019). Pomegranate peel extract exhibits a cytotoxic effect, reducing cell viability of human carcinoma cell lines of liver, breast, prostate, and lung cancer. The antioxidant and cytotoxic properties of pomegranate peel have been attributed to synergetic effects of phenolic phytochemicals, specifically to the large amounts of gallic and ellagic acid (Meselhy et al. 2018). Another study showed that pomegranate seed oil has good antioxidant and cytotoxic activities and is a good source of punicic acid γ-tocopherol and α-tocopherol. The best cytotoxic activity was obtained against the cervix adenocarcinoma cell line (HeLa), while moderate cytotoxic activity against colon and lung carcinoma cells was observed. Moreover, pomegranate seed oil did not have cytotoxic activity against normal cell lines, indicating that it is safe (Đurđević et al. 2018).

Antibacterial and Antifungal

Diseases caused by bacterial infections in clinical and food scenarios are one of the main threats to health and world development; bacterial infections are the main cause of chronic infections and mortality. For this reason, it is necessary to control and develop solutions to this type of disease. Frequently, antibiotics are the most common solution; however, incorrect and excessive use can cause changes in the bacteria, causing resistance and making treatment more difficult. The main mechanism generated by bacteria for resistance to antibiotics is the expression of enzymes that act on the antibiotic molecule by hydrolyzing the beta-lactam ring and inhibiting its mechanism of action (OMS, 2020, Wang et al. 2017). It is necessary to find other types of alternative compounds with antimicrobial activity that can be used in the pharmaceutical and food areas. Evaluation of new natural antimicrobials for the control of microbial growth is a viable option for many reasons. Natural antimicrobials are extracted from sources such as animals, plants, bacteria, mushrooms, algae, and viruses. They are organic compounds whose antimicrobial activity classifies them as chemical or biological. Plant sources have gained much importance among natural antimicrobials because of the wide variety of chemical compounds that can be obtained from plant leaves, roots, fruits, and seeds. These compounds are known as phytochemicals and the reported mechanisms of action by which they carry out their effect on microorganisms is through the inactivation of proteins, genes, signaling pathways, disruption of cell walls and biofilm formation (Abdelhamid and El-Dougdoug, 2020, Quinto et al. 2019).

Pomegranate antimicrobial activity has long been evaluated. Extracts from different fruit tissues, such as peel, seed, and juice, have shown a microbial growth-inhibiting effect, and the peel has a stronger antimicrobial effect than other pomegranate tissues (Fabrizio et al. 2017, Nozohour et al. 2018, Rongai et al. 2019). Analyses of extracts obtained from the peel have revealed the presence of compounds such as phenolic acids, anthocyanins, and tannins. Punicalagin and ellagic acid have been attributed to these compounds with antimicrobial activities. Among the reported mechanisms by which it exerts its antimicrobial activity is the inhibition of the catalytic activity of enzymes present in the bacterial cell wall, triggering an imbalance in the cell that leads to microbial cell death (Rongai et al. 2019, Singh et al. 2019).

Studies with ethanolic extracts of pomegranate peel and seed against *Staphylococcus aureus* and *P. aeruginosa* showed inhibition zones (18 to 27 mm) larger than those of the antibiotics used as controls. In addition, pomegranate peel extract had the strongest effect in terms of minimum inhibitory concentration (MIC) and minimum bactericidal concentration (MBC) (Nozohour et al. 2018). Recently, aqueous and methanolic extracts of pomegranate tannins showed an antibacterial effect against *Staphylococcus aureus*, *Escherichia coli*, and *Pseudomonas aeruginosa*, with inhibition zones of 20.6-30.3 mm for methanolic extracts and 10.8-15.3 mm for aqueous extracts (El moujahed et al. 2021).

Differences in the antimicrobial activity of other pomegranate tissues were also studied, in addition to the juice, seed, and peel. The activity of the mesocarp was analyzed, and the peel was the tissue with the highest activity against *Escherichia coli, Staphylococcus aureus, Pseudomonas aeruginosa, Enterococcus faecalis*. The inhibition zones for peel were mostly greater than 20 mm and the MIC less than 25 mg mL[1], suggesting the effectiveness of pomegranate tissues other than the juice for application in food for microbial control (Alsataf et al. 2021).

In food systems, antimicrobial applications have been carried out seeking objectives such as suppressing enteric infections, food preservation and gut health. Pomegranate extracts have been incorporated in the development of edible coatings for food grade packaging material or directly added to food matrices to extend shelf life. In all cases, incorporation of the extracts has resulted in a reduction in the microbial load or growth, showing it to be an option for the control of foodborne diseases (Licciardello et al. 2018, Chen et al. 2020).

Other work has focused on evaluating the antifungal effect of pomegranate extracts, using *in vitro* and *in vivo* methods and focusing mostly on peel extracts. The tests range from direct contact with microorganisms in previously inoculated fruit and its storage or in a food matrix.

Table 6. Antifungal effects of pomegranate tissues in test *in vitro* and *in vivo*

Test	Tissue	Microorganism	Reference
In vitro	Peels and seeds	*Candida albicans, Aspergillus brasilien*	Alsataf et al. 2021
In vitro	Peel	*Fusarium sambucinum*	Elsherbiny et al. 2016
In vitro	--	*F. oxysporum*	Rongai et al. 2019
In vivo	Peel	*Penicillium digitatum, Saccharomyces cerevisiae*	Kharchoufi et al. 2018
In vivo	Tannins extracts	*S. cerevisiae, C. tropicalis*	El moujahed et al. 2021
In vitro-vivo	Peel	*Monilinia laxa, M. fructigena*	El Khetabi et al. 2020
In vitro-vivo	Peel	*Botrytis cinerea, P. digitatum, P. expansum*	Destri et al. 2016
In vitro-vivo	Peel	*Xanthomonas campestris* pv. Campestris, *P. carotovorum* subsp. carotovorum	de Sousa et al. 2021

Some studies carried out on the antifungal activity of pomegranate extracts are shown in Table 6. All the studies show that pomegranate extract has antimicrobial activity and evaluate the possible application as an alternative to synthetic pesticides to improve the quality of the plant products, suggesting the use of the extract for plant disease control in preharvest or in postharvest to prevent fruit decomposition caused by pathogens during storage (Rongai et al. 2019, Belgacem et al. 2021).

In most studies, the extraction method is maceration and contact of solvents with pomegranate tissues for times ranging from minutes to hours. However, Alexandre et al. (2019) compared the effect of the extraction method on antimicrobial activity, comparing the effect of high pressure with that of enzymes (pectinase and cellulase) to obtain pomegranate peel extracts with antimicrobial compounds and growth inhibitors against *Bacillus cereus*, *Staphylococcus aureus*, *Listeria innocua*, *L. monocytogenes*, *Salmonella enteritidis*, *Pseudomonas aeruginosa* and *E. coli*, some of which were antibiotic resistant strains. The extract from enzymatic extraction did not have a significant impact on bacterial inhibition, while extracts from high-pressure extraction caused inhibition zones of around 22 mm, and in some cases, they were larger than the control extract from maceration. This result was possibly due to the punicalagin found in the high-pressure extract, whose concentrations were 4 times higher than in the enzymatic extract (Alexandre et al. 2019). Antimicrobial activity not only varies by the extraction method, other factors, such as the varieties, growing conditions, cultivation practices, and stages of development, can affect the presence and concentration of compounds with antimicrobial activity (Chen et al. 2020, Rosas-Burgos et al. 2017).

In general, pomegranate extracts (peel, juice, seeds and whole fruit) have shown antimicrobial effect against bacteria (gram-positive and gram-negative) and fungi. In most cases, the greatest inhibitory effect was against bacteria under different extraction conditions and pomegranate varieties. Therefore, the integral use of pomegranate is an alternative for control of pathogenic microorganisms in areas ranging from the clinic, food, and the environment.

Oral Health

The pericarp is the part of the fruit that contains more compounds used against oral pathogens, mainly those related to dental caries, such as *Streptococcus mutans*, *Streptococcus sanguinis*, and *Lactobacillus casei* (Millo et al. 2017). Hajifattahi et al. (2016) evaluated hydroalcoholic extract from pomegranate petals on common oral microorganisms achieving the maximum effect against *S. mutants*, the main microorganism responsible for dental plaque and caries. Also, Fabrizio et al. (2017) found effective inhibition against bacteria isolated from carious dentine, such as *Streptococcus mutans* and *Rothia dentocariosa*, using peel extract with MIC and MBC concentrations of 10 mg mL^{-1} and 15 mg mL^{-1}, respectively. However, inhibition by pomegranate juice was moderate. De Medeiros et al. (2015, 301) reported the use of a pomegranate mouthwash (6.25% juice) but found no differences in plaque index or bleeding on a probing index. However, it was effective in reducing *Streptococci*. Similar results were reported by Lalwani et al. (2014). Also, Millo et al. (2017) evaluated a dental gel (based on pomegranate juice) to prevent dental caries with promising results.

In other studies, pomegranate peel extract was reported to have antibacterial and antifungal effects against oral pathogens (Abdollahzadeh et al. 2011). Peel extract was also evaluated against oral candidiasis (Madugula et al. 2017); it was found that peel extract has an inhibitory effect *in vitro* like that of clotrimazole. The researchers concluded that peel extract may be used as a substitute for antifungal agents. Recently, a mouthwash (from peel extract) was evaluated for the control of dental caries pathogens; it was found to be highly effective relative to chlorhexidine (Benoy et al. 2021).

Skin Health Applications

The selection of suitable natural raw materials in cosmetic research and development is a key point, not only to obtain the expected results but also to avoid undesirable side effects (Karray et al. 2021). Pomegranate bioactive compounds (rind and seeds) can be integrated into skin health products, demonstrating the potential of biowaste converted into value-added products (Ko et al. 2021). Pomegranate phenolics are reported to have cutaneous benefits and to be effective in treating skin disorders, including hyperpigmentation. *In vitro*, pomegranate extract exhibited strong antioxidant activity with a tyrosinase inhibitory effect. Skin hyperpigmentation treating potency was confirmed by the suppression of cellular melanogenesis through inhibition of tyrosinase and TRP-2 as examined in the B16F10 melanoma cells (Kanlayavattanakul et al. 2020). Pomegranates have the potential as a novel approach to ameliorating the inflammation and pain associated with a range of skin conditions, including cold sores and herpetic stromal keratitis (Houston et al. 2017). Also, pomegranate extracts showed antifungal activities against *Candida glabrata* (Jayan et al. 2020). Additionally, pomegranate peel can be incorporated into cosmetic formulations as an excellent preservative due to its exceptional amount of phenolic compounds, powerful antioxidant activity, and antibacterial activity against pathogenic strains of Gram negative bacteria (*Salmonella enteric*, *Klebsiella pneumoniae*, *E. coli*, and *S. typhimurium*) and Gram-positive bacteria (*Staphylococcus aureus*, *Bacillus cereus*, and *Micrococcus luteus*) with inhibition zones between 16 and 25 mm (Karray et al. 2021).

Pomegranate juice, plant part extracts and phytochemicals were found to be effective in healing both acute and chronic wounds, such as burns, surgical incisions, excisions, gingival, oral lichen planus, aphthous stomatitis, and diabetic wounds, and gastric ulcers (Stefanou et al. 2021). Leong-Perng et al. (2021) probed fermented pomegranate as a functional supplement and a cosmetic ingredient. They found that the functional supplement improved moisture, brightness, elasticity, and collagen density of the skin of most subjects after 8 weeks relative to the baseline without treatment. Moreover, after 4 weeks of consumption of a cosmetic serum, moisture, brightness, elasticity, spots, UV spots, and collagen density of skin were slightly better than those at week 0. Therefore, daily fermented pomegranate extracts can protect the skin against oxidative stress and slow skin aging. Furthermore, pomegranate anthocyanins can be used as a safe, stable, homogeneous, non-irritating, effective topical anti-aging drug formulation for the aged (Kanlayavattanakul et al. 2020).

Another study reveals that a 10% powder of standardized pomegranate extract with 40% ellagic acid is promising for healing deep second-degree burn wounds. Pomegranate extract produced highly dense collagen with a good arrangement, accompanied by a complete and mature epithelium, a low number of inflammatory cells, and angiogenesis. This may be caused by the compounds in the pomegranate extract, which have antioxidant, anti-inflammatory, and anti-bacterial effects (Sektiari et al. 2019). Also, Lay et al. (2017) reported that topical formulation (hydrogel) containing a suspension of silibinin-loaded pomegranate oil-based nano-capsules is an alternative for the treatment of cutaneous UVB radiation-induced inflammation. It is important to mention that pomegranate extracts and phytochemicals have had results almost consistently better than or comparable to commercial medicines, with no side effects, while corticosteroids and antibiotics cause adverse effects that may be severe. These results indicate that pomegranate can be a safe, effective alternative wound healing agent (Stefanou et al. 2021).

Conclusion

Pomegranate waste is a rich source of natural bioactive constituents that can be used to improve the nutritional and functional properties of food products for animal or human consumption and as nutraceutical products. Peel extracts increase food stability upon processing, storage, or gastrointestinal digestion conditions by inhibiting oxidation as well as the growth of pathogenic microorganisms due to the large amounts of phenolic compounds with high antioxidant activity. Transformation of pomegranate by-products also reduces pollution caused by industrial waste, contributing to a circular bioeconomy through sustainable food waste management.

References

Abdelhamid, Ahmed G & Noha K El-Dougdoug. 2020. "Controlling foodborne pathogens with natural antimicrobials by biological control and antivirulence strategies," *Heliyon,* 6 :1-9, doi : 10.1016/j.heliyon.2020.e05020.

Abdollahzadeh, Sh., RY. Mashouf, H. Mortazavi, MH. Moghaddam, N. Roozbahani & M. Vahedi. 2011. "Antibacterial and antifungal activities of punica granatum peel extracts against oral pathogens." *Journal of Dentistry Tehran, University of Medical Science*, Iran) 8(1) :1-6.

Abid, Mouna, Héla Yaich, Hayfa Hidouri, Hamadi Attia, & MA. Ayadi. 2018. "Effect of substituted gelling agents from pomegranate peel on colour, textural and sensory properties of pomegranate jam." *Food chemistry,* 239 :1047-1054, doi: 10.1016/j.foodchem.2017.07.006.

Abou-Taleb, Mohamed (2022). "Fish burger quality treated by pomegranate peels powder during cold storage." *Egyptian Journal of Aquatic Biology and Fisheries*, 26(1) :201-215.

Ahmed, Mahesar Sarfaraz, Abdul Hameed Kori, Syed Tufail Hussain Sherazi, Aftab Ahmed Kandhro, and Zahid Husain Laghari. 2019b. Pomegranate (*Punica granatum*) seed oil. In : Fawzy R. M (editor), *Fruit oils: Chemistry and functionality* (pp. 691-709). Springer, Cham. Egypt.

Ahmed, Mohamed H, Hala M. Aldesouki & Farid A Badria. 2019a. "Effect of phenolic compounds from the rind of *Punica granatum* on the activity of three metabolism-related enzymes." *Biotechnology and Applied Biochemistry*, doi :10.1002/bab.1866.

Alexandre, Elisabete MC., Sara Silva, Sonia AO Santos, Armando JD Silvestre, Maria F Duarte, Jorge A Saraiva & Manuela Pintado. 2019. "Antimicrobial activity of pomegranate peel extracts performed by high pressure and enzymatic assisted extraction." *Food research international*, 115 :167-176.

Alsataf, Seba, Bulent Başyiğit, & Mehmet Karaaslan. 2021. "Multivariate analyses of the antioxidant, antidiabetic, antimicrobial activity of pomegranate tissues with respect to pomegranate juice." *Waste and Biomass Valorization*, 12(11) :5909-5921.

Andrade, Mariana A, Vasco Lima, Ana Sanchez Silva, Fernanda Vilarinho, Maria Conceição Castilho, Khaola Khwaldia, & Fernando Ramos. (2019). "Pomegranate and grape by-products and their active compounds : Are they a valuable source for food applications?" *Trends in Food Science & Technology,* 86:68-84.

Ayoubi, Azam, Mohammad Balvardi, Hamid-Reza Akhavan, Reza Hajimohammadi-Farimani1. 2022. "Fortified cake with pomegranate seed powder as a functional product." *Journal of Food Science and Technology*, 59 : 308-316.

Badr, Ahmed Noah, Hatem Salama Ali, Adel Gabr Abdel-Razek, Mohamed Gamal Shehata, & Najla A. Albaridi. 2020. "Bioactive components of pomegranate oil and their influence on mycotoxin secretion." *Toxins*, 12(12) :748, doi :10.3390/toxins12120748.

Belgacem, Imen, Maria G. Li Destri Nicosia, Sonia Pangallo, Ahmed Abdelfattah, Massimo Benuzzi, Giovanni E. Agosteo & Leonardo Schena. (2021). "Pomegranate peel extracts as safe natural treatments to control plant diseases and increase the shelf-life and safety of fresh fruits and vegetables." *Plants,* 10(3),453 :1-13.

Benoy, Jacob, Nivedhitha Malli Sureshbabu, Manish Ranjan, Aishwarya Ranganath & Riluwan Siddique. 2021. "The antimicrobial effect of pomegranate peel extract versus chlorhexidine in high caries risk individuals using quantitative real-time polymerase chain reaction: a randomized triple-blind controlled clinical trial." *International Journal of Dentistry,* 5563945 :1-14, doi :10.1155/2021/5563945

Bostami, Rubayet A. B. M., S.T. Ahmed, M. M. Islam, H. S. Mun, S. Y. Ko, S. S. Kim & C. J. Yang. 2015. "Growth performance, fecal noxious gas emission and economic efficacy in broilers fed fermented pomegranate by-products as residue of fruit industry." *Int. J. Adv. Res* 3(3) :102-114.

Bourekoua, Hayat, Renata Różyło, Urszula Gawlik-Dziki, Leila Benatallah, Mohammed Nasreddine Zidoune & Dariusz Dziki. 2018. "Pomegranate seed powder as a functional component of gluten-free bread

(Physical, sensorial and antioxidant evaluation." *International Journal of Food Science & Technology,* 53(8) :1906-1913, doi :10.1111/ijfs.13777.

Caldeira, Morzella Maresa, Jocelem Mastrodi Salgadob, Adna Prado Massariolib, Patricia Bachiegab, Alessandro de Oliveira Riosc, Severino Matias Alencarb, Andrés R. Schwemberd & Adriano Costa de Camargob. 2019. "Potential benefits of phenolics from pomegranate pulp and peel in Alzheimer's disease: Antioxidant activity and inhibition of acetylcholinesterase." *Journal of Food Bioactives,* 5 :136-141, doi: 10.31665/JFB.2019.5181.

Carocho, Márcio, Patricia Morales & Isabel CFR Ferreira. 2017. "Antioxidants: Reviewing the chemistry, food applications, legislation and role as preservatives." *Trends in Food Science & Technology,* 71 :107-120.

Castagna, Fabio, Cristian Piras, Ernesto Palma, Vincenzo Musolino, Carmine Lupia, Antonio Bosco, Laura Rinaldi, Giuseppe Cringoli, Vincenzo Musella & Domenico Britti. 2021. "Green veterinary pharmacology applied to parasite control : evaluation of *Punica granatum*, Artemisia campestris, Salix caprea aqueous macerates against gastrointestinal nematodes of sheep." *Veterinary Sciences,* 8(10) :237, doi :10.3390/vetsci8100237.

Chen, Jing, Chunling Liao, Xiaolu Ouyang, Ibrahim Kahramanoğlu, Yudi Gan & Mingxi Li. 2020. "Antimicrobial activity of pomegranate peel and its applications on food preservation." *Journal of Food Quality,* 8850339 :1-8.

Ciriminna, Rosaria, Norberto Chavarría-Hernández, Adriana Inés Rodríguez Hernández & Mario Pagliaro. 2015. "Pectin: A new perspective from the biorefi nery standpoint" *Biofuels, Bioproducts and Biorefining,* 9 :368-377, doi : 10.1002/bbb.1551.

Costa, AMM., LO. Silva, & AG Torres. 2019. "Chemical composition of commercial cold-pressed pomegranate (*Punica granatum*) seed oil from Turkey and Israel and the use of bioactive compounds for samples origin preliminary discrimination." *J. of Food Composition and Analysis,* 75 : 8-16, doi :10.1016/j.jfca.2018.09.004.

Daliu, Patricia, Antonello Santini & Ettore Novellino. 2019. "From pharmaceuticals to nutraceuticals: Bridging disease prevention and management." *Expert Review of Clinical Pharmacology,* 12(1) :1-7.

David, MJ, Joachim Bugert, Stephen P Denyer, & Charles M Heard. 2017. "Anti-inflammatory activity of *Punica granatum* L. (Pomegranate) rind extracts applied topically to ex vivo skin." *European Journal of Pharmaceutics and Biopharmaceutics*, 112, 30-37, doi :10.1016/j.ejpb.2016.11.014.

De Medeiros, Nóbrega Danúbia R., Ravely Lucena Santos, Renata de Souza Coelho Soares, Pollianna Muniz Alves, Ana Cláudia Dantas Medeiros & Jozinete Vieira Pereira. 2015. "A Randomized, Controlled Clinical Trial on the Clinical and Microbiological Efficacy of Punica granatum L. Mouthwash." *Pesquisa Brasileira em Odontopediatria e Clínica Integrada*, 15. 301-308.

De Sousa, Silva S., PCS Alves, DF Coutinho, TRSA Luz, GMG Fontoura, AA Berretta MC & Gonçalves Maciel. 2021. "*Punica granatum* L. extract contributes to phytopathogens control and enhances *Eruca vesicaria* (L.) Cav. germination in vitro and in vivo." *Advances in Horticultural Science,* 35(3) :217-231.

Deng, Yuanle, Yali Li, Fangfang Yang, Anqui Zeng, Shubuping Yang, Yi Luo, Yiwen Zhang, Yongmei Xie, Tinghong Ye, Yong Xia &Wenya Yin. 2017. "The extract from *Punica granatum* (pomegranate) peel induces apoptosis and impairs metastasis in prostate cancer cells." *Biomedicine & pharmacotherapy,* 93 : 976-984, doi :10.1016/j.biopha.2017.07.008.

Destri Nicoasia Li M, Sonia Pangallo, Ginat Raphae, Flora V. Romeo, Maria C. Strano, Paolo Rapisarda, Samir Droby & Leonardo Schena. 2016. "Control of postharvest fungal rots on citrus fruit and sweet cherries using a pomegranate peel extract." *Postharvest Biology and Technology*, 114 :54-61.

Dhineshkumar, V. & Ramasamy, D. 2016. "Pomegranate Processing and Value Addition : A Review." *Indian Horticulture Journal* 6(1) :1-12.

Durazzo, Alessandra, Massimo Lucarini & Antonello Santini. 2020. "*Nutraceuticals in human health. Foods*" 9(3) :370, doi :10.3390/foods9030370.

Đurđević, Sanja, katarina Šavikin, Jeleana Živković, Volker Böhm, Tatjana Stanojković, Ana Damjanović, & Slobodam Petrović. 2018. Antioxidant and cytotoxic activity of fatty oil isolated by supercritical fluid extraction from microwave pretreated seeds of wild growing Punica granatum L. *The Journal of Supercritical Fluids*, 133 :225-232, doi :10.1016/j.supflu.2017.10.021.

Echeverria, Francisca, Paula Andrea Jimenez Patino, Mauricio Castro-Sepulveda, Andres Bustamante, Paula Andrea Garcia Concha, Carlos Poblete-Aro, Rodrigo Valenzuela & Diego F. Garcia-Diaz. 2021.

"Microencapsulated pomegranate peel extract induces mitochondrial complex IV activity and prevents mitochondrial cristae alteration in brown adipose tissue in mice fed on a high-fat diet." *British Journal of Nutrition*, 126(6):825-836, doi :10.1017/S000711452000481X.

El Khetabi, Assia, Rachid Lahlali, Latifa Askarne, Said Ezraria, Lahsen El Ghadaroui, Abdessalem Tahiri, Jovana Hrustić, Said Amiri. 2020. "Efficacy assessment of pomegranate peel aqueous extract for brown rot (*Monilinia* spp.) disease control." *Physiological and Molecular Plant Pathology*, 110 :101482 :1-9.

El moujahed, Sara, Fouad Ouazzani Chahdi, Youssef Kandri Rodi, Lahsen El ghadraoui, Lamiaa Lemjallad, Faouzi Errachid. 2021. "The moroccan pomegranate: an underrated source of tannins extracts and natural antimicrobials from juice processing by-products." *Waste and Biomass Valorization*, 12(10) :5383-5399.

Elsherbiny, E. A., Basma H. Amin & Zakaria A. Baka. 2016. "Efficiency of pomegranate (*Punica granatum* L.) peels extract as a high potential natural tool towards Fusarium dry rot on potato tubers." *Postharvest Biology and Technology*, 111 :256-263.

Fabrizio, Ferrazano Giamaria, Elisa Scioscia, Daniela Sateriale, Gabiria Pastore, Roberta Colicchio, Chiara Pagliuca, Tiziana Cantile, Brunella Alcidi, Marco Coda, Aniello Ingenito, Elena Scaglione, Annunziata Gaetana Cicatiello, Maria Grazia Volpe, Michele Di Stasio, Paola Salvatore & Caterina Pagliarulo. 2017. "In vitro antibacterial activity of pomegranate juice and peel extracts on cariogenic bacteria." *BioMed research international*, doi :10.1155/2017/2152749.

Fabroni, Simona, Gabriele Ballistreri, Margherita Amenta, Flora V Romeo and Paolo Rapisarda. 2016. "Screening of the anthocyanin profile and in vitro pancreatic lipase inhibition by anthocyanin-containing extracts of fruits, vegetables, legumes and cereals." *Journal of the Science of Food and Agriculture*, 96(14) :4713-4723, doi :10.1002/jsfa.7708.

Faddladdeen, Khadija A, & Ahlam Abdulaziz Ojaimi. 2019. "Protective effect of pomegranate (*Punica granatum*) extract against diabetic changes in adult male rat liver: histological study." *Journal of microscopy and ultrastructure*, 7(4) :165-170.

Farag, Radwan S., Mohamed S. Abdel-Latif, Hanaa H. Abd El Baky & Layla S. Tawfeek. 2020. "Phytochemical screening and antioxidant activity of some medicinal plants" crude juices. *Biotechnology Reports*, 28, e00536. 1-7.

Fourati, Mariam, Hajer Ben Hlima, Khaoula Elhadef, Olfa Ben Braïek, Karim Ennouri, Ahlem Chakchouk Mtibaa & Lotfi Mellouli. 2020. "Bioactive compounds and pharmacological potential of pomegranate (*Punica granatum*) seeds-a review." *Plant Foods for Human Nutrition*, 75(4) :477-486.

Ghani, Usman, Muhammad Naeem, Hamza Rafeeq, Usman Imtiaz, Ayesha Amjad, Saqib Ullah, Asim Rehman & Fatima Qasim. 2019. "A novel approach towards nutraceuticals and biomedical applications." *Sch Int. J. Biochem*, 2 :245-252.

Ghimire, Arjun, Niraj Paudel, & Ranjana Poudel. 2022. "Effect of pomegranate peel extract on the storage stability of ground buffalo (*Bubalus bubalis*) meat." *LWT*, 154, 112690, doi :10.1016/j.lwt.2021.112690.

Ghorbani, Ehsan, Arasb Dabbagh Moghaddam, Anousheh Sharifan & Hossein Kiani. 2021. Emergency food product packaging by pectin-based antimicrobial coatings functionalized by pomegranate peel extracts. *Journal of Food Quality*, ID 6631021, 10 pages, doi :10.1155/2021/6631021.

Golmakani, Mohammad-Taghi, Mohammad Hadi Eskandari, Somayyeh Kooshesh & Mahboobeh Pishan. 2021. Investigation of the effects of pomegranate juice addition on physicochemical, microbiological, and functional properties of set and stirred yogurts. *Food Science & Nutrition*, 9(12) :6662-6671, doi :10.1002/fsn3.2615.

Grabež, Milkica, Ranko Škrbić, Miloš P. Stojiljković, Vesna Rudić-Grujića, Marija Paunović, Aleksandra Arsić, Snježana Petrović, Vesna Vučić, Bosa Mirjanić-Azarić, Katarina Šavikin, Nebojša Menković, Teodora Janković, & Nađa Vasiljević. 2020. "Beneficial effects of pomegranate peel extract on plasma lipid profile, fatty acids levels and blood pressure in patients with diabetes mellitus type-2 : A randomized, double-blind, placebo-controlled study." *Journal of Functional Foods*, 64, 103692, doi :10.1016/j.jff.2019.103692.

Gullón, Patricia, Gonzalo Astray, Beatriz Gullón, Igor Tomasevic & José M. Lorenzo. 2020. "Pomegranate peel as suitable source of high-added value bioactives: Tailored functionalized meat products." *Molecules*, 25(12) :2859, doi :10.3390/molecules25122859.

Gumienna, Małgorzata, Artur Szwengiel & Barbara Górna. 2016. "Bioactive components of pomegranate fruit and their transformation by fermentation processes." *European Food Research and Technology*, 242(5) :631-640.

Guzmán-Lorite, M, ML. Marina & MC. García. 2022."Pressurized liquids vs. high intensity focused ultrasounds for the extraction of proteins from a pomegranate seed waste." *Innovative Food Science & Emerging Technologies,* 77 :102958.

Hajifattahi, Farnaz, ElhamMoravej-Salehi, Maryam Taheri, ArashMahboubi & Mohammad Kamalinejad. 2016. "Antibacterial effect of hydroalcoholic extract of *Punica granatum* linn. petal on common oral microorganisms." *International journal of biomaterials*, 8098943 :1-6, doi :10.1155/2016/8098943.

Hanani, ZA Nur, F Cheng Yeea & MAR Nor-Khaizura. 2019. "Effect of pomegranate (*Punica granatum* L.) peel powder on the antioxidant and antimicrobial properties of fish gelatin films as active packaging." *Food Hydrocolloids*, 89 :253-259.

Hayrapetyan, Hasmik, Wilma C Hazeleger & Rijkelt R Beumer. 2012. "Inhibition of Listeria monocytogenes by pomegranate (*Punica granatum*) peel extract in meat paté at different temperatures." *Food Control*, 23(1) :66-72.

Hou, Chen, Weimin Zhang, Jianke Li, Lin Du, Ou Lv, Shengjuan Zhao & Jia Li 2019. "Beneficial effects of pomegranate on lipid metabolism in metabolic disorders." *Molecular nutrition & food research*, 63(16), 1800773, doi :10.1002/mnfr.201800773Houston,

Jauhar, Safiullah, Mohammad Rashedi Ismail-Fitry, Gun Hean Chong, Mahmud Ab Rashid Nor-Khaizura, & Wan Zunairah Wan Ibadullah. 2018. "Polyphenol compounds from Pomegranate (*Punica Granatum*) extracted via various methods and its application on meat and meat products: a review." *Journal of Advanced Research in Applied Sciences and Engineering Technology*, 12(1) :1-12.

Jayan, Lekshmy, N Priyadharsini, R Ramya & K Rajkumar. 2020. "Evaluation of antifungal activity of mint, pomegranate and coriander on fluconazole-resistant *Candida glabrata*." *J. of oral and maxillofacial pathology,* 24(3) :517-522, doi : 10.4103/jomfp.JOMFP_355_19.

Jeria, Nicolas, Cornejo Sebastian, Sebastian Cornejo, Gabriel Prado, Andres Bustamante, Diego F Garcia-Diaz, Paula Jimenez, Rodrigo Valenzuela, Carlos Poblete-Aro & Francisca Echeverria. 2022. "Beneficial effects of bioactive compounds obtained from agro-industrial by-products on obesity and metabolic syndrome components." *Food Reviews International*, 1-30, doi :10.1080/87559129.2021.2013498.

Kaderides, Kyriakos, Anastasia Kyriakoudi, Ioannis Mourtzinos, & Athanasia M. Goula. 2021. "Potential of pomegranate peel extract as a natural additive in foods." *Trends in Food Science & Technology*, 115 :380-390.

Kandylis, Panagiotis, & Evangelos Kokkinomagoulos. 2020. "Food applications and potential health benefits of pomegranate and its derivatives." *Foods,* 9(2):122, doi :10.3390/foods9020122.

Kanlayavattanakul, Mayuree, Wichayada Chongnativisit, Puxvadee Chaikul, & Nataya Lourith. 2020. "Phenolic-rich pomegranate peel extract: *in vitro,* cellular, and *in vivo* activities for skin hyperpigmentation treatment." *Planta Medica*, 86(11) :749-759, doi : 10.1055/a-1170-7785.

Karray, Aida, Najeh Krayem & Hajer Ben Saad & Adel Sayari. 2021. "Spirulina platensis, *Punica granatum* peel, and moringa leaves extracts in cosmetic formulations : an integrated approach of in vitro biological activities and acceptability studies." *Environmental Science and Pollution Research*, 28(7) :8802-8811, doi :10.1007/s11356-020-11156-6.

Khabeer, Trichur Shamsiya, Akila Prashant & Manonmani Haravey Krishnan. 2019. "Dietary fatty acids from pomegranate seeds (*Punica granatum*) inhibit adipogenesis and impact the expression of the obesity-associated mRNA transcripts in human adipose-derived mesenchymal stem cells." *J Food Biochem,* 43(3) : e12739, doi :10.1111/jfbc.12739.

Kharchoufi, Samira, Lucia Parafati, Fabio Licciardello, Giuseppe Muratore, Mokthar Hamdi, Gabriella Cirvilleri & Cristina Restucci. 2018. "Edible coatings incorporating pomegranate peel extract and biocontrol yeast to reduce *Penicillium digitatum* postharvest decay of oranges." *Food Microbiology*, 74 :107-112.

Ko, Katharine, Younas Dadmohammadi, & Alireza Abbaspourrad. 2021. "Nutritional and bioactive components of pomegranate waste used in food and cosmetic applications : A review." *Foods*, 10(3) :657.

Kumar Nishant, Pratibha, & Neeraj. 2020. "Functional properties of pomegranate peel in edible coating/film : A review." *International Journal of Postharvest Technology and Innovation*, 7(3) :205-216.

Kupnik, Kaja, Maja Leitgeb, Mateja Primoži˘c, Vesna Postružnik, Petra Kotnik, Nika Ku˘cuk, Željko Knez & Maša Knez Marevci. 2022. "Supercritical Fluid and Conventional Extractions of High Value-Added Compounds from Pomegranate Peels Waste : Production, Quantification and Antimicrobial Activity of Bioactive Constituents." *Plants,* 11(7) : 928.

Lalwani, Vidushi, Koneru Anila, Vanishree M, Vardendra M, Hunasgi Santosh, & Surekha R. 2014. "Antimicrobial activity of *Punica granatum* on streptococcus in dental caries patients and healthy individuals: A comparative study." *Journal of Advanced Clinical and Research Insights*, 1 :94-98.

Lay Marchiori Marila Crivellaro, Cristina Rigon, Camila Camponogara, Sara Marchesan Oliveira, Letícia Cruz. 2017. "Hydrogel containing silibinin-loaded pomegranate oil based nanocapsules exhibits anti-inflammatory effects on skin damage UVB radiation-induced in mice." *Journal of Photochemistry and Photobiology B : Biology*, 170 :25-32, doi :10.1016/j.jphotobiol.2017.03.015.

Leong-Perng Chan, Ya-Ping Tseng, Cheng Liu & Chia-Hua Liang 2021. "Fermented pomegranate extracts protect against oxidative stress and aging of skin." *Journal of Cosmetic Dermatology*, doi :10.1111/jocd.14379.

Licciardello, Fabio, Samira Kharchoufib, Giuseppe Muratorec & Cristina Restuccia. 2018. "Effect of edible coating combined with pomegranate peel extract on the quality maintenance of white shrimps (*Parapenaeus longirostris*) during refrigerated storage." *Food packaging and shelf life*, 17 :114-119.

Liu, Yi, Kin Weng Kong, Ding-Tao Wu, Hong-Yan Liu, Hua-Bin Li, Jia-Rong Zhang & Ren-You Gan. 2022. "Pomegranate peel-derived punicalagin: ultrasonic-assisted extraction, purification, and its α-glucosidase inhibitory mechanism." *Food Chem* 374 :131635, doi :10.1016/j.foodchem.2021.131635.

Lv, Ou, Jianke Li, Qianqian Ma & Wei Zhao. 2016. "Effects of pomegranate peel polyphenols on lipid accumulation and cholesterol metabolic transformation in L-02 human hepatic cells via the PPARγ-ABCA1/CYP7A1 pathway." *Food & function*, 7(12) :4976-4983, doi :10.1039/c6fo01261b.

Lytou, Anastasia E, George-John E Nychas & Efstathios Z Panagou. 2018. "Effect of pomegranate based marinades on the microbiological, chemical and sensory quality of chicken meat: A metabolomics approach." *International journal of food microbiology*, 267 :42-53.

Madugula Prethi, Sudhakara Reddy, Jyothirmai Koneru, Atla Srinivasa Rao, Rayapureddi Sruthi, & Divya Teja Dalli. 2017 "Rhetoric to Reality"- Efficacy of *Punica granatum* Peel Extract on Oral Candidiasis: An *in vitro* Study. *J Clin Diagn Res,* 11(1) :114-117.

Marić, Mirela, Antonela Ninčević Grassino, Zhenzhou Zhu, Francisco J. Barba, Mladen Brnčićb, & Suzana Rimac Brnčićb. 2018. "An overview of the traditional and innovative approaches for pectin extraction from plant food wastes and by-products: Ultrasound-, microwaves-, and enzyme-assisted extraction." *Trends in Food Science & Technology*, 76 :28-37.

Meselhy, Khaled M, Manal M. Shams, Noheir H. Sherif & Sawsan M. Elsonbaty. 2020. "Phytochemical study, potential cytotoxic and antioxidant activities of selected food by-products (pomegranate peel, rice bran, rice straw & mulberry bark)." *Natural Product Research*, 34(4) :530-533.

Micó-Vicent, Barbara, Marina Ramos, Valentin Viqueira, Francesca Luzi, Franco Dominici, Andrea Terenzi, Etienne Maron, Mahmoud Hamzaoui, Stephane Kohnen, Luigi Torre, Alfonso Jiménez, Debora Puglia, & María Carmen Garrigós. (2021). "Anthocyanin hybrid nanopigments from pomegranate waste: colour, thermomechanical stability and environmental impact of polyester-based bionanocomposites." *Polymers*, 13(12), 1966 :1-25.

Millo, Grazielle, Apa Juntavee, Ariya Ratanathongkam, Natsajee Nualkaew Jomjai Peerapattana & Supaporn Chatchiwiwattana. 2017. "Antibacterial inhibitory effects of *Punica granatum* gel on cariogenic bacteria: an *in vitro* study." *International journal of clinical pediatric dentistry*, 10(2) :152, doi :10.5005%2Fjp-journals-10005-1426.

Moazzen, Hossein & Mohammad Alizadeh. 2017. "Effects of pomegranate juice on cardiovascular risk factors in patients with metabolic syndrome : a double-blinded, randomized crossover controlled trial." *Plant foods for human nutrition*, 72(2) :126-133.

Momeni, Nafiseh & Hossein Ali Asadi-Gharneh. 2021. "Fatty acids composition of seed oils obtained from eight Iranian pomegranate cultivars." *Journal of Medicinal Plants*, 20(77) :26-36.

Morsy, Mohamed K., Enas Mekawi, & Rasha Elsabagh. 2018. "Impact of pomegranate peel nanoparticles on quality attributes of meatballs during refrigerated storage." *Food Science and Technology*. 89 :489-495.

Natalello, Antonio, Giuseppe Luciano, Luciano Morbidini, Bernardo Valenti, Mariano Pauselli, Pilar Frutos, Luisa Biondi, Pablo J. Rufino-Moya, Massimiliano Lanza, & Alessandro Priolo. 2019. "Effect of feeding pomegranate by-product on fatty acid composition of ruminal digesta, liver, and muscle in lambs." *Journal of agricultural and food chemistry*, 67(16) :4472-4482.

Nozohour, Yaser, Reza Golmohammadi, Reza Mirnejad & Majid Fartashvand. 2018. "Antibacterial activity of pomegranate (*Punica granatum* L.) seed and peel alcoholic extracts on *Staphylococcus aureus* and *Pseudomonas aeruginosa* isolated from health centers." *Journal of Applied Biotechnology Reports*, 5(1) :32-36.

OMS, 2020. "*Resistencia a los antibióticos*" ["*Antibiotic resistance*"] available In : https://www.who.int/news-room/fact-sheets/detail/antibiotic-resistance.

Panza, Olimpia, Conte Amalia, Matteo Alessandro Del Nobile. 2021. "Pomegranate by-products as natural preservative to prolong the shelf life of breaded cod stick." *Molecules*, 26(8),2385 :1-12.

Parafati, Lucia, Fabiola Pesce, Laura Siracusa, Biagio Fallico, Cristina Restuccia, & Rosa Palmeri. 2021. "Pomegranate by-product extracts as ingredients for producing experimental cheese with enhanced microbiological, functional, and physical characteristics." *Foods*, 10(11), 2669, doi:10.3390/foods10112669.

Pereira, Paulo Henrique F, Túlio Ítalo S Oliveira, Morsyleide F Rosa, Fabio Lima Cavalcante, & Graham K Moates, Nikolaus Wellner, Keith W Waldron, & Henriette MC Azeredoc. 2016. "Pectin extraction from pomegranate peels with citric acid." *International journal of biological macromolecules*, 88,373-379, doi :10.1016/j.ijbiomac.2016.03.07.

Pinnamaneni, Rajasekhar. 2020. "Cell viability studies and anti-cancerous activity evaluation of pomegranate (*Punica granatum* L) extract." *Research J. Pharm. and Tech,* 13(1) :303- 307, doi :10.5958/0974-360X.2020.00061. X.

Prabhakar, Rao Sreedevi, Vijayalakshmi Krishnamurthy. 2019. "Ethanol extract of *Punica granatum* L. leaf in combination with gallic acid ameliorates liver dysfunction in high-fat diet-induced obese rats." *Indian Journal of Pharmaceutical Sciences*, 81(4) : 673-680.

Quinto, Emiliano J, Irma Caro, Luz H. Villalobos-Delgado, Javier Mateo, Beatriz De-Mateo-Silleras & María P. Redondo-Del-Río. 2019. *Food safety through natural antimicrobials. Antibiotics*, 8(4), 208 :1-30.

Rashid, Rubiya, F. A. Masoodi, Sajad Mohd Wani, Shaziya Manzoor, & Amir Gull. 2022. "Ultrasound assisted extraction of bioactive compounds from pomegranate peel, their nanoencapsulation and application for improvement in shelf life extension of edible oils." *Food Chemistry,* 385 : 132608.

Rifna, E. J., & Madhuresh Dwivedi. 2022. "Effect of pulsed ultrasound assisted extraction and aqueous acetone mixture on total hydrolysable tannins from pomegranate peel." *Food Bioscience,* 45 : 101496.

Rongai, Domenico, Patrizio Pulcini, Giovanni Di Lernia, Paolo Nota, Pjerin Preka, & Filomena Milano. 2019. "Punicalagin content and antifungal activity of different pomegranate (*Punica ganatum* L.) genotypes." *Horticulturae*, 5(3), 52 :1-9.

Rosas-Burgos, Ema C., Armando Burgos-Hernández, Luis Noguera-Artiaga, Miroslava Kačániová, Francisca Hernández-García, Jose L Cárdenas-López, & Ángel A Carbonell-Barrachina. 2017. "Antimicrobial activity of pomegranate peel extracts as affected by cultivar." *Journal of the Science of Food and Agriculture*, 97(3) :802-810.

Sabraoui, Talal, Taleb Khider, Boubker Nasser, Rabiaa Eddoha, Abderrahman Moujahid, Maryam Benbachir, & Abdelkhalid Essamadi. 2020. "Determination of punicalagins content, metal chelating, and antioxidant properties of edible pomegranate (*Punica granatum* L) peels and seeds grown in Morocco." *International Journal of Food Science*, doi :10.1155/2020/8885889.

Sahebkar, Amirhossein, Claudio Ferri, Paolo Giorgini, Simona Bo, Petr Nachtigal, & David Grassi. 2017. "Effects of pomegranate juice on blood pressure : A systematic review and meta-analysis of randomized controlled trials." *Pharmacological Research*, 115 :149-161, doi :10.1016/j.phrs.2016.11.018.

Sanna, Cinzia, Arianna Marengo, Stefano Acquadro, Alessia Caredda, Roberta Lai, Angela Corona, Enzo Tramontano, Patrizia Rubiolo, & Francesca Esposito. 2021. "*In Vitro* Anti-HIV-1 reverse transcriptase and integrase properties of *Punica granatum* L. leaves, bark, and peel extracts and their main compounds." *Plants*, 10(10) :2124, doi :10.3390/plants10102124.

Sarojini, A, Raju CV, Lakshmisha IP, Amitha & Gajedra. 2019. "Effect of pomegranate (*Punica granatum*) peel extract on lipid oxidation in sardine fish oil." *J. Entomol. Zool. Stud*, 7 :140-144.

Sektiari, Lukiswanto B, Alya Miranti, Sri Agus Sudjarwo, Hardany Primarizky, & Misaco Wiwik Yuniarti. 2019. "Evaluation of wound healing potential of pomegranate (*Punica granatum*) whole fruit extract on skin burn wound in rats (*Rattus norvegicus*)." *Journal of Advanced Veterinary and Animal Research*, 6(2) :202-207.

Sharma, Minaxi, Zeba Usmani, Vijai Kumar Gupta & Rajeev Bhat. 2021. "Valorization of fruits and vegetable wastes and by-products to produce natural pigments." *Critical Reviews in Biotechnology*, 41(4) :535-563.

Singh, Balwinder, Jatinder Pal Singh, Amritpal Kaur & Narpinder Singh. 2019. "Antimicrobial potential of pomegranate peel: A review." *International Journal of Food Science & Technology*, 54(4) :959-965.

Stefanou, Valentina, Dimitris Timbis, Anastasia Kanellou, Dimitra Margari, Myrto Trianti, Ioannis Tsaknis, Anastasia Azar Naka & Vladimiros Lougovois. 2021. "Wound Healing Properties of Pomegranate." *Archives of Microbiology & Immunology*, 5(3) :263-291.

Talekar, Sachin, Antonio F. Patti, R. Vijayraghavan, & Amit Arora. 2019. "Recyclable enzymatic recovery of pectin and punicalagin rich phenolics from waste pomegranate peels using magnetic nanobiocatalyst." *Food Hydrocolloids*, 89 :468-480.

Télessy, István G. 2019. "Nutraceuticals." In : Singh Ram B, Ross Watson Ronald, Takahashi Toru (eds). *The role of functional food security in global health* (pp. 409-421). Academic Press. San Diego Ca.

Toda, Keisuke, Mai Ueyama, Shomu Tanaka, Izumi Tsukayama, Takuto Mega, Yuka Konoike, Asako Tamenobu, Februadi Bastian, Iria Akai, Hideyuki Ito, Yuki Kawakami, Yoshitaka Takahashi, & Toshiko Suzuki-Yamamoto. 2020. "Ellagitannins from *Punica granatum* leaves suppress microsomal prostaglandin E synthase-1 expression and induce lung cancer cells to undergo apoptosis." *Bioscience, biotechnology, and biochemistry*, 84(4) :757-763.

Trigo, João P, Elisabete M. C. Alexandre, Sara Silva, Eduardo Costa, Jorge A. Saraiva, & Manuela Pintado. 2020. "Study of viability of high pressure extract from pomegranate peel to improve carrot juice characteristics." *Food & Function*, 11(4) :3410-3419.

USDA. 2019. "*National nutrient database of standard.*" United States Department of Agriculture https://fdc.nal.usda.gov/fdc-app.html#/ NDB numer raw 9286, and NDB numer juice bottled 9442. Acceded 11/04/2022.

Vaithiyanathan, S, BM Naveena, M Muthukumar, PS Girish & N Kondaiah. 2011. "Effect of dipping in pomegranate (*Punica granatum*) fruit juice phenolic solution on the shelf life of chicken meat under refrigerated storage (4°C)." *Meat science*, 88(3) :409-414.

Valdés, Arantzazu, Esther Garcia-Serna, Antonio Martínez-Abad, Francisco Vilaplana, Alfonso Jimenez, & María Carmen Garrigós. 2019. "Gelatin-based antimicrobial films incorporating pomegranate (*Punica granatum* L.) seed juice by-product." *Molecules*, 25(1), 166 :1-20, doi :10.3390/molecules25010166.

Vallarino, Giulia, Annalisa Salis, Elena Lucarini, Federica Turrini, Guendalina Olivero, Alessandra Roggeri, Gianluca Damonte, Raffaella Boggia, Mannelli Lorenzo Di Cesare, Carla Ghelardini, & Anna Pittaluga. 2022. "Healthy Properties of a New Formulation of Pomegranate-Peel Extract in Mice Suffering from Experimental Autoimmune Encephalomyelitis." *Molecules,* 27, 3 :914.

Wang, Dongdong, Cigdem Özen, Ibrahim M Abu-Reidah, Sridevi Chigurupati, Jayanta kumar Patra, Jaroslaw O Horbanczuk, Artur Jóźwik, Nikolay T Tzvetkov, Pavel Urrin, & Atanas G Atanasov. 2018. "Vasculoprotective effects of pomegranate (*Punica granatum* L.)." *Frontiers in pharmacology*, 9, 544, doi :10.3389/fphar.2018.00544.

Wang, Linlin, Chen Hu, Longquan Shao. 2017. "The antimicrobial activity of nanoparticles: present situation and prospects for the future." *International journal of nanomedicine*, 12 :1227-1249.

Witt, Elaine de Paula, Gracielle Johann, Carina Contini Triques, Edson Antonio da Silva, Camila da Silva, Dionisio Borsato & Tatiana Rodrigues da Silva Baumgartner. "Potential application of supercritical extract of pomegranate peel as bio-antioxidant for biodiesel." *Environmental Progress & Sustainable Energy,* 41, no. 2 (2022) : e13755.

Xiang, Qiwen, Meifeng Li, Jiayu Wen, Fajian Ren, Zhou Yang, Xingyue Jiang & Yan Chen. 2022. "The bioactivity and applications of pomegranate peel extract: A review." *Journal of Food Biochemistry*, e14105, doi :10.1111/jfbc.14105.

Yan, Linlin, Xiaojing Zhou, Lingling Shi, Dilinuer Shalimu, Chao Ma, & Yujun Liu. 2017. "Phenolic profiles and antioxidant activities of six Chinese pomegranate (*Punica granatum* L.) cultivars." *International journal of food properties*, 20(sup1), S94-S107.

Zhao, Shenjuan, Jianke Li, Lifang Wanga & Xiaoxia Wu. 2016. "Pomegranate peel polyphenols inhibit lipid accumulation and enhance cholesterol efflux in raw264. 7 macrophages." *Food & function*, 7(7) :3201-3210, doi :10.1039/c6fo00347h.

Chapter 7

Nut Byproducts: An Opportunity to Decrease the Environmental Impact and Produce Functional Ingredients

M. de L. González-Muñoz
M. Rodríguez-Rodríguez
G. A. Miranda-Mejía
and M. Morales-de la Peña*
Tecnologico de Monterrey, Escuela de Ingeniería y Ciencias, Bio-Foods Research Lab, Querétaro, Mexico

Abstract

Nuts, such as pecans, peanuts and almonds, are highly consumed worldwide as basic snacks or as main ingredients in different dishes and processed foods. Their high nutritional value, due to their content of proteins and fats, and characteristic taste, make them very attractive for consumers. However, their production results in the generation of different byproducts; mainly, shells, leaves, branches, skin, and cakes. Currently, most of these byproducts are used as cattle feed or incinerated, contributing in a negative way to the environmental impact. In these sense, researchers are looking for alternative ways to exploit them as funcional ingredients in the development of new proucts. Several studies report that nut byproducts posses an elevated concentration of essencital nutrients, such as amino acids, fatty acids, phenolic compounds and flavonoids, among others. As well, some of them have good technological funcionalities such as water and oil absorption. This chapter presents the most recent information related to the evaluation of pecan, peanut and almonds byproducts as funcional ingredients, their advantages, chalenges and opportunitines.

Keywords: byproducts, pecans, peanuts, almonds, functional ingredients

* Corresponding Author's Email: mariana.moralesdlp@tec.mx.

In: Food Byproducts
Editors: S. Teresita Martín-del-Campo, A. Cardador-Martínez and J. del Pilar Ramírez-Anaya
ISBN: 979-8-88697-612-0

Introduction

Pecans, peanuts, and almonds are some of the most widespread edible nuts worldwide. Due to their high concentration of nutrients, especially protein and fat, they are considered energy-dense foods with health-related properties. Overall, they have been characterized as being rich in essential amino acids, unsaturated fatty acids, dietary fiber, minerals, vitamins, phenolic acids, and flavonoids which can provide a wide range of health benefits. Furthermore, their characteristic taste makes these nuts very attractive to consumers, representing an excellent alternative as snacks, oil sources, and also, as main ingredients for bakery and confectionary products, ice creams, milk substitutes, and dressings, among others.

The transformation and preservation processes of nuts include different stages, from the harvest up to the packaging and storage. Through these processes and depending on the desired final characteristics of the nuts to be commercialized for human consumption, a substantial amount of different byproducts is generated, such as shells or hulls, leaves, branches, skin, and cakes. Today, some of these byproducts are used as feed for livestock or, in the worst case, they are incinerated, resulting in pollution and environmental impact. In this sense, in order to decrease the amount of waste, different research groups worldwide have focused on nut byproducts characterization in terms of technological properties as well as nutrient and bioactive compounds profile.

Currently, the obtained results from these studies indicate that nut byproducts, depending on the matrix, have a great potential to be used as functional ingredients in food formulation and as excellent additives to improve some technological characteristics of foods such as water and oil retention capacity, or emulsification properties, among others. However, further studies are required to complete specific information about toxicity or allergenic problems in order to avoid any inconvenient. This chapter reviews the most recent information about the main characteristics of pecan, peanut, and almond byproducts and their potential use as functional ingredients or additives.

Pecan

Nuts

Pecan nut, *Carya illinoinensis*, is one of the principal nuts consumed worldwide and one of the most economically valuable horticultural crops in North America (Poletto et al., 2019; Badenes and Byrne, 2012). The pecan tree is originally from California, United States; however, the biggest production comes from Mexico, with 124,000 tons per year (de Prá Andrade et al., 2021). The geographic distribution of pecan trees in Mexico has been found in the northern part of the country, in Chihuahua, with the highest local production in the following municipalities: Jiménez, Camargo, Allende, Buenaventura, and Aldama (del Carmen and Urrea López, 2016).

Pecan tree measures around 30 m in height and has a trunk diameter of 2-2.5 m. Its bark is light-gray or brown, covered with leaves of 25 cm long that may have between 9 and 17 leaflets with a serrated edge; the color can vary between yellow-green. These trees have an ideal development in drained clay soils, although the seedlings can hold short periods of flooding (del Carmen and Urrea López, 2016). The harvest of pecan nuts lasts approximately from the last week of September to the third week of November.

Pecan nut is characterized by being a large fruit with fleshy mesocarp and hard endocarp, separated in two brown valves. The seed is divided into a hard shell, middle septum, and kernels, and it has two or four lobes and holes. They can be classified into two varieties, mainly based on their size and shell thickness: the first ones are "creoles", characterized by small size, hard shell, and high oil content. On the other hand, the "fine or paper-shelled" variety, which has been improved by grafts, has a larger size and a thinner shell (del Carmen and Urrea López, 2016). Overall, pecan nuts are an excellent source of lipids, with approximately 70% of its composition also containing carbohydrates, dietary fiber, and protein. The protein profile of pecan nuts is made up of 18 amino acids, of which 10 are essential such as arginine, phenylalanine, histidine, isoleucine, leucine, lysine, methionine, threonine, tryptophan, and valine. Among them, glutamic acid and arginine are the most abundant, with a concentration of more than 1 g/100 grams (del Carmen and Urrea López, 2016). Regarding its fatty acid profile, monounsaturated fatty acids are present at the highest concentration (66%) with a lower content of polyunsaturated (26%) and saturated (8%) fatty acids (Alves et al., 2019; Maciel et al., 2020; McKay et al., 2022; Reckziegel et al., 2011). Oleic acid is the major fatty acid (56.4%), followed by linoleic acid (18.9%).

Additionally, pecan nuts are a good source of dietary fiber (11%) and contains a significant concentration of bioactive compounds such as carotenoids (β-carotene, vitamin E, lutein) (McKay et al., 2022), phenolic compounds (proanthocyanidins, condensed tannins, and ellagic acid) (Marchetti et al., 2018), vitamins, like choline and gamma-tocopherol; and minerals, standing out potassium, phosphorus, and magnesium (del Carmen and Urrea López, 2016). In this sense, it has been proven that continuous consumption of pecan nuts decreases the risk of cancer and cardio-metabolic diseases, could improve lung functions, has anti-inflammatory properties, and a positive effect on the skin by inducing pro-collagen synthesis and inhibiting pro-inflammatory molecules involved in aging (del Carmen and Urrea López, 2016; Maciel et al., 2020; McKay et al., 2022; Reckziegel et al., 2011; Marchetti et al., 2018). Among all nuts, pecans have the highest antioxidant capacity (del Carmen and Urrea López, 2016).

The high lipid concentration of pecans makes them semi-perishable foods that can be easily oxidized. Usually, when pecans are harvested they present moisture of 12%. They are collected, sanitized with chlorine to avoid bacteria such as *E. Coli*, and dried to diminish the moisture content up to 4% with warm air (25 °C, 60% Relative Humidity). To avoid oxidation during storage, they have to be protected from light, high temperature, and oxygen. The pecan shell color is a key factor that determines quality. It should be a golden brown or amber color. In fact, processors suggest that high-quality pecans should be lighter than golden. The darkening of the shell may imply inappropriate storage conditions (McKay et al., 2022).

Pecans can be commercialized shelled, or unshelled. After drying, the cracking process removes the outer shell maintaining the kernel intact. The cracking process is very important since it determines the value and retail price. During cracking, pecans are pressed to mechanically break the shell, reducing it by half their size. Later, they passed through different hole sizes to classify them into various categories. Finally, obtained kernels are blown to remove the lighter shell fragments. Intact kernels have the highest value, including an extended shelf-life compared to smaller pieces, due to less exposure to deteriorating factors. The yield in processing plants ranges from 50–80% (del Carmen and Urrea López, 2016; McKay et al., 2022).

Usually, pecan kernels, which constitute less than 50% of the whole nut, are consumed fresh, as pecan oil, or used as ingredients in bakery, confectionery, desserts, candies, and snacks, among other products (Idowu et al., 2017; Cervantes Vázquez et al., 2018; SAGARPA, 2017) (Figure 1). In Mexico, pecans are also used for traditional dishes or drinks like mole and pulque (SAGARPA, 2017). Due to their complete nutrient profile and antioxidant compounds, different pecan-based products have been recently developed. Namely, a "Dulce de leche" (a rich, creamy, caramel-like milk-based sauce or spread) was Prepared with pre-emulsified pecan oil as fat source due to the content of oleic and linoleic acids (Ranalli et al., 2017).

The transformation of pecan nuts into other derivative products like those mentioned before generates a significant number of byproducts such as branches, leaves, shells, and cake, which can be considered as an alternative source of macronutrients and phytochemical compounds (Maciel et al., 2020; Reckziegel et al., 2011). Some research groups worldwide have been working on the use of these byproducts, especially pecan shells and cake, to generate foods with functional properties and also to decrease the environmental impact caused by the pecan industry. Only few studies have been done on branches and leaves.

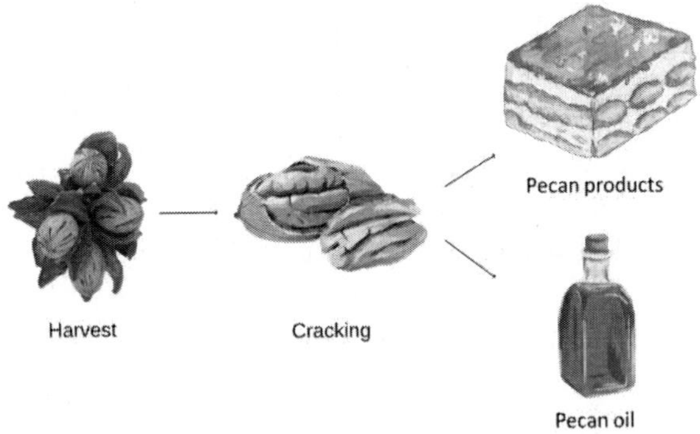

Figure 1. General pecan transformation process.

Shells

Shells are an abundant residue from pecan production (Aldana et al., 2015). The shell corresponds to 50% of the whole pecan nut weight. Since it is considered an agricultural residue, pecan shells are commonly used as residual biomass, allowing them to be a cheap renewable energy source. The bio residues can be processed through chemical routes to produce heat, electricity, fuels, and chemical products (Aldana et al., 2015). Also, because of their nature, they are characterized as good adsorbent material for water treatments (de Prá Andrade et al., 2021).

Other studies have demonstrated that pecan shells are characterized by having a high concentration of insoluble fiber (cellulose, lignin, and hemicellulose) and phenolic compounds such as tannins and proanthocyanidins (del Carmen and Urrea López, 2016). These compounds could be extracted from shells and used as nutraceuticals or functional ingredients. Traditionally, the extraction of these compounds from pecan shells has been done by infusions in hot water and ethanol, hydro distillation, solvent extraction, spray drying, and lyophilization. Ethanolic extracts could function as antimicrobials against foodborne illness-causing microorganisms such as *Staphylococcus aureus* and *Listeria monocytogenes* (del Carmen and Urrea López, 2016).

New extraction technologies can enhance the extraction of bioactive compounds from different food matrices, such as pecan shells. Namely, enzyme-assisted extraction (EAE), microwave-assisted extraction (EAM), ultrasound-assisted extraction (UAE), pressurized liquid extraction (ELP), and supercritical fluid extraction (SFE) are some of the most effective techniques; reaching extraction yield up to 98% with SFE technology. Once the bioactive compounds are obtained, it is necessary to quantify them using sensitive and selective methods, like spectrophotometry and chromatography tests.

Cake

Pecan cake, also known as okara or bagasse, is the residue from the oil or aqueous extraction from pecan kernels to produce pecan oil or pecan-based beverages. This byproduct is considered an important source of fiber, proteins, and carbohydrates with low energy value, and also, it contains phenolic compounds, tannins, and minerals such as Cu, Mg, and Mn (del Carmen and Urrea López, 2016). Although its nutritional properties, sensory characteristics, and functional attributes, pecan cake is mostly commercialized at a low cost for animal feed. Nonetheless, during the last decade, researchers have demonstrated that it is a potential ingredient to develop products such as desserts, jelly, pasta, yogurts, cheese, bakery, and confectionery. Furthermore, bioactive compounds can be extracted from pecan cake and used as nutraceutical or additive ingredients in the food, pharmaceutical, and cosmetic industry (Maciel et al., 2020).

Commonly, pecan cake remains with significant amounts of pecan oil. Its extraction from this residual cake is an alternative to increasing profits in the pecan oil industry (del Carmen and Urrea López, 2016). Oil extraction yields from pecan cake can be optimized by using different solvents, modifying processing temperature, or using emerging extraction methods, such as supercritical fluids, subcritical water, microwave, high hydrostatic pressure, pulsed electric fields, or ultrasound (Maciel et al., 2020).

It has been proved that pecan cake can replace wheat flour in baked products, improving the lipid profile of monounsaturated fatty acids and fiber content. Also, the mineral content (Mn, K, Ca, Fe, Mg, Zn, and Se) significantly increases in pecan cake-based products (Marchetti et al., 2018). According to Marchetti et al. (2018) and Maciel et al. (2020), pecan nut cake has excellent techno-functional properties like water and oil absorption, which could improve moistness, texture, appearance, and taste of bakery products, snacks, pasta, sausages, and cheeses.

Branches and Leaves

Branches and leaves are the byproducts generated during harvesting, gathering, and selecting pecans for commercialization. These byproducts, especially the leaves, are good sources of bioactive compounds, mainly phenols and flavonoids. However, it has been reported that pecan leaves contain a higher concentration of flavonoids, quantified as the sum of individual flavonoids determined by high performance liquid chromatography (HPLC) in methanol extract, than those found in shells, contrary to the concentration of phenols which are lower in leaves than in shells (Hawary et al., 2016). On the other hand, it has been reported that triterpenoids and phytosterols found in pecan leaves possess antimicrobial activity against Mycobacterium tuberculosis H37Rv. According to Cruz-Vega et al. (2008), pecan leaf extract (125 mg/mL) inactivated around 50–80% of the bacteria. Likewise, some phenols (gallic and ellagic acids) and flavonoids (rutin, catechin, and epicatechin) present in pecan leaves may have antimicrobial activity against Gram-positive and Gram-negative bacteria, as well as yeasts (Bottari et al. 2017).

As seen, pecan nuts contain different functional properties in their byproducts: leaves, branches, shells, and cake, which represent a great opportunity to use as raw materials to extract different compounds with different health benefits and techno-functional properties. In this sense, it is important to continuously study their attributes, as well as to find the right extraction procedures in order to use the whole gamma of pecan byproducts in the food, cosmetic and pharmaceutical industries.

Peanut

Peanut (*Arachis hypogaea L.*) is a relevant oilseed crop, characterized by a high content of oil, protein, and fiber, grown in different parts of the world but originated in South America (Zhao et al. 2012; Akhtar et al., 2014). Peanuts are technically legumes but are widely identified as part of the nut's food group (Ros, 2010). They are recognized as highly nutritious foods compared with foods from animal sources, including meat. Their content of carbohydrates, lipids, proteins, vitamins B_1, C, E, niacin, and minerals such as calcium, magnesium, and potassium, are similar to or higher than that in meat or eggs. Furthermore, their consumption does not affect cholesterol or saturated fatty acids blood levels (Arya et al., 2016).

According to literature, the fat content in peanuts is around 40%, making them energy-dense foods. Their lipid profile is characterized by a high fraction of unsaturated fatty acids, approximately 40% oleic acid (monounsaturated) and 40% linoleic acid (polyunsaturated), and only a trace amount of linolenic acid, contributing to the benefits for cardiovascular health (Akhtar et al., 2014). Due to its fatty acid profile, peanut oil has convenient oxidative stability and has been considered high-quality cooking and frying oil (Carrin et al., 2010).

Their nutritional and unique flavor make peanuts highly attractive to consumers worldwide. Usually, peanuts can be consumed as peanut oil, peanut butter, roasted peanuts, peanut flour, and as the main ingredients in energy bars, snacks, and candies, among others (Settaluri et al., 2012). The average production of crushed peanuts worldwide was 14.09 million metric tons from 2000 to 2010 (Zhao et al., 2012). During the peanut transformation processing (pre- and post-harvest) into commercial products, they pass through important steps such as removal, threshing, sun drying, and cleaning to remove dirt, stones, plant residues, and other sewage. Also, depending on the desired final product characteristics, they can be peeled, classified, and roasted (Larrauri et al., 2016). Along these transformation processes, different byproducts are generated, including shells, skins, peanut cake, and peanut vine, which can be considered potential pollutants; however, they are also recognized to be a source of valuable nutrients such as protein, fiber, antioxidant compounds and antimicrobials (Lorenzo et al., 2018). Hence, taking into account the environmental awareness and limited landfill space, the use of byproducts has become an attractive alternative to disposal. Particularly, peanut byproducts may be used in the food processing industry to develop other edible products or to be incorporated into processed foods as functional ingredients (Zhao et al., 2011; Sorita et al., 2020).

Skins

Skin removal from peanut kernel is one of the primary steps in peanut processing. Peanut skin is the nontoxic pink-red layer that covers peanuts, and it is the main residue of peanut processing with low commercial value. It is estimated that 35–45 g of peanuts skins are produced per kg of a shelled peanut kernel. Approximately more than 0.93 million metric tons of peanut skins are produced each year globally (Davis and Dean, 2016). Unfortunately, most of them are considered wastes and are discarded, while only a small amount is utilized to extract phenolic compounds or as a minor component of animal feed (Sobolev and Cole 2003). Furthermore, peanut skins have been characterized by a high content of potent antioxidants such as proanthocyanidins (Win et al., 2011; Lorenzo et al., 2018). Hence, they can be exploited as an inexpensive source and renewable raw material of antioxidants for use as functional ingredients in foods or dietary supplements (Yu et al. 2006).

Peanut skin is commonly used in China to treat hemorrhage and bronchitis (Lou et al., 1999); however, some studies have shown that their application as a functional ingredient in food product development at 5-10% level of concentration resulted in bitter and astringent flavor, indicating its limited application (Hathorn and Sanders, 2012). In addition, one of the main reasons for the scarce use of peanut byproducts, such as peanut skin, is the lack of effective extraction methods to obtain high-quality compounds required to serve as functional ingredients (Lorenzo et al., 2018). Interestingly, de Camargo et al. (2015) evaluated the effect of gamma irradiation (2.5 and 5 kGy) on peanut skin phenolic composition. Their results demonstrated that treatments increased the free phenolic and proanthocyanidin content, as well as the antioxidant potential. Proanthocyanidins are considered the second most abundant type of polyphenols in the human diet, particularly due to the consumption of chocolate, wine, and peanuts with skin (Appeldoorn et al., 2009). Recently, Christman et al. (2019) demonstrated that phenolic extracts from peanut skins attenuate the decrease in cell viability in high glucose treated HepG2 cells, showing a protective effect against hyperglycemia-induced cell death and antidiabetic effects.

Although phenolic compounds are recognized mainly for their role in oxidative reactions, they have also been known for their antimicrobial activity (Cueva et al. 2010). de Camargo et al. (2017) evaluated the antimicrobial potential of peanuts skins against different microorganisms, including bacteria, yeasts, and molds, and conclude that phenolic compounds contained in them can inhibit the development of pathogenic and spoilage microorganisms; however, the effectiveness depends on the target microorganism. This determination could be the key to allowing the application of peanut skin phenolic compounds to increase the shelf-life and guarantee the safety of foods. Some applications of peanut skin phenolic extracts have been conducted in meat products. A methanol extract (200 ppm) of peanut skin prevented lipid oxidation in cooked ground beef without any effect on color, aroma, and microbial growth during storage (14 days) (O'Keefe and Wang, 2006). On the other hand, peanut skin phenolic extracts have also been used to prevent the oxidative processes of soybean oil during storage with great success (Franco et al., 2018).

Although some research about peanut skin properties, functionality, and characteristics has been conducted during the last decades, further studies are required to utilize peanut skin benefits efficiently. The exploitation of peanut skins can potentially protect and enhance the environment and might improve some product characteristics by using natural additives.

Shells

Peanut shells are usually obtained when peanuts are passed through the shelling machines resulting in peanut kernels and hulls or shells. Shell production is estimated to be around 230–300 g per kg of peanut (Wang and Xu, 2008), representing a significant waste disposal problem around the area where peanuts are cultivated or treated. Their annual production accounts for about 46 million tons (Carneiro et al., 2017). Currently, some bioactive compounds, such as flavonoids with high antioxidant capacity, have been identified from peanut shells water extracts (Yen and Duh, 1993; Bertioli et al., 2011; Yang et al., 2007). Furthermore, this byproduct has been characterized as an excellent source of fiber (47.3%) (Zhao et al., 2012; Collins and Post, 1981). Today, most of the uses for peanut shells are applied in the chemical industry as absorbents for chemical and environmental clean-ups, landscaping mulch, kitty litter, shipping packing materials, and alternative charcoal (National Peanut Board, 2018).

In a recent study (Albergamo et al., 2021), peanut shells and skins were used to elaborate a snack bar. It was demonstrated that peanut shells contributed to a greater extent than skins to boost the antioxidant capacity of the final product. From the sensory point of view, their incorporation caused a slightly astringent taste and bitter notes; however, it was reported that the panelists emphasized that the use of such natural additives provides good desirability to the product. Hence, the use of this kind of peanut byproducts represents an excellent alternative to obtaining natural additives for food production with high antioxidant capacity.

Cake

Peanut cake is obtained from the peanut oil extraction process. This byproduct is available at peanut processing plants in all production regions, but currently, it is not sold at a competitive price. The global average peanut-cake flour production was 5.78 million metric tons between 2000 and 2010 (Zhao et al., 2012). According to Zhao et al. (2012), the yield of peanut cake obtained after oil extraction depends on the procedure followed, being hot or cold extraction. Approximately per 1000 kg of peanuts subjected to a cold extraction process, 700 kg of peanut cake are generated, while after the hot extraction process, only 500 kg of peanut cake are obtained.

Peanut cake has high protein content (50%), fat (9%), and other functional components, such as lectin and resveratrol (Sales and Resurreccion 2009; Aletor and Ojelabi, 2006; Ezekiel et al., 2011). Feedstuffs (2001) compared the ash, crude protein, and total digestible nutrients concentration of peanut-cake (6.3%, 52.3%, and 77.0%, respectively), cottonseed cake (7.0%, 45.6%, 76.0%, respectively) and soybean cake (6.5%, 55.1%, 87.0%, respectively). When comparing the bioactive profile of peanut cake with soybean cake, results revealed higher concentrations of niacin, riboflavin, thiamin, and pantothenic acid in peanut cake but lower concentrations of the essential amino acids lysine, methionine, and tryptophan (Hill, 2002). Regarding the fatty acid profile, peanut cake has a greater proportion of palmitic, oleic, and myristic acids, 38.74%, 23.10%, and 15.35%, respectively, compared to soybean cake (Oliveira et al., 2012).

The use of peanut cake has been mostly implemented in animal nutrition because of its similarity to soybean meal (Abdalla et al., 2008; Bezzerra et al. 2016). Silva et al. (2015) and Correia et al. (2016) evaluated the replacement of soybean cake with peanut cake in the diet of goats and young bulls, respectively, obtaining interesting results and good acceptability from the animals. On the other hand, defatted peanut flour obtained from the peanut cake can be used as a protein source for the fortification of different products, such as bread, pasta, cookies, breakfast cereals, and popular drinks (Badwaik et al., 2012). Namely, Badwaik et al. (2012) used defatted peanut flour combined with carrot and semolina to produce a protein and fiber-rich pasta. The authors concluded that the elaborated pasta could decrease the problem of protein-calorie malnutrition and provide a high concentration of dietary fiber, solving the problem of constipation. Hence, it can be said that peanut cake byproduct could be a potential alternative ingredient to develop foods with high protein and fiber concentration; however, research on the technological functionality of the peanut-cake protein is necessary to complement this information.

Vine

The peanut vines include roots, stems, leaves, and flowers (Figure 2). The production of peanut vine from harvested peanuts is estimated to be around 60–65%. The yield of peanut vine generated annually worldwide as a byproduct of the peanut industry is much more than the peanut kernel, skin, and hull. However, up today, very limited information on functional components from peanut vines is available in the literature. Romero et al. (1987) characterized peanut vine and stated that it has a high feeding value due to its elevated concentration of crude protein, minerals, and vitamins. Du and Fu (2008) reported that peanut vines are rich in dietary fibers and flavonoid compounds, such as rutin, quercetin, and luteolin, providing a wide range of health benefits. Hence, this byproduct represents an interesting area to be studied and evaluated.

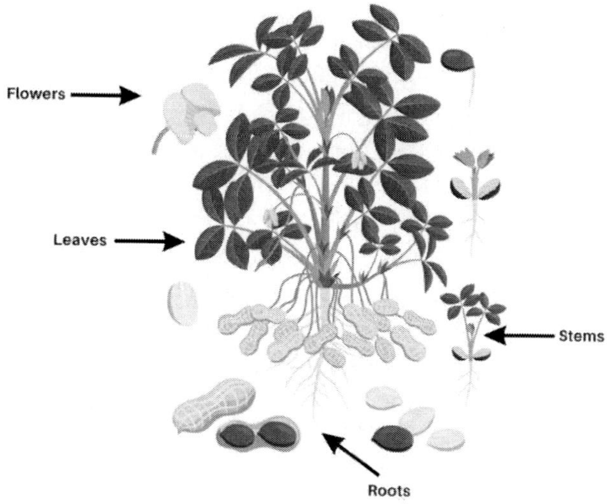

Figure 2. Peanut vine.

Almond

Almond (*Prunus amygdalus*, syn. *Prunus dulcis*) is a tree from the Rosaceae Family (apples, pears, prunes, raspberries), which is very well known for its edible seed. The fruit of the almond is a drupe with 3 distinct parts: (1) inner kernel, (2) middle shell portion, and (3) outer green shell cover or hull (Esfahlan et al., 2010). The inner kernel, also known as almond, is the most consumed nut in the world, just behind peanuts (Ros, 2010). Commercially, almonds can be found in the shell, shelled, or peeled, being peeled almonds the most consumed (Smeriglio et al., 2016). There are several almond varieties that vary mainly in shell texture and phytochemicals concentration, being the Nonpareil, Mission, California, Neplus ultra, and Peerless the major varieties produced and commercialized.

Global almond production in 2021-2022 was estimated at 1.6 million metric tons shelled basis, being the United States, specifically California, the largest almond producer in the world, followed by a modest contribution from Australia, Spain, Italy, China, and India (USDA, 2021b, 2022). Just California produces 77-83% of the world's commercial almonds (Garcia-Perez et al., 2021; Kendall et al., 2015; USDA, 2022). At present, almonds are considered the most consumed tree nut in the world, with 30% of the total consumption of nuts worldwide, followed by walnuts with 20% (King et al., 2022). Per capita consumption of almonds has had the highest growth among nuts, with an average annual rate of 5.5% since 2000 and estimated consumption of 1.075 kg per capita in 2018 (USDA, 2021a; Widmar, 2021).

Almonds are known to be a rich source of nutrients, phytochemicals, and antioxidants. Phenolic compounds, tocopherols, sterols, resveratrol, monounsaturated and polyunsaturated fatty acids, arginine, and potassium are the main health-related compounds found in almonds (Gupta et al., 2020; Mandalari et al., 2010; Ros, 2010). Thus, almond consumption has been recommended to reduce and prevent disease factors such as high triacylglycerols, high LDL cholesterol, cardiovascular, and other chronic diseases (Wijeratne et al., 2006). Almonds with or without the brown skin are consumed mainly as snacks and as ingredients in a variety of processed foods: ready-to-eat cereals, energy and granola bars, baked goods, almond butter, almond snack mixes, chocolate and non-chocolate confectionary, frozen novelties, ice cream, and milk substitutes, among others (Sumner et al., 2014).

The almond industry comprises four segments: almond growing, almond hulling and shelling, almond handling, and the manufacture of almond products. In general, the almonds production process involves four steps, orchard establishment (1-2 years), orchard production and harvest (3-25 years), and orchard removal (Kendall et al., 2015). Almond trees achieve a steady almond production from year 7 to 25, and harvest starts each year when almonds drupes are mature and partly dried (Esfahlan et al., 2010). After harvest, in-shell almonds are transported to hulling and shelling facilities, where hulls and shells are removed. Approximately two million tons of hulls are collected annually after being separated from the almond meat (Sumner et al., 2014). The resulting almond kernel or meat (11% of the original almond fruit) can be further processed for retail and fresh consumption or skin removal, where almonds are blanched in hot water and peeled (Garcia-Perez et al., 2021; Smeriglio et al., 2016). Usually, almonds are commercialized with or without skins whole, sliced, slivered, chopped, diced, roasted, or raw (Sumner et al., 2014).

The high demand for almonds worldwide leads to an even higher production of byproducts which are generally discarded or used as low-value products. Hulls, shells, brown skins, and blanching water are the main byproducts generated during almond production, as shown in

Figure 3. The annual volume of byproducts from almond industrial processing is estimated as 70–80% of the total almond fruits produced (Smeriglio et al., 2016). When almond byproducts are not further processed, they are usually discarded as waste. The destination and volume of almond byproducts are shown in Table 1 (Kendall et al., 2015).

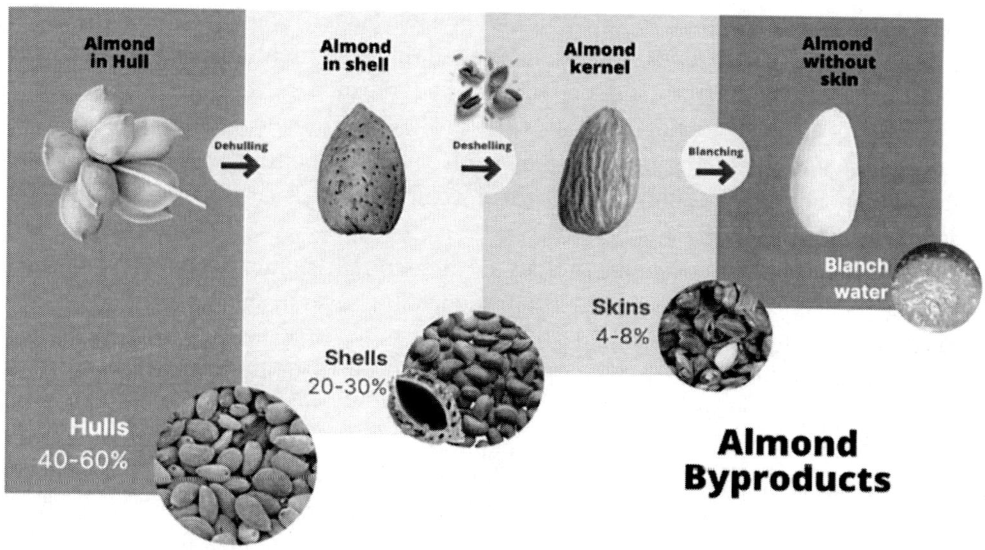

Figure 3. Almond byproducts generated during the transformation process.

Table 1. Destination and volume of almond byproducts

Byproduct	Destination	Volume
Brown skin almonds	Animal feed and energy	4-8%
Hulls and hash	Livestock feed	40-60%
Shell	Energy and livestock bedding	20-30%
Blanching water	Waste	Unknown

Hull

Almond hull, also known as almond green shell cover, is the mesocarp of the almond drupe and represents 40–60% of the total almond drupe (Garcia-Perez et al., 2021). Worldwide annual production of almond hulls is between 0.8 to 1.7 million tons (Barral-Martinez et al., 2021). Hulls are leathery, astringent, and with a low moisture content (8–20%), but with a high energy value for ruminants, and therefore they are mainly used for livestock feed, reducing the need for alfalfa and the cost of almond processing (De Peters et al., 2020; Sumner et al., 2014). The almond hull mainly consists of 28–32% neutral detergent fiber, 18–25% acid detergent fiber, 9–14% lignin, 2–3% protein, 5–6% nonfibrous carbohydrate, 2–2.5% total tannin, and 3–3.5% total phenolic compounds, being phenolic acids present at a higher concentration than flavonoids (Najari et al., 2022; Rubilar et al., 2007). Additionally, almond hulls are considered a rich source of triterpenoids (betulinic, urosolic, and oleanolic acids), which have a critical role in scavenging free radicals and inflammatory inhibition (Takeoka et al., 2000; Wijeratne et al., 2006). The content of phenolic compounds in almond hulls differs depending on the almond variety and the extraction method used. Some authors have even explored the use of emerging technologies such as ultrasound and pulsed electric fields to increase extraction efficiency. For example, Kahlaoui et al. (2019) evaluated the extraction yield efficiency of total phenolic compounds, flavonoids, and tannins, as well as the radical scavenging activity using ultrasound in different hull varieties, and results showed extraction yields up to 2.7–8.2, 5.2–19.0, and 3.5–10.8 times higher, for phenolic compounds, tannins, and flavonoids, respectively. The main phenolic compounds found in almond hulls are chlorogenic acid, catechin, hydroxybenzoic and cinnamic acid derivatives, and flavan-3-ols and flavonol glycosides (Garcia-Perez et al., 2021; Kahlaoui et al., 2019). These phenolic compounds are present in lower concentrations than skins, but the almond hull can still be considered a source of multiple bioactive compounds that could be used for developing novel and functional foods or as ingredients for pharmaceutical applications.

The high free radical scavenging activity of almond hull extract, which is related to the presence of phenolic compounds, has been studied by several authors. Meshkini (2016) evaluated the polyphenol content and antioxidant activity of almond hull extracts in the protection of human erythrocyte membrane function from oxidative stress, and the results showed that almond hull extracts have the potential to provide protection against induced oxidative damage and membrane protein degradation in human erythrocytes caused by hydrogen peroxide. Similarly, An et al. (2020) tested the bioavailability of the phenolic-rich extracts from almond hulls and their protective effect on oxidative stressed Caco-2 cells induced by tert-butylhydroperoxide. Results demonstrated that almond hull extracts were able to meliorate the oxidative stressed Caco-2 cells and protect cells from oxidative stress-induced toxicity. These experimental observations show that almond green hull can be used as a potent source of natural antioxidants in the food and pharmaceutic industry to limit the risk of various oxidative-associated diseases.

Pectin and other valuable products can also be recovered from almond hulls. Najari et al. (2022) analyzed the potential to recover pectin and phenolic compounds and to produce pullulan (an exopolysaccharide) and single-cell protein (edible biomass) from almond hulls. In their study, the authors were able to achieve a maximum extraction yield of 26.32% for pectin and 6.97% for phenolic compounds; from the remained residues, the authors were able to produce 34.29–24.56g/L of pullulan and biomass containing 13.44-19.31% of single-cell proteins, confirming that quality value-added compounds can be obtained from almond hulls.

Furthermore, almond hulls contain 28.9% of lignin and 56.1% of polysaccharides, especially celluloses and hemicelluloses, so this byproduct has been evaluated as a potential source of xylooligosaccharides to produce prebiotic and sweetener agents; as a source of lignocellulosic biomass to produce bioethanol, biobutanol, and other fermentation products, and as a source of pure cellulose to produce nanofibers (Garcia-Perez et al., 2021; Sillero et al., 2021; Urruzola et al., 2014). Other possible applications that have been evaluated for almond hulls is their use as heavy metal adsorbent (Bulut and Tez, 2007; Mehrasbi et al., 2009), dyes adsorbent (Doulati Ardejani et al., 2008), growing media for agricultural products (Urrestarazu et al., 2005), and production of activated carbon (Toles et al., 2000).

Skin

The almond skin or seed coat is the leathery brown coat that protects the almond from oxidation and microbial contamination. They represent 4–8% of the total almond kernel weight. The positive health benefits associated with the consumption of almonds are partly due to the almond skin fiber concentration and bioactive compound composition. It has been reported that almond skins contain significant amounts of phenolics and flavonoids; and high antioxidant scavenging, metal chelating, cytoprotective, and anti-bacterial activities, which allows them to protect the almond from oxidation reactions, mainly lipid oxidation (Esfahlan et al., 2010; Mandalari et al., 2010; Milbury et al., 2006; Smeriglio et al., 2016; Wijeratne et al., 2006). However, since many food applications require almonds without skin, it is very common to remove them through the blanching process. During blanching, a substantial amount of phenolics (>60%) is transferred to the blanch water; therefore, some biological and health properties are lost or transferred to blanch water (Smeriglio et al., 2016; Esfahlan et al., 2010; Mandalari et al., 2010).

Mandalari et al. (2010) estimated the content of phenolics, lipids, proteins, fiber content, and antioxidant activity of natural almond skins vs. blanched skins and blanch water. Results showed that total phenolic content and antioxidant activity were higher in natural skins. Blanched skins and blanch water contained only a fraction, 11% and 0.6%, respectively, of the total flavonoids and phenolics acids present in natural skins. Similar results were found by Smeriglio et al. (2016), where natural skins presented the highest phenolic content and antioxidant activity, followed by blanched skins and blanch water. However, the proportion of total phenolic compounds that remained in blanched skins (39%) and blanching water (11%) was higher. Lack of standardization of industrial processes and diversity among almond varieties may result in fluctuations in the polyphenolic content and profile of these byproducts. However, studies conducted so far show clear evidence that the blanching process causes a significant decrease in polyphenol content. However, the remaining compounds in blanched skins and blanching water can be valuable for the food, pharmaceutical, and cosmetic industries if further processed.

Flavonols, flavanols, and flavanones are the most abundant phenolic compounds in natural skins, blanched skins, and blanch water. The main phenolic acids identified are hydroxybenzoic acid, vanillic acid, and t-ferulic acid; and the major flavonoids are (+)-catechin, (–)-epicatechin, kaempferol, kaempferol-3-O-rutinoside, kaempferol-3-O-glucoside, isorhamnetin-3-O-rutinoside, and naringenin. Among these, isorhamnetin-3-O-rutinoside, naringenin, kaempferol-3-O-glucoside, and kaempferol-3-O-rutinoside represent the predominant compounds (Mandalari et al., 2010; Monagas et al., 2007; Smeriglio et al., 2016).

In addition, almond skins are known to be a good source of dietary fiber as their cell wall material is mainly composed of pectic substances. Mandalari et al. (2010) analyzed the total dietary fiber and carbohydrate composition in natural skins, blanched skins, and blanching water and found that glucose and galacturonic acid were the major sugars, followed by arabinose, xylose, and galactose. The concentration of carbohydrates was similar between natural and blanched skins, but in blanched water, glucose (>70%) was the main carbohydrate present as a result of the hydrolysis of other sugars during the blanching process. In terms of total dietary fiber (TDF), no significant differences were found between natural and blanched skins (≈ 45% TDF), where 96% was identified as insoluble dietary fiber and 3–4% as soluble.

Potential applications of blanched skins and blanch water in the food industry have been explored. As a good source of dietary fiber, blanched skins have been tested successfully for the elaboration of biscuits with increased fiber and phenolic content (Pasqualone et al., 2020). Similarly, Pasqualone et al. (2018) tested the rheological properties of dough in combination with blanched almond skin (powder) or blanch water. Findings suggest that due to the hygroscopic effect of fiber, bread is perhaps not the best bakery product to be formulated with almond skins. However, the use of blanch water did not interfere with any rheological property and, therefore, can be used in any kind of bakery.

The use of blanched skins also represents an interesting strategy to formulate functional food products with prebiotic properties, as there is clinical evidence that the consumption of blanched almond skins may lead to the improvement of the intestinal microbiota profile and the modification of the intestinal bacterial activities, due to the abundance of dietary fiber and polyphenols. Liu et al. (2014) observed significant increases in the populations of *Bifidobacterium* spp. and *Lactobacillus* spp. and the inhibition of *Clostridum perfringens* in the fecal samples of human patients as a consequence of almond skin supplementation.

Blanch Water

This byproduct results from the rapid treatment of almonds with boiling water (100 °C for 10 min) to remove the skins. Almond blanch water is one of the less studied almond byproducts, which, as mentioned before, might be a source of soluble phytochemicals since some skin bioactive compounds are transferred to the water during the process (Pérez-Jiménez and Torres, 2012). Like blanched skins, phenolic compounds and flavonoids are present in blanched water but in smaller concentrations.

The most representative compounds identified in blanch water are naringenin-7-O-glucoside and kaempferol-7-O-rutinoside, followed by catechin and 5-hydroxybenzoic acid (Mandalari et al., 2013). Among flavonoids, Pérez-Jiménez and Torres (2012) studied the presence of proanthocyanidins, which are polymers of flavan-3-ols associated with different health effects (antioxidant, anticancer, antidiabetic, neuroprotective, and antimicrobial). There is an increasing interest in identifying new sources of these compounds, especially in agricultural and food processing wastes, where they are abundant. Through a combination of mass spectrophotometric techniques, Pérez-Jiménez and Torres (2012) identified about 70 proanthocyanidins in almond blanch water, including procyanidins, prodelphinidins, and propelargonidins, meaning that blanch water is a rich source of soluble proanthocyanidins. Then, it can be considered a sustainable source of functional ingredients.

In terms of pharmaceutical applications, blanch water extracts have been tested as a potential photoprotector. Mandalari et al. (2013) analyzed the antioxidant capacity of blanch water extracts. They tested the photoprotective effect of a formulation with 2% of blanched water extracts in vivo against photooxidative damage, obtaining a 50.58% inhibition of erythema, demonstrating the potential of blanch water extracts as a source of highly valuable compounds.

Conclusion

More than a global concern, nut byproducts, shells, skins, blanch water, cake, leaves, and branches, are becoming an economic niche of interest for functional ingredients, and the demand continues to rise exponentially. The research generated so far has shown clear evidence that nut byproducts are potential sources of bioactive compounds to produce new functional foods and additives. Each byproduct features unique characteristics which make them good alternatives for a wide range of applications. Although research in nut byproducts has mainly focused on their characterization, with all the information generated so far, research is now moving forward towards finding new applications and the development of innovative products, processes, and ingredients. Further research in nut byproducts should encourage their revalorization, increase economic value, reduce the environmental impact, and promote sustainability. Also, the evaluation of the possible toxicity and allergenic factors must be conducted to assure their use in food, cosmetic, and pharmaceutic industries. Fortunately, these industries are eager to incorporate new active ingredients and develop more accessible functional products to meet the growing demand for products with health-promoting benefits. Hence, the appropriate reuse of industrial low-value products with high bioactive compounds, such as nut byproducts, represents a feasible alternative to be widely exploited, reducing the environmental impact.

References

Abdalla, A. L., Silva Filho, J. C. D., Godoi, A. R. D., Carmo, C. D. A., and Eduardo, J. L. D. P. 2008. "Use of biodiesel industry byproduts in rumiants feed." *Revista Brasileira de Zootecnia* 37(spe):260–268. doi:10.1590/s1516-35982008001300030.

Akhtar, S., Khalid, N., Ahmed, I., Shahzad, A., and Suleria, H. A. R. 2014. "Physicochemical characteristics, functional properties, and nutritional benefits of peanut oil: a review." *Critical Reviews in Food Science and Nutrition 54*(12): 1562–1575. doi:10.1080/10408398.2011.644353.

Albergamo, A., Salvo, A., Carabetta, S., Arrigo, S., Di Sanzo, R., Costa, R., ... and Russo, M. 2021. "Development of an antioxidant formula based on peanut byproducts and effects on sensory properties and aroma stability of fortified peanut snacks during storage." *Journal of the Science of Food and Agriculture* 101(2): 638-647. doi: 10.1002/jsfa.10676.

Aldana, H., Lozano, F. J., Acevedo, J., and Mendoza, A. 2015. "Thermogravimetric characterization and gasification of pecan nut shells." *Bioresource Technology* 198:634–641. doi:10.1016/j.biortech. 2015.09.069.

Aletor, O., and Ojelabi, A. 2006. "Comparative evaluation of the nutritive and functional attributes of some traditional nigerian snacks and oil seed cakes." *Pakistan Journal of Nutrition* 6(1):99–103. doi:10.3923/ pjn.2007.99.103.

Alves, J. dos S., Confortin, T. C., Todero, I., Rodriguez, A. S., Ribeiro, S. R., Boeira, C. P., Wagner, R., Mazutti, M. A., and da Rosa, C. S. 2019. "Simultaneous extraction of oil and bioactive compounds from pecan nut using pressurized solvents." *Journal of Supercritical Fluids* 153. doi:10.1016/j.supflu. 2019.104598.

An, J., Liu, J., Liang, Y., Ma, Y., Chen, C., Cheng, Y., Peng, P., Zhou, N., Zhang, R., Addy, M., Chen, P., Liu, Y., Huang, G., Ren, D., and Ruan, R. 2020. "Characterization, bioavailability and protective effects of phenolic-rich extracts from almond hulls against pro-oxidant induced toxicity in Caco-2 cells." *Food Chemistry* 322:126742. doi:10.1016/J.FOODCHEM.2020.126742

Appeldoorn, M. M., Vincken, J. P., Sanders, M., Hollman, P. C., and Gruppen, H. 2009. "Combined normal-phase and reversed-phase liquid chromatography/ESI-MS as a tool to determine the molecular diversity of A-type procyanidins in peanut skins." *Journal of Agricultural and Food Chemistry* 57(14): 6007–6013. doi: 10.1021/jf9010092.

Arya, S. S., Salve, A. R., and Chauhan, S. 2016. "Peanuts as functional food: a review." *Journal of food science and technology* 53(1):31–41. doi:10.1007/s13197-015-2007-9

Badenes, M. L. and Byrne, D. H. (Eds). 2012. *Fruit Breeding*. Springer, New York.

Badwaik, L. S., Prasad, K., and Seth, D. 2014. "Optimization of ingredient levels for the development of peanut based fiber rich pasta." *Journal of Food Science and technology* 51(10): 2713-2719. doi: 10.1007/s13197-012-0779-8.

Barral-Martinez, M., Fraga-Corral, M., Garcia-Perez, P., Simal-Gandara, J., and Prieto, M. A. 2021. "Almond by-products: valorization for sustainability and competitiveness of the industry." *Foods* 10(8): 1793. doi:10.3390/FOODS10081793

Bertioli, D. J., Seijo, G., Freitas, F. O., Valls, J. F. M., Leal-Bertioli, S. C. M., and Moretzsohn, M. C. 2011. "An overview of peanut and its wild relatives." *Plant Genetic Resources* 9(01):134–149. doi:10.1017/s1479262110000444.

Birima, A. H., Hammad, H. A., Desa, M. N. M., and Muda, Z. C. 2013. "Extraction of natural coagulant from peanut seeds for treatment of turbid water." *IOP Conference Series: Earth and Environmental Science* 16:012065. doi:10.1088/1755-1315/16/1/012065

Bottari, N. B., Lopes, L. Q. S., Pizzuti, K., Filippi dos Santos Alves, C., Corrêa, M. S., Bolzan, L. P., Zago, A., de Almeida Vaucher, R. Boligon, A. A., Giongo, J. L., Baldissera, M. D., and Santos, R. C. V. 2017. "Antimicrobial activity and phytochemical characterization of Carya illinoensis." *Microbiology Pathogenesis* 104:190–195. doi: 10.1016/j.micpath.2017.01.037.

Bulut, Y., and Tez, Z. 2007. "Adsorption studies on ground shells of hazelnut and almond." *Journal of Hazardous Materials* 149(1): 35–41. doi.org/10.1016/J.JHAZMAT.2007.03.044.

Carneiro, A. P., Rodríguez, O., and Macedo, E. A. 2017. "Dissolution and fractionation of nut shells in ionic liquids." *Bioresource Technology* 227: 188–196. doi:10.1016/j.biortech.2016.11.112

Carrín, M. E., and Carelli, A. A. 2010. "Peanut oil: Compositional data." *European Journal of Lipid Scicence and Technology* 112:697-707. doi:10.1002/ejlt.200900176

Cervantes Vázquez, M. G., Orona Castillo, I., Vázqez Vázquez, C., Fortis Hernández, M., and de Jesus Espinoza Arellano, J. 2018. "Comparative analysis of pecan orchards (*Carya illinoensis* Koch) in the Comarca Lagunera." *Revista Mexicana de Ciencias Agrícolas* 9 (1):25 – 35. doi:10.29312/remexca.v9i1.845

Christman, L. M., Dean, L. L., Allen, J. C., Godinez, S. F., and Toomer, O. T. 2019. "Peanut skin phenolic extract attenuates hyperglycemic responses *in vivo* and *in vitro*." *PLoS One* 14(3): e0214591. doi: 10.1371/journal.pone.0214591.

Correia, B., Carvalho, G., Oliveira, R., Pires, A., Ribeiro, O., Silva, R., Leão, A., Simionato, J., and Carvalho, B. 2016. "Production and quality of beef from young bulls fed diets supplemented with peanut cake." *Meat Science* 118:157–163. doi:10.1016/j.meatsci.2016.03.017

Collins, J. L., and A. R. Post. 1981. "Peanut hull flour as a potential source of dietary fiber." *Journal of Food Science* 46 (2):445–8. doi:10.1111/j.1365-2621.1981.tb04881.x

Cruz-Vega, D. E., Verde-Star, M. J., Salinas-González, N., Rosales-Hernández, B., Estrada-García, I., Mendez-Aragón, P., Carranza-Rosales, P., González-Garza, M. T., and Castro-Garza, J. 2008. "Antimycobacterial activity of *Juglans regia*, *Juglans mollis*, *Carya illinoensis* and *Bocconia frutescens*." *Phytotherarpy Research* 22:557–559. doi: 10.1002/ptr.2343.

Cueva, C., Moreno-Arribas, M. V., Martín-Álvarez, P. J., Bills, G., Vicente, M. F., Basilio, A…and Bartolomé, B. 2010. "Antimicrobial activity of phenolic acids against commensal, probiotic and pathogenic bacteria." *Research in Microbiology* 161(5):372–382. doi:10.1016/j.resmic.2010.04.006

Davis, J. P., and Dean, L. L. 2016. "Peanut composition, flavor and nutrition." In *Peanuts*. Oxford, UK: Academic Press and AOCS Press

de Camargo, A. C., Regitano-d'Arce, M. A. B., Gallo, C. R., and Shahidi, F. 2015. "Gamma irradiation induced changes in microbiological status, phenolic profile and antioxidant activity of peanut skin." *Journal of Functional Foods* 12:129–143. doi:10.1016/j.jff.2014.10.034

de Camargo, A. C., Regitano-d'Arce, M. A. B., Rasera, G. B., Canniatti-Brazaca, S. G., do Prado Silva, L., Alvarenga, V. O…Shahidi, F. 2017. "Phenolic acids and flavonoids of peanut byproducts: antioxidant capacity and antimicrobial effects." *Food Chemistry* 237:538–544. doi:10.1016/j.foodchem.2017.05.046.

De Peters, E. J., Swanson, K. L., Bill, H. M., Asmus, J., and Heguy, J. M. 2020. "Nutritional composition of almond hulls." *Applied Animal Science* 36(6):761–770. doi:10.15232/AAS.2020-02035.

de Prá Andrade, M., Piazza, D., and Poletto, M. 2021. "Pecan nutshell: morphological, chemical and thermal characterization." *Journal of Materials Research and Technology* 13:2229–2238. doi:10.1016/j.jmrt.2021.05.106.

del Carmen, N., and Urrea López, R. 2016. *"Challenges and opportunities for the exploitation of pecan nut in Mexico"*. www.ciatej.mx

Doulati Ardejani, F., Badii, K., Limaee, N. Y., Shafaei, S. Z., and Mirhabibi, A. R. 2008. "Adsorption of Direct Red 80 dye from aqueous solution onto almond shells: Effect of pH, initial concentration and shell type." *Journal of Hazardous Materials* 151(2–3): 730–737. doi:10.1016/J.JHAZMAT.2007.06.048.

Du F. Y., and Fu K. Q. 2008. "Study on total flavonoids contents in peanut vines of different plant organs." *Food Science* 1:137–140.

Esfahlan, A. J., Jamei, R., and Esfahlan, R. J. 2010. "The importance of almond (*Prunus amygdalus* L.) and its byproducts." *Food Chemistry* 120(2):349–360. doi:10.1016/J.FOODCHEM.2009.09.063.

Ezekiel, C. N., Alabi, O. A., Anokwuru, C. P., and Oginni, O. 2011. "Studies on dietary aflatoxin-induced genotoxicity using two In vivo bioassays." *Scholars Research Library* 3(2):97–106.

Feedstuffs. *Feedstuffs reference issue & buyers guide*. Vol. 73(29). Carol Stream (IL): Feedstuffs; 2001.

Franco, D., Rodríguez-Amado, I., Agregán, R., Munekata, P. E., Vázquez, J. A., Barba, F. J., and Lorenzo, J. M. 2018. "Optimization of antioxidants extraction from peanut skin to prevent oxidative processes during soybean oil storage." *LWT*, 88,1-8. doi:10.1016/j.lwt.2017.09.027.

Garcia-Perez, P., Xiao, J., Munekata, P. E. S., Lorenzo, J. M., Barba, F. J., Rajoka, M. S. R., Barros, L., Mascoloti Sprea, R., Amaral, J. S., Prieto, M. A., and Simal-Gandara, J. 2021. "Revalorization of almond by-products for the design of novel functional foods: an updated review." *Foods* 10(8):1823. doi:10.3390/foods10081823

Gupta, A., Sharma, R., and Sharma, S. 2020. "Almond." In *Antioxidants in Vegetables and Nuts - Properties and Health Benefits*. 423–452. Srpinger, Singapore. doi:10.1007/978-981-15-7470-2_21.

Hathorn, C. S., and Sanders, T. H. 2012. "Flavor and antioxidant capacity of peanut paste and peanut butter supplemented with peanut skins." *Journal of Food Science* 77:S407–S411. doi10.1111/j.1750-3841.2012.02953.x.

Hawary, S. S., Saad, S., El Halawany, A. M., Ali, Z. Y., and El Bishbishy, M. 2016. "Phenolic content and anti-hyperglycemic activity of pecan cultivars from Egypt." *Pharmaceutical biology* 54(5):788–198. doi: 10.3109/13880209.2015.1080732.

Hill, G. M. 2002. "Peanut byproducts fed to cattle." *Veterinary Clinics of North America: Food Animal Practice* 18(2):295–315. doi:10.1016/s0749-0720(02)00019-1

Idowu, O. J., Sanogo, S., abd Brewer, C. E. 2017. "Short term impacts of pecan waste byproducts on soil quality in texturally different arid soils." *Communication in Soil Science and Plant Analysis* 48(15):1781–1791. doi:10.1080/00103624.2017.1395448

Kahlaoui, M., Vecchia, S. B. D., Giovine, F., Kbaier, H. B. H., Bouzouita, N., Pereira, L. B., and Zeppa, G. 2019. "Characterization of polyphenolic compounds extracted from different varieties of almond hulls (Prunus dulcis L.)." *Antioxidants* 8(12):647. doi:10.3390/ANTIOX8120647.

Kendall, A., Marvinney, E., Brodt, S., and Zhu, W. 2015. "Life cycle–based assessment of energy use and greenhouse gas emissions in almond production, part I: analytical framework and baseline results." *Journal of Industrial Ecology* 19(6):1008–1018. doi:10.1111/JIEC.12332.

King, E. S., Chapman, D. M., Luo, K., Ferris, S., Huang, G., and Mitchell, A. E. 2022. "Progress in almond quality and sensory assessment: an overview". *Agriculture* 12(5):710. doi:10.3390/AGRICULTURE12050710.

Larrauri, M., Zunino, M. P., Zygadlo, J. A., Grosso, N. R., and Nepote, V. 2016. "Chemical characterization and antioxidant properties of fractions separated from extract of peanut skin derived from different industrial processes." *Industrial Crops and Products* 94: 964–971. doi:10.1016/j.indcrop.2016.09.066.

List, G. R. 2016. "Processing and Food Uses of Peanut Oil and Protein." *Peanuts*, 405–428:10.1016/b978-1-63067-038-2.00015-0.

Liu, Z., Lin, X., Huang, G., Zhang, W., Rao, P., and Ni, L. 2014. "Prebiotic effects of almonds and almond skins on intestinal microbiota in healthy adult humans." *Anaerobe* 26:1–6. doi:10.1016/J. ANAEROBE.2013.11.007.

López Rivilli, P., Alarcón, R., Isasmendi, G. L., and Pérez, J. D. 2012. "Stepwise isothermal fast pyrolysis (SIFP): Part II. SIFP of peanut shells –Antifungal properties of phenolic fractions." *BioResources* 7:112–117.

Lorenzo, J. M., Munekata, P. E., Sant'Ana, A. S., Carvalho, R. B., Barba, F. J., Toldrá, F., ... and Trindade, M. A. 2018. "Main characteristics of peanut skin and its role for the preservation of meat products." *Trends in Food Science & Technology* 77:1-10. doi:10.1016/j.foodres.2019.108718.

Lou, H., Yamazaki, Y., Sasaki, T., Uchida, M., Tanaka, H., and Oka, S. 1999. "A-type proanthocyanidins from peanut skins." *Phytochemistry* 51(2):297–308. doi:10.1016/S0031-9422(98)00736-5.

Maciel, L. G., Ribeiro, F. L., Teixeira, G. L., Molognoni, L., Nascimento dos Santos, J., Larroza Nunes, I., and Mara Block, J. 2020. "The potential of the pecan nut cake as an ingredient for the food industry." *Food Research International*, 127. doi:10.1016/j.foodres.2019.108718.

Mandalari, G., Tomaino, A., Arcoraci, T., Martorana, M., Turco, V. lo, Cacciola, F., Rich, G. T., Bisignano, C., Saija, A., Dugo, P., Cross, K. L., Parker, M. L., Waldron, K. W., and Wickham, M. S. J. 2010. "Characterization of polyphenols, lipids and dietary fibre from almond skins (*Amygdalus communis* L.)." *Journal of Food Composition and Analysis* 23(2):166–174. doi:10.1016/J.JFCA.2009.08.015.

Mandalari, G., Arcoraci, T., Martorana, M., Bisignano, C., Rizza, L., Bonina, F. P., Trombetta, D., and Tomaino, A. 2013. "Antioxidant and photoprotective effects of blanch water, a byproduct of the almond processing industry." *Molecules* 18(10), 12426–12440. Doi: 10.3390/MOLECULES181012426.

Marchetti, L., Califano, A. N., and Andrés, S. C. 2018. "Partial replacement of wheat flour by pecan nut expeller meal on bakery products. Effect on muffins quality." *LWT-Food Science and Technology* 95:85–91. doi:10.1016/j.lwt.2018.04.050.

McKay, A. M., Kerr, W. L., Dorick, J. M., and Dunn, L. L. 2022. "Conditions for optimal shelling, microbial reduction, and kernel quality in pecans." *Postharvest Biology and Technology* 191:111966. doi:10.1016/j.postharvbio.2022.111966.

Mehrasbi, M. R., Farahmandkia, Z., Taghibeigloo, B., and Taromi, A. 2009. "Adsorption of lead and cadmium from aqueous solution by using almond shells." *Water, Air, and Soil Pollution* 199(1–4),343–351.

Meshkini, A. 2016. "Acetone extract of almond aulls provides protection against oxidative damage and membrane protein degradation." *Journal of Acupuncture and Meridian Studies* 9(3):134–142. doi:10.1007/S11270-008-9883-9/TABLES/3.

Milbury, P. E., Chen, C. Y., Dolnikowski, G. G., and Blumberg, J. B. 2006. "Determination of flavonoids and phenolics and their distribution in almonds." *Journal of Agricultural and Food Chemistry* 54(14):5027–5033. doi:10.1021/JF0603937/ASSET/IMAGES/LARGE/JF0603937F00002.JPEG.

Monagas, M., Garrido, I., Lebrón-Aguilar, R., Bartolome, B., and Gómez-Cordovés, C. 2007. "Almond (*Prunus dulcis* (Mill.) D. A. Webb) skins as a potential source of bioactive polyphenols." *Journal of Agricultural and Food Chemistry* 55(21):8498–8507. doi:10.1021/JF071780Z/ASSET/IMAGES/LARGE/JF-2007-01780Z_0003.JPEG.

Najari, Z., Khodaiyan, F., Yarmand, M. S., and Hosseini, S. S. 2022. "Almond hulls waste valorization towards sustainable agricultural development: Production of pectin, phenolics, pullulan, and single cell protein." *Waste Management* 141:208–219. doi:10.1016/J.WASMAN.2022.01.007.

National Peanut Board. 2018. Accessed June 21, 2018. http://www.nationalpeanutboard.org/more/sustainability/5-surprising-uses-forpeanut-shells.htm

O'Keefe, S. F., and Wang, H. 2006. "Effects of peanut skin extract on quality and storage stability of beef products." *Meat Science* 73(2):278–286. doi:10.1016/j.meatsci.2005.12.001.

Oliveira, R. L., Leão, A. G., Ribeiro, O. L., Borja, M. S., Pinheiro, A. A., Oliveira, R. L., and Santana, M. C. A. 2012. "Biodiesel industry byproducts used for ruminant feed." *Scopus*, 25(4):625–638.

Pasqualone, A., Laddomada, B., Spina, A., Todaro, A., Guzmàn, C., Summo, C., Mita, G., and Giannone, V. 2018. "Almond byproducts: Extraction and characterization of phenolic compounds and evaluation of their potential use in composite dough with wheat flour." *LWT* 89:299–306. doi:10.1016/J.LWT.2017.10.066.

Pasqualone, A., Laddomada, B., Boukid, F., de Angelis, D., and Summo, C. 2020. "Use of almond skins to improve nutritional and functional properties of biscuits: an example of upcycling." *Foods* 9(11):1705. doi:10.3390/FOODS9111705.

Pérez-Jiménez, J., and Torres, J. L. 2012. "Analysis of proanthocyanidins in almond blanch water by HPLC–ESI–QqQ–MS/MS and MALDI–TOF/TOF MS." *Food Research International* 49(2):798–806. doi:10.1016/J.FOODRES.2012.09.005.

Poletto, T., Poletto, I., Moraes Silva, L. M., Brião Muniz, M. F., Silveira Reinger, L. R., Richards, N., and Stefenon, V. M. 2019. "Morphological, chemical and genetic analysis of southern Brazilian pecan accessions." *Scientia Horticulturae* 261:108863. doi:10.1016/j.scienta.2019.108863.

Ranalli, N., Andres, S. C., and Califano, A. N. 2017. "Dulce de leche-like product enriched with emulsified pecan oil: assessment of physicochemical characteristics, quality attributes, and shelf-life." *European Journal of Lipid Science and Technology* 119(7): 1600377. doi:10.1002/ejlt.201600377.

Reckziegel, P., Boufleur, N., Barcelos, R. C. S., Benvegnú, D. M., Pase, C. S., Muller, L. G., Teixeira, A. M., Zanella, R., Prado, A. C. P., Fett, R., Block, J. M., and Burger, M. E. 2011. "Oxidative stress and anxiety-like symptoms related to withdrawal of passive cigarette smoke in mice: Beneficial effects of pecan nut shells extract, a byproduct of the nut industry." *Ecotoxicology and Environmental Safety* 74(6): 1770–1778. doi:10.1016/j.ecoenv.2011.04.022.

Romero, F., van Horn, H. H., Prine, G. M., and French, E. C. 1987. "Effect of cutting interval upon yield, composition, and digestibility of Florida 77 alfalfa and florigraze rhizoma peanut." *Journal of Animal Science* 65:786-791. doi:10.2527/jas1987.653786x.

Ros, E. 2010. "Health Benefits of Nut Consumption." *Nutrients* 2(7):652–682. doi:10.3390/NU2070652.

Rubilar, M., Pinelo, M., Shene, C., Sineiro, J., and Nuñez, M. J. 2007. "Separation and HPLC-MS identification of phenolic antioxidants from agricultural residues: Almond hulls and grape pomace." *Journal of Agricultural and Food Chemistry* 55(25):10101–10109. doi:10.1021/JF0721996/ASSET/IMAGES/LARGE/JF-2007-021996_0006.JPEG.

Sabliov, C., Boldor, D., Coronel, P., and Sanders, T. 2008. "Continuous microwave processing of peanut beverages." *Journal of Food Processing and Preservation* 32(6):935-945. doi:10.1111/j.1745-4549.2008.00223.x.

SAGARPA. 2017. *National Agricultural Planning: Mexican pecan nut.*

Sales J. M., and Resurreccion A. V. A. 2009. "Maximising resveratrol and piceid contents in UV and ultrasound treated peanuts." *Food Chemistry* 117:674–680. doi:10.1016/j.foodchem.2009.04.075.

Settaluri, V. S., Kandala, C. V. K., Puppala, N., and Sundaram, J. 2012. "Peanuts and their nutritional aspects-a review." *Food and Nutrition Sciences* 3(12):1644–1650. doi:10.4236/fns.2012.312215.

Sfahlan, A. J., Mahmoodzadeh, A., Hasanzadeh, A., Heidari, R., and Jamei, R. 2009. "Antioxidants and antiradicals in almond hull and shell (*Amygdalus communis* L.) as a function of genotype." *Food Chemistry* 115(2):529–533. doi:10.1016/J.FOODCHEM.2008.12.049.

Silva, T. M., Medeiros, A. D., Oliveira, R. L., Neto, S. G., Ribeiro, M. D., Bagaldo, A. R., and Ribeiro, O. L. 2015. "Peanut cake as a substitute for soybean meal in the diet of goats." *Journal of Animal Science* 93(6):2998–3005. doi:10.2527/jas.2014-8548.

Sillero, L., Morales, A., Fernández-Marín, R., Hernández-Ramos, F., Dávila, I., Erdocia, X., and Labidi, J. 2021. "Life cycle asessment of various biorefinery approaches for the valorisation of almond shells." *Sustainable Production and Consumption* 28:749–759. doi:10.1016/J.SPC.2021.07.004.

Smeriglio, A., Mandalari, G., Bisignano, C., Filocamo, A., Barreca, D., Bellocco, E., and Trombetta, D. 2016. "Polyphenolic content and biological properties of Avola almond (*Prunus dulcis Mill.* D. A. Webb) skin and its industrial byproducts." *Industrial Crops and Products* 83:283–293. doi:10.1016/J.INDCROP.2015.11.089.

Sobolev V. S., and Cole R. J. 2003. "Note on utilization of peanut seed test." *Journal of the Science of Food and Agriculture* 84:105–111. doi:10.1002/jsfa.1593.

Sorita, G. D., Leimann, F. V., and Ferreira, S. R. S. 2020. "Biorefinery approach: Is it an upgrade opportunity for peanut byproducts?" *Trends in Food Science & Technology* 105:56–69. doi:10.1016/j.tifs.2020.08.011.

Sumner, D. A., Matthews, W. A., Medellín-Azuara, J., and Bradley, A. 2014. "The economic impacts of the California almond industry." *A Report Prepared for the Almond Board of California*.

Takeoka, G., Dao, L., Teranishi, R., Wong, R., Flessa, S., Harden, L., and Edwards, R. 2000. "Identification of three triterpenoids in almond hulls." *Journal of Agricultural and Food Chemistry* 48(8):3437–3439. doi:10.1021/JF9908289.

Toles, C. A., Marshall, W. E., Johns, M. M., Wartelle, L. H., and McAloon, A. 2000. "Acid-activated carbons from almond shells: physical, chemical and adsorptive properties and estimated cost of production." *Bioresource Technology* 71(1):87–92. doi:10.1016/S0960-8524(99)00029-2.

Urrestarazu, M., Martínez, G. A., and Salas, M. D. C. 2005. "Almond shell waste: possible local rockwool substitute in soilless crop culture." *Scientia Horticulturae* 103(4):453–460. doi:10.1016/J.SCIENTA.2004.06.011.

Urruzola, I., Robles, E., Serrano, L., and Labidi, J. 2014. "Nanopaper from almond (*Prunus dulcis*) shell." *Cellulose* 21(3):1619–1629. doi:10.1007/S10570-014-0238-Y/FIGURES/9.

USDA. 2021a. *"Food availability (Per Capita) data system"*. https://www.ers.usda.gov/data-products/food-availability-per-capita-data-system/

USDA. 2021b. *"Tree Nuts: World Markets and Trade."* https://downloads.usda.library.cornell.edu/usda-smis/files/tm70mv16z/n296xz43s/6682z4061/TreeNuts.pdf

USDA. 2022. *"Tree Nuts - Production for United States."* https://apps.fas.usda.gov/psdonline/app/index.html#/app/home/statsByCountry

Vázquez-Ovando, A., Betancur-Ancona, D., and Chel-Guerrero, L. 2013. "Physicochemical and functional properties of a protein-rich fraction produced by dry fractionation of chia seeds (*Salvia hispanica* L.)." *CyTA - Journal of Food* 11(1):75–80. doi:10.1080/19476337.2012.692123.

Wang D., and Xu H. 2008. "Biological activity components and functionality in peanut hull." *Food and Nutrition in China* 12:23–25

Widmar, D. 2021. "U.S. Almond Production and Consumption Trends." *Agricultural Economic Insights*. https://aei.ag/2021/05/17/united-states-almond-production-consumption-trends/

Wijeratne, S. S. K., Abou-Zaid, M. M., and Shahidi, F. 2006. "Antioxidant polyphenols in almond and its coproducts." *Journal of Agricultural and Food Chemistry* 54(2):312–318. doi.:10.1021/JF051692J/ASSET/IMAGES/LARGE/JF051692JF00001.JPEG.

Win, M. M., Abdul-Hamid, A., Baharin, B. S., Anwar, F., Sabu, M. C., and Pak-Dek, M. S. 2011. "Phenolic compounds and antioxidant activity of peanuts skin, hull, raw kernel and roasted kernel flour." *Pakistan Journal of Botany* 43(3): 1635–1642. doi:10.1007/s00217-011-1544-3.

Yang, C. H., Yang, S. F., and Liu, W. H. 2007. "Production of Xylooligosaccharides from Xylans by Extracellular Xylanases from *Thermobifida fusca*." *Journal of Agricultural and Food Chemistry* 55(10):3955–3959. doi:10.1021/jf0635964.

Ye, L., Liao, Y., Sun, W., and Zhao, M. 2014. "Effect of protein oxidation on the stability of peanut beverage." *CyTA - Journal of Food* 13(1):49–55. doi:10.1080/19476337.2014.910556.

Yen, G. C., and Duh, P. D. 1993. "Antioxidative properties of methanolic extracts from peanut hulls." *Journal of the American Oil Chemists Society* 70:383–386.

Yu, J.M, Ahmedna, M., Goktepe, I., and Dai, J. 2006. "Peanut skin procyanidins: Composition and antioxidant activities as affected by processing." *Journal of Food Composition and Analysis* 19(4):364–371. doi:10.1016/j.jfca.2005.08.003.

Yu, J. M., Aahmedna, M., and Goktepe, I. 2007. "Peanut protein concentrate: Production and functional properties as affected by processing." *Food Chemistry* 103:121–129. doi:10.1016/j.foodchem.2006.08.012.

Yu, M., Liu, H., Shi, A., Liu, L., and Wang, Q. 2016. "Preparation of resveratrol-enriched and poor allergic protein peanut sprout from ultrasound treated peanut seeds." *Ultrasonics Sonochemistry* 28:334-340. doi:10.1016/j.ultsonch.2015.08.008.

Zhao, X., J. Chen, and F. Du. 2012. "Potential use of peanut byproducts in food processing: a review." *Journal of Food Science and Technology* 49 (5):521–9.

Chapter 8

Olive Byproducts

Beatriz Martín-García[1,2]
Ana María Gómez-Caravaca[1,3]
and Vito Verardo[2,3,*]

[1]Department of Analytical Chemistry, Faculty of Sciences, University of Granada, Granada, Spain
[2]Department of Nutrition and Food Science, University of Granada, Campus of Cartuja, Granada, Spain
[3]Institute of Nutrition and Food Technology 'José Mataix', Biomedical Research Center, University of Granada, Armilla, Granada, Spain

Abstract

In the last few years, the consumption of olive oil has been increasing due to its organoleptic and beneficial properties. As a result, the production of olive waste (wood, branches, leaves) and byproducts (olive pomace, olive mill wastewater, olive stones) derived from the olive oil Industry has increased and, consequently, this fact is generating undesirable effects in terms of sustainability and environmental impact and high costs for management and disposal. However, these wastes and byproducts are a rich source of bioactive compounds such as triterpenic acids and phenolic compounds, which can be reused for several purposes following the circular economy. The recovery of these compounds from olive byproducts can be carried out by green extraction technologies, including ultrasound-assisted extraction, supercritical fluid extraction, pressurized liquid extraction and microwave-assisted extraction. As well as this, the pulsed electric field, radiofrequency drying, high voltage electrical discharge and high-hydrostatic pressure. These techniques are, furthermore, environmentally friendly and provide high-efficiency short extraction time, which improves the monitoring process and has low energy consumption. In this sense, numerous processes about the purification of phenolic compounds from olive byproducts with membrane treatment have been used, such as microfiltration, ultrafiltration, nanofiltration, and reverse osmosis. Results showed that these purification processes concentrate phenolic compounds with a low molecular weight, which can be used in food, pharmaceutical and cosmetic industries. Therefore, the development of eco-friendly processes could be a suitable strategy to recover the bioactive compounds from olive oil byproducts with applications in the innovative functional foods sector.

---

* Corresponding Author's Email: vitoverardo@ugr.es.

In: Food Byproducts
Editors: S. Teresita Martín-del-Campo, A. Cardador-Martínez and J. del Pilar Ramírez-Anaya
ISBN: 979-8-88697-612-0

Keywords: byproducts, bioactive compounds, green extraction technologies, functional foods, olive process

Introduction

Olive industries generate a large quantity of solid and liquid wastes. Concretely, during olive oil production, one ton of olives produces around 0.40 tons of olive pomace, which is composed mainly of fragments of olive skin, pulp, and bone (Nunes et al. 2016; Tapia-Quirós et al. 2020), and 1200 L of olive mill waste water, depending on the oil extraction technologies employed (Dermeche et al. 2013). These wastes are harmful to the environment due to part of this olive mill wastewater is often still thrown into the nearby environment, causing soil contamination, natural stream pollution, or other serious effects on the aquatic fauna (Komnitsas et al. 2016). Furthermore, the management and transport of the wet olive pomace is difficult because this possesses a strong odor and a pasty texture (Souilem et al. 2017). Generally, the treatment of olive mill effluents is difficult mainly for the following grounds: it is high organic loading, the seasonal operation of olive mills, it is high territorial scattering, the high costs of installing a waste treatment facility and the presence of organic compounds which are hardly biodegradable (Vlyssides, Lamprou, and Vlysidis 2017).

Nevertheless, these olive byproducts possess bioactive compounds, such as phenolic compounds (Tapia-Quirós et al. 2020). Phenolic compounds are secondary metabolites of plants that possess more than one phenol group per molecule. These molecules possess a strong antioxidant potential; thus, their use could be interesting in the application in food, cosmetic, and pharmaceutical industries because of their benefits to human health. As olive oil production are major economic activities in Southern Europe, it is well worth delving deeply into the recovery of these active compounds from production wastes and byproducts. For instance, Spain produced about 5000 kton/year of olive pomace with an average total phenolic content of 900 g/ton (Panzella et al. 2020). Thus, the recovery of phenolic compounds could be a good option for valorizing these byproducts and would support a more sustainable bioeconomy, reducing environmental problems produced by these wastes(Tapia-Quirós et al. 2020). Extraction is the first step in the recovery of phenolic compounds. Several techniques can be applied to olive byproducts, such as ultrasound-assisted extraction (UAE), microwave-assisted extraction (MAE), or pressurized liquid extraction (PLE) (Gómez-Cruz et al. 2021; Xie et al. 2019; Martiny et al. 2021; Martín-García et al. 2022). UAE consists of the application of high-frequency waves ($\geq$2 MHz) that allow the production of cavitation bubbles and the rupture of the cell membranes in the sample, improving the mass transfer of phenolic compounds from the solid matrix to the liquid phase. UAE can be done in an ultrasonic bath or with an ultrasonic probe (Barba et al. 2016; Ameer, Shahbaz, and Kwon 2017a). MAE uses microwave energy, which can be absorbed by polar molecules. PLE consists of the use of solvents at high pressure (Araújo et al. 2015). High pressure combined with high temperature improves both the kinetics of the mass transfer, and the efficiency of the extraction system (Vazquez-Roig and Picó 2015). In addition, pressure-driven membrane technologies have been used in olive mill wastewater including microfiltration (MF), ultrafiltration (UF), nanofiltration (NF), and reverse osmosis (RO) which are effective in the phenolic recovery and are environmentally friendly.

This chapter is focused on different sections: (a) byproducts generated in processing and/or products discarded during harvesting, transport due to non-accomplishment to quality standards, concretely during the olive oil production, (b) quality evaluation of byproducts and their implications economic, environmental, social, etc., as well as their added-value in food and feed ingredients, industrial bio-based materials, bio-fertilizers and bioenergy, (c) bioactive molecules present or that could be generated from the olive byproducts and (d) recovery and uses systems employed in olive leaves, olive pomace and olive waste mill water. In this last section, the yields of phenolic compounds recovered from these olive byproducts have been evaluated by different extraction techniques which are influenced by different parameters such as solvent, temperature, ratio of solid to solvent and time among other specific factors employed in specific extraction process.

Methods

Byproducts Generated in Processing and/or Products Discarded during Harvesting, Transport Due to Non-accomplishment to Quality Standards

Byproducts and waste from olive production (agriculture) and the olive oil industry (mills and refineries) are important environmental issues in Mediterranean areas. Olive oil production involves different processes to the olives, including washing, grinding, beating, and the extraction itself, which represent the basic stage of the whole process (Rodrigues, Pimentel, and Oliveira 2015; Roselló-Soto et al. 2015a). Different byproducts are obtained depending on the process employed for the olive oil production. Specifically, on the type of horizontal centrifugation (two-phase or three-phase), the process most used to separate olive oil from the olive paste obtained after malaxation is shown in the Figure 1.

When the two-phase decanter process is used, water is not added to the olive paste resulting from malaxation, providing a humid semi-solid pomace byproduct (wet olive pomace, wet OP, or alperujo), which is constituted of olive pulp and husk, crushed olive stone and aqueous solution. Two byproducts are obtained from the case of three phase decanter, solid OP and olive oil mill wastewater (OMWW) or alpechin. Particularly, the modern two-phase centrifugal extraction technology merges OP with OMWW to produce a single byproduct named olive mill waste (OMW) (Bhatnagar et al. 2014). OMW shows a wide spectrum of toxicity, antimicrobial activity, high acidity and phytotoxicity, that make its biological degradation challenging and its disposal problematic for major producing countries, given the significant amounts of OMW generated every year (Abbattista et al. 2021). In the European Union producer countries, 9.6 million tonnes/year of byproducts subject to valorization (olive pomace, leaves and stone) are generated from olive mills, while 11.8 million tons of additional biomass is obtained from olive tree pruning biomass (OTPB). Nevertheless, only 30% of this biomass can be economically used due to part of the production takes place on remote mountainous farms (Berbel and Posadillo 2018b; Donner and Radić 2021). This gives an estimate for the European union of 13.1 million tonnes (9.6 tonnes of waste from mills and +30% of the OTPB). Olive leaves are generated both at the farm during the pruning of the olive trees as well as during the harvesting and cleaning of olives prior to oil extraction in the processing plant (Berbel and Posadillo 2018b). Olive leaves represent 10% of the total weight from the olives harvest and 25 kg/tree during olive tree pruning (Khwaldia et al. 2022). The olive pomace constitutes around 35-40% of the total weight of olives processed in the mill (Khwaldia et al. 2022; Akay et al. 2015).

OMWW is made up of 90% water with a low number of organic compounds and mineral salts, which is produced by the three-phase olive-oil production process. The annual world production of OMWW is around 10 and more than 30 million meter cubes (Genethliou, Kornaros, and Dailianis 2020). Olive stones comprises regarding 10% of the weight of olives and are mostly made up of cellulose (20.9%), hemicellulose (26%), xylose (26.6%), galactose (1.4%), arabinose (1.3%), and lignin (35.6%) (Padilla-Rascón et al. 2020).

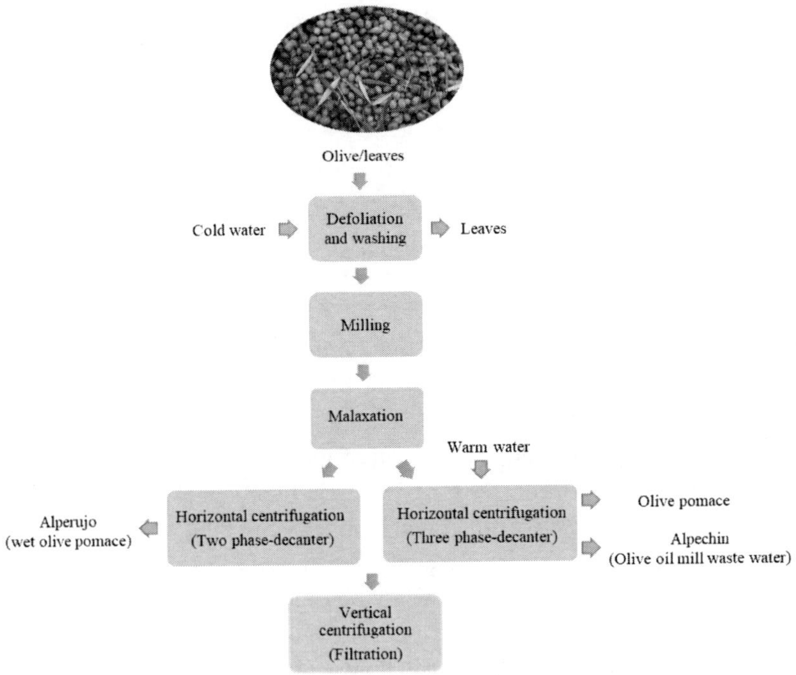

Figure 1. The main steps for the olive oil production (Abbattista et al. 2021).

Quality Evaluation of Byproducts and Their Implications (Economic, Environmental, Social, etc.)

Solutions for olive waste and byproducts need to be found to avoid negative environmental impacts in combination with an effort to preserve scarce natural resources (energy, water, soil, raw material) locally. In this sense, the treatment of olive mill waste and olive byproducts is not common European Union legislation. There are national regulations in several countries such as Spain, Italy, or Portugal for olive byproduct disposal, but not in all olive oil-producing countries and also not for all types of olive waste (e.g., not for leaves and branches). Nevertheless, waste and byproducts from olive oil production should not only be considered from a mere 'management or treatment' point of view but can provide opportunities to be valorized, such as these can be converted into new ingredients and products with a high value-added. The different conversion strategies based on a broader bioeconomy perspective can be separated into diverse categories of biomass upgrading (Berbel and Posadillo 2018b), such as specialty molecules, food and feed ingredients, industrial bio-based materials, bio-fertilizers and bioenergy. Using olive byproducts as biomass for energy needs the highest volume but has the lowest priority. Bio-fertilizers and biomaterials have more value, followed by food and feed ingredients. The highest value-added from the lowest volume is reached within the fine chemical or pharmaceutical sector.

Low-Value Uses

Direct release of the olive mill to the environment is not recommended due to the phytotoxicity of olive mill waste (Majbar et al. 2018). In addition, the olive mill waste possesses a strong odor and a doughy texture that makes it difficult to manage and transport. The use of olive seeds as a biofuel for heating in olive oil production regions is very common, concretely in agro-industries, animal farms, greenhouses, and household heating systems. The conversion of waste to solid (pellets) or liquid (biodiesel) biofuel incorporates more value than direct energy use or composting. Other research papers have reported the conversion of byproducts to biodiesel (Yücel 2011; Hernández et al. 2014). The use of olive waste and byproducts is a profitable solution in terms of employment and value generation. Furthermore, the savings in greenhouse gas emissions for Andalusia, which produces around 50% of the EU's olive oil, are predicted to be 1.6 million tonnes of CO2-eq annually (AGAPA 2015).

High-Value Uses

Olive biomass is a source of bioactive compounds; thus, their recovery by using the extraction techniques may be marketable in the pharmaceutical and cosmetic industries. Galanakis et al. (2017) reported the use of different technologies for the recovery of these compounds from olive oil processing byproducts and provide an integrated strategy for ensuring the process's long-term viability. There are several studies about the utilization of waste products with promising future applications. There are two options for using: the recovery of valuable components and bioconversion into valuable products. Olive mill wastewater is a rich source of phenolic compounds such as hydroxytyrosol with considerable antioxidant activities that potentially affect the blood lipid levels. In addition, these phenolic compounds recovered from olive mill wastewater are used in cosmetics due to these active UV filters (Galanakis, Tsatalas, and Galanakis 2018). In addition, these phenolic compounds possess an inhibitory capacity in sugar fermentation. Therefore, the disposal of these compounds from the aqueous extract could enhance ethanol production from the glucose contained in the extractive fraction, improving biogas or bioethanol production (Manzanares et al. 2017).

Medium-Value Uses

In some industries is well-established the use of olive byproducts for animal feed. Nevertheless, olive biomass possesses little available protein, contains phytosterols and is a good energy source. Nevertheless, its high energy content may, in fact, reduce the total feed intake of animals and thus might be counterproductive. Therefore, olive byproducts are a good supplement for a varied diet, but fat-rich byproducts should be reduced to a certain percentage (5% is recommended). The most used byproduct is olive pomace; for every 100 kg of olives,

40–80 kg of olive pomace. The use of olive byproducts as feed for aquaculture and livestock has shown an improvement in the fatty acids profile of animal products by decreasing the saturated acids and increasing unsaturated components in both, meat and milk (Berbel and Posadillo 2018b, 2018a).

Bioactive Molecules Present or That Could Be Generated from the Byproducts

Olive byproducts are inexpensive raw materials containing bioactive compounds, which have the potential to improve the economic value of nutraceuticals, functional food, dietary supplements, packaging materials, and pharmaceutical formulations. Olive oil byproducts are characterized by a large variability of the bioactive compounds due to several factors, including olive cultivar, agronomical conditions, climate, level of ripening, a production process that affects their profile, and their biological activities (Khwaldia et al. 2022).

Olive byproducts possess a wide range of bioactive compounds, which possess beneficial activities in the health of humans. Olive byproducts, such as leaves and pruning biomass, seeds, pomace, cake, or OMWW, contain a wide range of bioactive compounds which present beneficial effects on the health of humans. For this reason, the utilization of these compounds could be beneficial to obtain a yield added-value products and ingredients (Khwaldia et al. 2022). The main bioactive compounds present in olive residues are fatty acids, phytosterols, triterpenoids, phenolic compounds (phenolic acids and derivatives, flavonoids, secoiridoids, phenylpropanoids and lignans) and tocopherols, whereas various other bioactive components such as carotenoids or chlorophylls have been found in low quantities (Hannachi et al. 2020). The content of these bioactive compounds depends on the type of byproduct, and also other agronomic factors affect the chemical composition of olives, such as cultivar, ripeness, climatic conditions, growth conditions, etc. (Romero et al. 2017). The quantity of recovered bioactive compounds from the main olive byproducts, including olive pomace, leaves OMWW and seeds, appears in Table 1.

Fatty Acids

Olive fruits are rich in oil, composed mainly of triglycerides containing fatty acids, concretely oleic, palmitic and linoleic acids. The majority of these fatty acids are obtained from olive oil production, whereas around 20% of fatty acids are lost in the olive oil byproducts, including olive pomace, olive mill wastewater and seeds. The most concentrated fatty acids in these olive residues are oleic acid and linoleic acid. Olive seeds are the byproduct with the highest fatty acid content (14000 and 4500 mg/100 g d.w.), which represent around 91.7% of its bioactive composition (Maestri et al. 2019; Hannachi et al. 2020). Olive pomace contain 2000 and 350 mg/100 g of oleic and linoleic acid, its oleic acid content represents around 60% of the total fatty acid content (Wedyan et al. 2017; Antónia Nunes et al. 2018). Lastly, the content of oleic acid in olive mill wastewater represents around 92.9% of its total bioactive composition (Otero et al. 2021; Martins, Martins, and Braga 2021).

Table 1. Bioactive compounds recovered from the main olive byproducts

Bioactive compounds		Olive byproducts				Reference
Group	Main compounds	Olive pomace	Leaves	Olive mill wastewater	Seeds	
Fatty acids	Oleic acid	2000		3000	14000	(Hannachi et al. 2020; Maestri et al. 2019; Martins, Martins, and Braga 2021; Antónia Nunes et al. 2018)
	Linoleic acid	350			4500	
Phytosterols	β-sitosterol	148			167.49-355.5	(Maestri et al. 2019; Ranalli et al. 2002)
	Campesterol	4.34			7.27-19.23	
	Stigmasterol	6.37			5.39-6.45	
Triterpenoids	Squalene	300	-	25	300	(Xie et al. 2021; Mwakalukwa et al. 2020; Otero et al. 2021; Guinda et al. 2010)
	Maslinic acid	55.1-400	243.4-730	18	-	
	Oleanolic acid	36.7-350	1188-3450	8.21	-	
	Ursolic acid	73.8	200-633.4	-	733.8	
	Erythrodiol	n.d.	80-321.9	-	271.7	
Tocopherols	α-tocopherol	2.63-7.64	1.12-28.46	1.56-3.91	7.98-40.1	(Botsoglou et al. 2012; Tarchoune et al. 2019; Maestri et al. 2019; Antónia Nunes et al. 2018; Aggoun et al. 2016; González-Hidalgo, Bañón, and Ros 2012)
	β-tocopherol	0.03-0.09			0.905	
	γ-tocopherol	0.04-0.13		0.26-0.56	3.90	
	δ-tocopherol				0.99	
Phenolic acid and derivatives	Gallic cid	10	0.74-5.58	-	3	(Benincasa et al. 2022; Chanioti and Tzia 2018; Talhaoui, Taamalli, et al. 2015a; Otero et al. 2021)
	Caffeic acid	0.1-5	0.14-0.45	9	140	
	p-coumaric acid	10	-	0.5-5	300	
	Hydroxybenzoic acid	10	0.06-2.38	-	5	
	Ferulic acid	20	0.7-9.14	-	30	
	Vanillic acid	0.038-30	1.28-11	0.136-20	10	
Phenyl alcohols	Hydroxytyrosol	4.71-89	0.21-112	11.0-148.1	84.2	(Benincasa et al. 2022; Chanioti and Tzia 2018; Talhaoui et al. 2016, 2014; Talhaoui, Taamalli, et al. 2015a; De Bruno et al. 2021; González-Hidalgo, Bañón, and Ros 2012)
	Tyrosol	0.63-2.15	9-66	0.95-204.3	29.9	
	Hydroxytyrosol hexose	-	80.9-237	-	-	
	Tyrosol glucoside	-	76-144	-	-	
	Verbascoside	8	30-1860	15-70	-	

Table 1. (Continued)

Bioactive compounds		Olive byproducts				Reference
Group	Main compounds	Olive pomace	Leaves	Olive mill wastewater	Seeds	
Flavonoids	Luteolin Apigenin Rutin	3-73 0.053-4 0.23-74	0.6-560 0.46-33.95 18-29	6.24-20 0.955-4 0.72-5	60 30 70	(Benincasa et al. 2022; Chanioti and Tzia 2018; Talhaoui et al. 2016; Talhaoui, Taamalli, et al. 2015a; Otero et al. 2021; De Bruno et al. 2021)
Secoiridoids	Oleuropein Oleocanthal Oleacein Ligstroside	0.089-3500 30 20 50	987-6644 - - 78-99	0.15-1.61 1 70 9	56.87 - - -	(Chanioti and Tzia 2018; Talhaoui et al. 2014; Talhaoui, Gómez-Caravaca, et al. 2015; Talhaoui et al. 2016; Otero et al. 2021; De Bruno et al. 2021; González-Hidalgo, Bañón, and Ros 2012)
Lignans	Pinoresinol 1-acetoxypinoresinol	0.59-3 2	- -	1.92-30.3 0.113	3 -	(Mwakalukwa et al. 2020; De Bruno et al. 2021)

Phytosterols

Phytosterols have been found in olive pomace and in seeds. The sterol content in olive pomace is around 183.2 mg/100 g d.w., whereas its content in olive seeds was 412.13 mg/100 g respectively. Among them, the most concentrated sterols in these byproducts are β-sitosterol, campesterol and stigmasterol (Table 1) (Maestri et al. 2019; Ranalli et al. 2002). Phytosterols, these olive byproducts (olive pomace and seeds), correspond to around 8.9% and 2.7% of their total bioactive composition.

Triterpenoids

Squalene, maslinic and oleanolic acids are the most representative triterpenoid compounds in olive byproducts. The presence of squalene has been found in both olive pomace and seeds was 300 mg/100 g d.w., whereas its content in olive mill waste water is lower 25 mg/100 g d.w. (Xie et al. 2021; Mwakalukwa et al. 2020; Otero et al. 2021). Maslinic and oleanolic acids have been reported in olive pomace, leaves and olive mill waste water, but not in seeds (Table 1). Olive leaves contain important amounts of oleanolic acid, followed by significant concentrations of maslinic acid. The content of oleanolic and maslinic acids in olive leaves is about 1188-3450 and 243.4-730 mg/100 g d.w. in olive leaves, respectively (Xie et al. 2021; Guinda et al. 2010). This content is higher than in olive pomace, with a content of 36.7-350 and 55.1-400, respectively, while lower levels have been reported in OMWW (8 and 18 mg/100 g of oleanoic and maslinic acid (Xie et al. 2021; Mwakalukwa et al. 2020; Otero et al. 2021). Other triterpenoids, such as ursolic acid and erytrodiol, have been found mainly in leaves and seeds (Xie et al. 2021).

Tocopherols

Olives are rich in tocopherols, which are isomers of vitamin E; these might be present as α-, β-, γ- and δ-tocopherol isoforms. The content of tocopherols in olive byproducts is around 77.5% higher than the present in olive oil. Concretely, α-tocopherol is the most concentrated isoform in the olive byproducts. In olive leaves and seeds, this content reaches around 30-40 mg/100 g d.w., while its lowest content is observed in olive mill wastewater (2-4 mg/100 g d.w.) (Botsoglou et al. 2012; Tarchoune et al. 2019; Maestri et al. 2019; Antónia Nunes et al. 2018; Aggoun et al. 2016; González-Hidalgo, Bañón, and Ros 2012).

Phenolic Compounds

The content of phenolic compounds in olive byproducts depends mainly on the olive cultivars and maturation stage. These phenolic components are classified according to their structure in phenolic acid and derivatives, phenyl alcohols, flavonoids, secoiridoids and lignans (Table 1). The most abundant phenolic compounds in olive byproducts are the secoiridoids, including oleuropein and ligstroside (Otero et al. 2021). In addition, phenyl alcohols such as hydroxytyrosol are found in high quantity in the olive leaves byproducts. Specifically, hydroxytyrosol is mainly obtained from the hydrolysis of oleuropein, whereas tyrosol is obtained from ligstroside degradation (Wani et al. 2018).

Phenolic Acid Derivatives

Comparing the distribution of phenolic acid derivatives, only 4.4% of these components are present in olive oil, whereas the remaining 95.6% can be found in its residues (Otero et al. 2021). The main phenolic acid derivatives in olive byproducts are gallic acid, caffeic acid, p-coumaric acid, hydroxybenzoic acid, ferulic acid and vanillic acid. Among the olive residues, seeds possess the highest content in phenolic acids with a total content of 488 mg/100 g d.w, which represent around 2.4% of the main bioactive compounds), followed by the olive pomace, with 85 mg/100 g d.w., which represent around 1.1% of its total bioactive composition. Olive mill wastewater and olive leaves possess a total content of 9.64-34 mg/100g d.w. and 2.92-28.55 mg/100g d.w. (Benincasa et al. 2022; Chanioti & Tzia 2018a; Talhaoui, Taamalli, et al. 2015a; Otero et al. 2021).

Phenyl Alcohols

The main phenyl alcohols present in the olive byproducts are hydroxytyrosol, tyrosol and verbascoside. Among them, the most concentrated in most of olive byproducts were hydroxytyrosol and tyrosol, which highest content has been found in the olive mill waste water, followed by the olive leaves, seeds and olive pomace. Hydroxytyrosol and tyrosol have been found mainly in their glucosylated form in the olive leaves, which have been found in a quantity of 80.9-237 and 76-144 mg/100 g d.w., respectively. In addition, the highest content of verbascoside has been reported in olive leaves, reaching concentrations of 1860 mg per 100 g d.w., followed by olive mill wastewater (70 mg/100g d.w.) and olive pomace (8 mg/100 g d.w.) (Talhaoui, Taamalli, et al. 2015b).

Flavonoids

With regard to the flavonoid, the most identified in olive residues, are flavones such as luteolin, apigenin and the flavonol rutin. Luteolin is the most abundant flavonoid in pomace, leaves and OMWW, whereas rutin is the most concentrated flavonoid in olive seeds. Olive pomace, leaves, OMWW and seeds contain up to 151, 622.95, 29 and 160 mg of these flavonoids per 100 g d.w. (Benincasa et al. 2022; Chanioti and Tzia 2018; Talhaoui et al. 2016; Talhaoui, Taamalli, et al. 2015a; Otero et al. 2021; De Bruno et al. 2021). This content represents the 2, 4, 0.9 and 0.8% of the bioactive compounds found in these matrices.

Secoiridoids

Secoiridoids are the most concentrated phenolic compounds present in olive byproducts. Oleuropein is the main secoiridois, biosynthetically derived from the esterification of hydroxytyrosol with elenolic acid, and ligstroside, obtained from the esterification of tyrosol with oleoside 11-methyl ester, a methylated derivative of elenoic acid (Czerwińska, Kiss, and Naruszewicz 2012). The quantity of these compounds is higher in the different residues, which represent 96.9% of total secoiridoids, compared with 3.1% found in Olive oil. According to previous studies, secoiridoids are more abundant in olive leaves (1065-6700 mg/100g d.w.) and in olive pomace (100-3600 mg/100 g d.w.), whereas this content is much lower in olive mill wastewater and seeds (81.61 and 56.87 mg/100 g) (Chanioti and Tzia 2018; Talhaoui et al. 2014; Talhaoui, Gómez-Caravaca, et al. 2015; Talhaoui et al. 2016; Otero et al. 2021; De Bruno et al. 2021; González-Hidalgo, Bañón, and Ros 2012). Among them, the most concentrated secoiridoid was oleuropein in olive pomace and in leaves, which content could represent around 97 and 98% of their total secoiridoid content.

Considering the bioactive composition of olive residues, the quantity of secoiridoids represents 15.2, 0.7 and 3.7% of the bioactive fraction of olive pomace, seeds and OMWW, respectively (Otero et al. 2021).

Lignans

The two main lignans found in olive byproducts are (+)-pinoresinol and (+)-1-acetoxypinoresinol. Both lignans have been identified in olive pomace (~3 and ~2 mg/100 g d.w., respectively) and OMWW (30.3 and 0.1 mg/100 g d.w., respectively), but only (+)-pinoresinol was detected in seeds (3 mg/100 g d.w.). Concretely, lignans represent a 0.1, 0.1 and 0.01% of the bioactive compounds of olive pomace, OMWW and seeds (Mwakalukwa et al. 2020; De Bruno et al. 2021).

Recovery and Uses Systems

Recovery of phenolic compounds from olive wastes is a multi-stage process that consists of the following steps: extraction of the phenolic compounds from the solid matrix; post-treatment of the generated extracts by membrane technologies; separation of phenolic compounds or individual components; and final treatment stages of drying to produce powder forms. Several techniques have been used to recover phenolic compounds from olive byproducts. Conventional extraction with solvents, also called solid-liquid extraction (SLE) or maceration, has been a widely used technique for years to extract bioactive compounds from matrices of vegetable origin (Tapia-Quirós et al. 2022). Numerous solvents have been studied for the extraction of phenolic compounds. Nevertheless, the chosen systems for food, pharmaceutical, or cosmetic applications are those based on water and ethanol (Tapia-Quirós et al. 2020).

New non-conventional extraction technologies have been used recently due to the increasing need for efficient recovery procedures. These processes aim to recover specific phenol (HT) in pure form or the recovery of the mixture of phenolic compounds as a crude product. The application of these emerging techniques to different food components (lipids, minerals, vitamins, polyphenols, aroma compounds, and enzymes) keeps their effectiveness and bioactive content and bioavailability since these technologies don't employ high temperatures avoiding the degradation of these compounds. These new techniques can reduce the extraction time, process temperature, and solvent consumption (Otero et al. 2018; Sarfarazi et al. 2020; Tapia-Quirós et al. 2020; Pappas, Lakka, Palaiogiannis, Athanasiadis, et al. 2021). Therefore, this produces a higher extraction efficiency and lower energy consumption. The most relevant technologies are the following: ultrasound-assisted extraction (UAE), microwave-assisted extraction (MAE), supercritical fluid extraction (SFE), pressurized liquid extraction (PLE), pulsed electrical fields (PEF), High Voltage electrical discharge (HVED) and High hydrostatic pressure (HHPP) (Galanakis 2021).

Ultrasound Assisted Extraction

Ultrasound-assisted extraction (UAE) consists of the use of high-frequency waves that produce negative pressure. This creates cavitation of small bubbles in the solvent. This phenomenon causes the rupture of the cell membranes in the samples, providing a contact between the solvent and the matrix. This fact improves the mass transfer of phenolic compounds from the sample to the solvent. UAE can be done using an ultrasonic bath or more efficiently with an ultrasonic probe (Barba et al. 2016; Ameer, Shahbaz, and Kwon 2017a). UAE is mainly influenced by different processing parameters such as time, temperature, liquid:solid ratio, power, and frequency (Kumar, Srivastav, and Sharanagat 2021). Therefore, their optimization is significant to achieve extracts with a high concentration in phenolic compounds. UAE with probe achieves high TPC in a short extraction time; this technique is a good choice for the extraction of phenolic compounds from plant sources. Regarding olive pomace, Gómez-Cruz et al. (2021) and Xie et al. (2019) have shown that UAE was a very effective method for the extraction of phenolic compounds from olive pomace resulting in high yields in a short time of 5 and 12 min, which employed 40% acetone and 90% ethanol and a frequency of 40 and 20 kHz with a ratio of 8.5% (w/v) and 30/1 mL/g (v/w). The UAE treatment was confirmed to be particularly effective for the recovery of hydroxytyrosol, maslinic acid, and oleanolic acid from olive pomace. Higher extraction times of 75 and 30 min have been reported in the case of US bath in comparison with US probe in olive pomace (Tapia-Quirós et al. 2020; Goldsmith et al. 2018), which employed water and 50% EtOH at 20 °C and 30 min with solvent ratios of 50/1 and 20/1 mL/g (v/w) respectively. Ultrasonic assisted extraction has shown a promising extraction technique combined with freeze-dried OMWW at the conditions of 100% methanol as solvent, and three times of 20 min (1.5 g/25 mL, w/v) offered the highest content in phenol (1521.3 μg GAE/mL), which was 61.07 and 162.92% higher than the obtained by liquid-liquid extraction and solid-phase extraction (Jerman Klen and Mozetič Vodopivec 2011).

Microwave Assisted Extraction

Microwave-assisted extraction (MAE) is an extraction technique that consists of the use of the microwave energy, which can be absorbed by polar molecules. Microwave energy heats the polar solvents in contact with solid samples, increasing the internal pressure inside the cell. This allows the rupture of the cell wall and the release of bioactive compounds to the solvent.

MAE is mostly performed with closed vessels (Barba et al. 2016; Roselló-Soto et al. 2015b). Different studies have been resported the application of MAE for the recovery of phenolic compounds from olive byproducts (Table 2). These studies have evaluated the application of MAE for the recovery of phenolic compounds, such as hydroxytyrosol, maslinic acid, and oleanolic acid from olive pomace and oleuropein and hydroxytyrosol in olive leaves. According to these studies, the microwave treatment employed short extraction times from 1.5 to 10 min and extraction mixture solvent of water and ethanol, power from 600 to 1000 W and temperatures between 50 °C and 100 °C. In the case of olive pomace, a similar concentration in hydroxytyrosol, maslinic acid and oleanoic acid were obtained by MAE at the conditions of 90% EtOH, 50 °C, 600 W, 2.45 GHz, 5 min and 30:1 mL/g (v/w) in comparison with those obtained by US probe at the same conditions of solvent, temperature and ratio with a frequency of 20 kHz and 3 min (Xie et al. 2019).

Supercritical Fluid Extraction (SFE)

Supercritical fluid extraction (SFE) is a technology that uses organic solvents containing carbon dioxide in its supercritical state. The increase of the pressure and the temperature of the liquid/gas over the critical point provides fluids in the supercritical state. CO_2 is the most used supercritical fluid due to its availability, non-toxicity, and flammability, low cost in comparison with liquid organic solvents, and the ability to manage critical temperature and pressure of 31.1°C and 7.38 MPa. Compared with liquid organic solvents, supercritical CO_2 has relatively low viscosity, high molecular diffusivity and low system surface tension. All these properties result in better mass transfer. The non-polar nature of CO_2 makes it particularly suitable for the extraction of non-polar compounds. Adding a co-solvent, such as ethanol, methanol, or acetone, increases the polarity of CO_2 and enables the extraction of polar compounds (Roselló-Soto et al. 2015b; Tapia-Quirós et al. 2022). A previous study reported a total phenolic content obtained in olive pomace at the SFE optimum conditions, which was 89% higher than the obtained with a conventional extraction method obtained at the conditions of 50% EtOH, 0.5 g OP/mL of solid to solvent ratio (S/L), 70°C and 60 min (180 ± 11 mgGAE/g DE) (Katsinas et al. 2021). Therefore, SFE employed an extraction time and solvent consumption, which were 67% and 38% lower compared to conventional extraction. Nevertheless, another study reported employed SFE in the recovery of phenolic compounds from olive leaves, and which highest yield of oleuropein was obtained by the Soxhlet method (37.84 mg/g d.w.) in comparison with SFE (14.26 mg/g d.w.) by using CO_2 modified by methanol at 300 bar and 100 °C in the SFE method (Şahin, Bilgin, and Dramur 2011). In addition, another study reported a low total phenolic content by SFE at the conditions of 16.5 MPa, 20°C and 210 min (11.49 mg/g d.w.) in comparison with the obtained by PLE at 60°C, 110 min and 80%EtOH (386.42 mg GAE/g extract) (A. D. Rosa et al. 2019). A previous study applied a Supercritical fluid extraction in OMWW, which was performed at 40°C, 350 bar, and for 60 min. This extract was compared with conventional solvent extraction (methanol, ethanol, mixture of ethanol to water 1:1, n-propanol, isopropanol, and ethyl acetate), and SC-CO_2 shown to be an efficient solvent for the recovery of phenolic compounds with high antioxidant activity in OMWW, but its phenolic yield was quite low (Lafka et al. 2011).

Pressurized Liquid Extraction (PLE)

Pressurized liquid extraction consists of using the extraction solvent at high temperatures and pressure. This technique possesses three operating modes including dynamic, static, and dynamic-static. In the dynamic mode, the solvent is constantly driven through the cell during extraction, whereas in the static mode, the solvent is pumped only between the extraction cycles; the dynamic-static mode represents a combination of the two (Araújo et al. 2015). pressurized liquid extraction (PLE) is considered an innovative extraction procedure for recovering phenolic compounds from olive byproducts, and this is a great alternative to conventional extraction procedures because of its high selectivity, low solvent consumption, short extraction times and non-harmful solvents with lower toxicity causing the minimum environmental impact after their usage (Ameer, Shahbaz, and Kwon 2017b). PLE runs under constant pressure and varying factors values including temperature, extraction time, the composition of extraction solvent, etc. (Xynos et al. 2014). PLE uses high temperatures; nevertheless, the application of high pressure makes the solvents keep in their liquid state, thus providing an improvement in the extraction efficiency (Xynos et al. 2012). Previous studies reported the same pressure of 10.34MPa in the PLE for the extraction of phenolic compounds in olive leaves and pomace, varying factors such as time, temperature and % of solvent. In olive pomace, the optimum PLE conditions were 20 min, 136.5ºC and 52.3% EtOH showed a content in phenolic compounds of 1659 mg/kg d.w., which was 588.9% higher than the obtained by a conventional technique that employed 120 min and 80% EtOH (281.7 mg/kg d.w., respectively) (Pavez et al. 2019). The most concentrated phenolic compounds in this olive pomace extract by PLE were hydroxytyrosol and luteolin. Another study employed 105°C, 100% EtOH and 5 min in olive leaves which the sum of phenolic compounds by HPLC-MS was 158.7 ± 0.4 mg/g d.w. In addition, another study optimized different extraction solvents such as ethyl/acetate/ethanol, acetone and mixture of ethanol/water, temperature and extraction time in PLE, where the optimal conditions were 60ºC, 110 min and 80%EtOH to obtain 386.42 mg GAE/g extract, which was lower than the reported by SFE at 16.5 MPa, 20°C and 210 min (11.49 mg/g d.w.), respectively. One of the advantages of extraction with PLE is the lower solvent consumption and higher extraction capacity when compared to the extractions with SFE and carbon dioxide, where the presence of oleuropein in the olive leaf extract was not identified through HPLC analyses (A. D. Rosa et al. 2019).

Pulsed Electric Field (PEF)

Pulsed electric fields (PEF) use high voltage, for an extremely short duration, in the range of a few μs, producing a strong electric field. PEF produces cell membrane electroporation, and therefore permeation through the membrane increases, and hence extraction improves (Maroun et al. 2018). The effectiveness of PEF treatment depends on the parameters, including electric field strength, pulse shape, pulse width, number of pulses, pulse-specific energy, and frequency (Martín-García et al. 2019). This technique has been applied in the olive leaves, where the PEF parameters were 10 μs of pulse duration, 1000 μs of pulse period, electric field strength1 kV/cm, 30 min of induction time and 25% EtOH obtaining a total phenolic content of 20.75 mg GAE/g d.w. extract. The phenolic recovery by PEF treatment was 31.85% higher in comparison with the control (15.74 mg GAE/g d.w.). In conclusion, PEF treatment increased the recovery of phenolic compounds in olive pomace. This indicates that PEF possesses an excellent potential for "green" selective extraction of biofunctional components from olive leaves that can be used for functional food manufacturing in a sustainable way, providing high-quality products that present several public health benefits (Pappas, Lakka, Palaiogiannis, Bozinou, et al. 2021). The effect of PEF (3 kV/cm, 15 μs and 20 Hz) at selected conditions after 10 min, showed a phenolic recovery in olive pomace, which was 35.08% higher in comparison with the control (1550.92 mg/100g) (Andreou et al. 2020).

High Voltage Electrical Discharge (HVED)

High Voltage Electrical discharge (HVED) is a technology based on the application of a high voltage electrical discharge between two electrodes submerged in an aqueous solution (Barišić, Flanjak, Kopjar, et al. 2020). Several reactions are produced by the application of HVED, including the generation of localized high-temperature plasmas, the emission of high-intensity UV light, the generation of shock waves, and the generation of free radicals produced by the photodissociation of water. Increasing shock waves and explosions of cavitation bubbles can reduce the particle size and influence cell membranes fragmentation (Barišić, Flanjak, Križić, et al. 2020; Deng, Ju, and Xi 2018). In addition, reactive species like molecular oxygen and hydrogen, hydrogen peroxide, hydroxyl radicals, and oxygen radical ions were reported (Banožić et al. 2021). All these processes lead to cell disruption, improve mass transfer and raise the release of intracellular compounds into the solvent. A recent study reported the use of HVED in olive leaves achieving a phenolic recovery at 9 min, 20 kV, 50% EtOH, 100 Hz and 400 μs of pulse width, which was three times higher than the obtained by a conventional extraction at the conditions of 50% EtOH and 9 min of stirring (20.61 ± 0.14 mg GAE/g) (Žuntar et al. 2019).

High Hydrostatic Pressure (HHPP)

High hydrostatic pressure (HHPP) is cold isostatic super high hydraulic pressure that ranges from 100 to 800 MPa (Pimenta Inada et al. 2020). This equipment generates an increase in the temperature; thus, some of this equipment uses cooling systems. During HHPP, the pressurized cells show an increase in their permeability due to the mass transfer theory. Under HHPP, a rapid permeation is observed due to the high differential pressure between the cell interior and the exterior of cell membranes (Ergin M. Altuner et al. 2012). A previous study reported the use of HPPP (200 MPa, 10 min, 6.41 kJ/kg) as a pretreatment for the recovery of individual phenolics, in which the sum of phenolic compounds was 61.48% higher than the control (1550.92 mg/100g) (Andreou et al. 2020).

The extraction techniques more used in the case of olive mill wastewater are liquid-liquid and solid-liquid extraction. In those methods, solvent selection is the most important factor. In addition, n-hexane and petroleum ether are the most common solvents used to remove the lipid fraction, whereas ethyl acetate was previously used for the phenolic recovery in olive mill wastewater (Papaphilippou et al. 2013; Martins, Martins, and Braga 2021). Other organic solvents could be used to recover compounds with different polarities. However, solvent recycling is important to guarantee the competitiveness of the liquid-liquid extraction process.

Table 2. Extraction conditions and yields of the main phenolic compounds from olive oil production byproducts

Technique	Olive byproduct	Conditions	Bioactive compounds (mg/g d.w.)	Concentration	Reference
US bath	Leaves	Solvent= 80%MeOH Time = 10 min Ratio = 1/20 g/mL (w/v)	SPC	52.13-60.64 mg/g d.w.	(Talhaoui et al. 2014)
		Solvent= 80%MeOH Time = 10 min Ratio = 1/20 g/mL (w/v)	SPC	46.04-83.30 mg/g d.w.	(Talhaoui et al. 2016)
	Olive pomace	Solvent = Water Temperature = 30ºC Power = 250 W Frequency = 50 Hz Extraction time = 75 min Ratio = 50/1 mL/g	TPC	19.71 mg GAE/g	(Goldsmith et al. 2018)
		Solvent = 50% EtOH Temperature = 20ºC Time = 30 min Ratio = 20/1 mL/g (v/w)	TPC	8.05 mg GA/g	(Tapia-Quirós et al. 2020)
US probe	Leaves	Solvent = 55% EtOH Time = 8 min Amplitude = 100%	SPC Oleuropein Hydroxytyrosol	40.9 ± 0.2 mg/g d.w. 24.6 ± 0.2 mg/g d.w. 1.01 ± 0.02 mg/g d.w.	(Martín-García et al. 2022)
		Temperature= 40 ºC Solvent= 60% EtOH Time= 17.91 min Ratio= 1/13 g/mL (w/v) Amplitude= 30%	TPC Oleuropein Luteolin-7-glucoside	210.25 mmol GAE/kg d.w. 69.91 mg/g d.w. 1.82 mg/g d.w.	(Lama-Muñoz et al. 2019)
	Olive pomace	Power= 100 W Frequency = 40 kHz Time = 12 min Amplitude = 70% Solvent = 40% acetone Solid load= 8.5% w/v	TPC	45.41 mg GAE/g d.w.	(Gómez-Cruz et al. 2021)
		Solvent = 90% EtOH Temperature = 50ºC Frequency = 20 kHz Time = 3 min Ratio = 30/1 mL/g (v/w)	Hydroxytyrosol Maslinic acid oleonolic acid	55.11 mg/g 381.20 mg/g 29.80 mg/g	(Xie et al. 2019)

Technique	Olive byproduct	Conditions	Bioactive compounds (mg/g d.w.)	Concentration	Reference
MAE	OMWW	Ratio = Freeze-dried OMWW (1.5 g/25 mL, w/v) Time = 3 x 20 min Solvent = 100% MeOH	SPC p-HPEA-DEDA Hydroxytyrosol Demethylligstroside	1521.3 μg GAE/mL 1065.9 μg GAE/mL 100.3 μg GAE/mL 244.8 μg/mL	(Jerman Klen and Mozetič Vodopivec 2011)
	Leaves	Power = 1000W Time = 3 min Temperature = 86°C Solvent = 100% H_2O	TPC Oleuropein Hydroxytyrosol	104.22 mg GAE/g d.w. 14.46 mg/ g d.w. 0.59 mg/g d.w.	(Da Rosa et al. 2021)
		Power = 800 W Temperature = 97.4ºC Solvent = 59.8%EtOH Ratio = 7.5 solid/solvent ratio	TPC	37.98 mg GAE/g d.w.	(Darvishzadeh and Orsat 2022)
		Power = 1000 W pH = 6 Time = 2 min Temperature = 100°C 100% H_2O	TPC Oleuropein	103.87 mg GAE/g d.b 11.59 mg/g d.b.	(Martiny et al. 2021)
	Olive pomace	Solvent = 90% EtOH Temperature = 50ºC Power = 600 W Frequency = 2.45 GHz Time = 5 min Ratio = 30:1 mL/g (v/w)	Hydroxytyrosol Maslinic acid Oleonolic acid	53.2 mg/g 356 mg/g 26.30 mg/g	(Xie et al. 2019)
		Solvent = 20% EtOH Power = 700 W Time = 10 min Ratio= 50/1 mL/g (v/w)	TPC Oleuropein Hydroxytyrosol Tyrosol	50.18 mg GA/g d.w. 0.03 mg/g d.w. 1.22 mg/g d.w. 0.13 mg/g d.w.	(Jurmanović et al. 2019)
PLE	Leaves	Pressure = 10.34 MPa Temperature = 138°C Time = 5 min 100%EtOH	SPC	158.7 mg/g d.w.	(Martín-García et al. 2020)
		Pressure = 10.3 MPa Temperature = 60ºC Time = 110 min 80%EtOH	TPC Oleuropein	386.42 mg GAE/g extract 73.65 mg/g extract	(Rosa et al. 2019)

Table 2. (Continued)

Technique	Olive byproduct	Conditions	Bioactive compounds (mg/g d.w.)	Concentration	Reference
PLE (continued)	Olive pomace	Pressure = 10.34 MPa Time = 20 min Temperature = 136.5°C Solvent = 52.3% EtOH	SPC Oxidized hydroxytyrosol Luteolin	1659 mg/kg d.w. 638 mg/Kg d.w. 221 mg/Kg d.w.	(Pavez et al. 2019)
SFE	Olive pomace	Temperature = 183.9°C Solvent = 84.7% EtOH Rario = 0.8 g/mL	TPC	336 mg GAE/ g d.e.	(Katsinas et al. 2021)
	Leaves	CO_2 modified with 20% methanol Presurre = 300 bar Temperature = 100°C Time = 2h	SPC	14.26 mg/g dried leaf	(Şahin, Bilgin, and Dramur 2011)
		Pressure = 16.5 MPa Temperature = 20°C Time = 210 min	TPC	11.49 mg GAE /g d.w.	(Rosa et al. 2019)
PEF	Olive leaves	Pulse duration = 10 µs Pulse period = 1000 µs Electric field strength = 1 kV/cm Induction time = 30 min Solvent = 25% EtOH	TPC	20.75 mg GAE/g d.w. extract	(Pappas, Lakka, Palaiogiannis, Bozinou, et al. 2021)
	Olive pomace	Electric field strength = 3 kV/cm n° pulses = 3,000 pulses Time = 10 min Pulse width = 15 µs Frequency = 20 Hz	TPC Oleuropein Hydroxytyrosol Tyrosol Rutin Luteolin	2095.06 mg/100g 91.82 mg/100g 1556.78 mg/100g 134.48 mg/100g 15.94 mg/100g 43.15 mg/100g	(Andreou et al. 2020)
HVED	Olive leaves	Time = 9 min Voltage = 20 kV Solvent = 50% EtOH Frequency =100 Hz Pulse width was 400 µs	TPC	65.99 mg GAE/ g d.w.	(Žuntar et al. 2019)

Technique	Olive byproduct	Conditions	Bioactive compounds (mg/g d.w.)	Concentration	Reference
HHPP	Olive pomace	Solvent = 50% EtOH Pressure = 200 MPa Time = 10 min Ratio = 10/1 mL/g (v/w)	SPC Oleuropein Hydroxytyrosol Tyrosol Rutin Luteolin	2504.5 mg/100g 84.65 mg/100g 2001.56 mg/100g 124.88 mg/100g 17.59 mg/100g 39.27 mg/100g	(Andreou et al. 2020)
		Solvent = Choline chloride-lactic acid Pressure = 600 MPa Time = 10 min Ratio = 12.5:1 mL/g (v/w)	TPC Oleuropein Hydroxytyrosol Rutin SPC	25.96 mg GAE/g dw 1.94 mg/g dw. 2.57 mg/g d.w. 0.66 mg/g dw 5.31 mg/g dw	(Chanioti and Tzia 2018)

SPC= Sum of phenolic compounds, TPC= Total phenolic content by Folin-Ciocalteu expressed by equivalent of gallic acid (GAE).

A recent study reported the content of the phenolic recovery presented in the olive wastewater fractions 2POMW (two-phase system olive mill wastewater from Spain) and POMW (press extraction olive mill wastewater from Portugal) extracted by ethyl acetate and analyzed by using HPLC-MS. The content of tyrosol, syringic acid and p-coumaric acid in POMW (13.98, 14.00 and 9.00 µg/g), which was 50-72% than the obtained in the fraction (Martins, Martins, and Braga 2021). Another study used two different solvent solutions: ethanol (99.8%) and ethanol:diethyl ether (1:2 v/v), to maximize the recovery of the phenolic compounds from wastewater collected during physical refining. In particular, the wastewater extract obtained with 99.8% EtOH showed the highest total phenolic content (0.430 g GAE/L), which content was 47.26% higher than the obtained by ethanol:diethyl ether (1:2 v/v) (0.292 g GAE/L), respectively (Venturi et al. 2017). Furthermore, Cloud point extraction (CPE) is a non-conventional technique that has been used in OMWW. This technique uses surfactants at temperatures over the cloud point. The surfactant forms micelles that become dehydrated, and then the micelles aggregate, separating organic molecules from the solvent. The use of Tween 80 as a surfactant at a concentration of 5% and a double-step CPE using 5% of Tween 80, the recovery of up to 94.4% of the total phenols from OMWW (Gómez-Caravaca et al. 2017). Another study reported another surfactant named Triton X-100, which has been used at concentrations between 1% and 10% (w/w) at three temperatures (70, 80, and 90°C) above the cloud point temperature of the surfactant (Şahin, Bilgin, and Dramur 2011). The efficiency of the extraction was evaluated by the determination of the concentration of phenolic compounds and the volume of both phases. CPE technique is a simple, fast, low-cost, sensitive, and selective procedure that allows obtaining natural antioxidants that can be used for different purposes, such as pharmaceutical, cosmetic, or food applications.

Treatment by Membrane Technologies

Pressure-driven membrane technologies, including microfiltration (MF), ultrafiltration (UF), nanofiltration (NF), and reverse osmosis (RO), are very interesting separation processes and environmentally friendly to the recovery of bioactive compounds from raw streams, such as phenolic compounds fractionation and remediation in olive byproducts. The advantages of these technologies could be the following: no chemical reagents, such as solvents, are indispensable to perform the separation and concentration; lower capital and operational costs and lively necessities than the tremendous larger part of classic separation technologies forms are inherent (Gómez-Caravaca et al. 2017).

Membrane operations can be considered a valid approach for the selective recovery of phenolic compounds in OMWW (Table 3). Several microfiltration (MF) and ultrafiltration (UF) membranes have been used to concentrate the recovered phenolic compounds from vegetation waters, using a final reverse osmosis (RO) consisting of a polymeric hydranautics membrane (composite polyamide). In addition, a previous study has demonstrated the possibility to concentrate OMW by means of direct contact membrane distillation (DCMD) process by using polytetrafluoroethylene membranes with a good polyphenol separation factor, which was close to 100% during 9 h of operating time (El-Abbassi et al. 2012). Garcia-Castello et al. (2010) used a system that includes MF and NF, osmotic distillation (OD) and vacuum membrane distillation (VMD) to recover, purify and concentrate phenolic compounds from OMW, obtaining a 78% of the initial content of phenolic compounds in the permeate stream

(Garcia-Castello et al. 2010). El-Abbassi et al. (2012). reported the separation of phenolic compounds from OMW by 100% after operating DCMD for 8 h. Nanofiltration (NF) and reverse osmosis (RO) processes have been proposed alternatively to concentrate specific phenol compounds such as hydroxytyrosol, protocatechuic acid, tyrosol, caffeic acid and oleuropein, although the corresponding separation of phenolic classes was not so successful as in the case of UF membranes. Depending on the wastewater and its phenolic content recovery obtained ranges from 0.5-19.3 g/L of phenolic compounds (Mudimu et al. 2012). Cassano et al. (2013) used two ultrafiltration (UF) processes followed by a nanofiltration (NF) step on laboratory scale operating in selected process parameters. The produced fractions were analyzed for their total content of polyphenols, total antioxidant activity (TAA), free low molecular weight phenolic compounds and total organic carbon (TOC). NF permeate produced at 5 bar is depleted in low molecular weight phenolic compounds and its TOC value was 95 mg/L. These results were obtained avoiding an additional step of RO (Cassano et al. 2013). A recent study reported the use of a combination of microfiltration (MF), nanofiltration (NF), and reverse osmosis (RO) in a sequential design to provide phenolic-enriched fractions in olive mill wastewater that have been investigated for their chemical profile using ultra-high-performance liquid chromatography (UHPLC). This process allowed to obtain a concentrated fraction (RO retentate) of 3 g/L of phenolic compounds, with hydroxytyrosol, tyrosol, and oleuropein, the most abundant compounds (Tundis et al. 2020).

Table 3. Membrane operations in OMWW

Membrane operations	Results	Ref
NF/ MF/OD/VMD	Recovery of 78% the initial content of polyphenols	(Garcia-Castello et al. 2010)
DCMD/ PTFE	100% phenol separation	(El-Abbassi et al. 2012)
UF with regenerated cellulose membranes/NF/RO	0.5 and 19.3 g/L total polyphenols concentrated	(Mudimu et al. 2012)
UF membranes in series followed by NF	NF permeate produced at 5 bar is depleted in low molecular weight polyphenols and its TOC value is 95 mg/L.	(Cassano et al. 2013)
MF/ NF and RO	3 g/L of phenolic compounds	(Tundis et al. 2020)
NF using the NF270 membrane.	The optimum results were obtained at 0.5 m/s and 15 bar. The rejections of the chemical oxygen demand (COD) were above 78%, whereas phenolic compounds were barely retained.	(Sánchez-Arévalo et al. 2021)

Another study has been evaluated the efficiency of a nanofiltration process to separate low-molecular-weight phenolic compounds from the sugars present in olive mill wastewaters (Sánchez-Arévalo et al. 2021). It can be noted that interest of the recovery of phenolic compounds from OMWW, and the use of different systems of membrane filtrations could be a good choice to counterbalance the wastewater treatment process costs. However, more research is needed to comprehend, modelize, minimize, and control the fouling problems associated (Gómez-Caravaca et al. 2017).

Conclusion

Olive byproducts have shown to be a rich source of bioactive compounds including phenolic compounds that possess beneficial properties due to their antioxidant activities. Different technologies for their recovery have been used, among them the application of ultrasound-assisted extraction by probe could be a good option since it requires less energy consumption than emerging technologies such as MAE, PLE, SFE. In addition, other non-thermal technologies such as PEF, HVED and HHPP have been shown effective in olive pomace and leaves with short extraction times. Furthermore, separation processes based on pressure-driven membrane technologies are very interesting to the recovery of bioactive compounds from raw streams, such as phenolic compounds fractionation and remediation in olive byproducts. These recovered bioactive compounds could be used in food products, to obtain innovative natural additives to improve food properties. For the moment, further research is needed in this field, aiming the utilization of olive byproducts to improve food nutritional profile, properties, and technological functions.

Acknowledgment

This study is part of the national project RTI2018-099835-A-I00 financed by MCIN/AEI/ 10.13039/ 501100011033/ FEDER "Una manera de hacer Europa"

References

Abbattista, Ramona, Giovanni Ventura, Cosima Damiana Calvano, Tommaso R.I. Cataldi, and Ilario Losito. 2021. "Bioactive Compounds in Waste By-Products from Olive Oil Production: Applications and Structural Characterization by Mass Spectrometry Techniques." *Foods* 10 (6): 1–28. https://doi.org/10.3390/foods10061236.

AGAPA. 2015. "Evaluación de La Producción y Usos de Los Subproductos de Las Agroindustrias Del Olivar En Andalucía [Evaluation of the Production and Uses of the By-products of the Olive Grove Agroindustries in Andalusia]." *Consejería de Agricultura, Pesca y Desarrollo.*

Aggoun, Moufida, Rabah Arhab, Agnès Cornu, Josiane Portelli, Malika Barkat, and Benoît Graulet. 2016. "Olive Mill Wastewater Microconstituents Composition According to Olive Variety and Extraction Process." *Food Chemistry* 209: 72–80. https://doi.org/10.1016/j.foodchem.2016.04.034.

Akay, Fazilet, Aslihan Kazan, Melih Soner Celiktas, and Ozlem Yesil-Celiktas. 2015. "A Holistic Engineering Approach for Utilization of Olive Pomace." *Journal of Supercritical Fluids* 99: 1–7. https://doi.org/10.1016/j.supflu.2015.01.025.

Ameer, Kashif, Hafiz Muhammad Shahbaz, and Joong-Ho Ho Kwon. 2017a. "Green Extraction Methods for Polyphenols from Plant Matrices and Their Byproducts: A Review." *Comprehensive Reviews in Food Science and Food Safety* 16 (2): 295–315. https://doi.org/10.1111/1541-4337.12253.

Ameer, Kashif, Hafiz Muhammad Shahbaz, and Joong Ho Kwon. 2017b. "Green Extraction Methods for Polyphenols from Plant Matrices and Their Byproducts: A Review." *Comprehensive Reviews in Food Science and Food Safety* 16: 295–315. https://doi.org/10.1111/1541-4337.12253.

Andreou, Varvara, Marios Psarianos, George Dimopoulos, Dimitrios Tsimogiannis, and Petros Taoukis. 2020. "Effect of Pulsed Electric Fields and High Pressure on Improved Recovery of High-Added-Value Compounds from Olive Pomace." *Journal of Food Science* 85 (5): 1500–1512. https://doi.org/10.1111/1750-3841.15122.

Antónia Nunes, M., Anabela S. G. Costa, Sílvia Bessada, Joana Santos, Helder Puga, Rita C. Alves, Vitor Freitas, and M. Beatriz P. P. Oliveira. 2018. "Olive Pomace as a Valuable Source of Bioactive Compounds: A Study Regarding Its Lipid- and Water-Soluble Components." *Science of the Total Environment* 644: 229–36. https://doi.org/10.1016/j.scitotenv.2018.06.350.

Araújo, Mariana, Filipa B. Pimentel, Rita C. Alves, and M. Beatriz P.P. Oliveira. 2015. "Phenolic Compounds from Olive Mill Wastes: Health Effects, Analytical Approach and Application as Food Antioxidants." *Trends in Food Science and Technology* 45 (2): 200–211. https://doi.org/10.1016/j.tifs.2015.06.010.

Banožić, Marija, Antun Jozinović, Jovana Grgić, Borislav Miličević, and Stela Jokić. 2021. "High Voltage Electric Discharge for Recovery of Chlorogenic Acid from Tobacco Waste." *Sustainability (Switzerland)* 13 (8). https://doi.org/10.3390/su13084481.

Barba, Francisco J., Zhenzhou Zhu, Mohamed Koubaa, Anderson S. Sant'Ana, and Vibeke Orlien. 2016. "Green Alternative Methods for the Extraction of Antioxidant Bioactive Compounds from Winery Wastes and Byproducts: A Review." *Trends in Food Science and Technology* 49: 96–109. https://doi.org/10.1016/j.tifs.2016.01.006.

Barišić, V., Flanjak, I., Kopjar, M., Benšić, M., Jozinović, A., Babić, J., Šubarić, D., Miličević, B., Doko, K., Jašić, M., & Ačkar, Đ. 2020. "Does High Voltage Electrical Discharge Treatment Induce Changes in Tannin and Fiber Properties of Cocoa Shell?" *Foods* 9 (6). https://doi.org/10.3390/foods9060810.

Barišić, Veronika, Ivana Flanjak, Ivana Križić, Antun Jozinović, Drago Šubarić, Jurislav Babić, Borislav Miličević, and Đurđica Ačkar. 2020. "Impact of High-Voltage Electric Discharge Treatment on Cocoa Shell Phenolic Components and Methylxanthines." *Journal of Food Process Engineering* 43 (1): 1–10. https://doi.org/10.1111/jfpe.13057.

Benincasa, Cinzia, Massimiliano Pellegrino, Elvira Romano, Salvatore Claps, Carmelo Fallara, and Enzo Perri. 2022. "Qualitative and Quantitative Analysis of Phenolic Compounds in Spray-Dried Olive Mill Wastewater." *Frontiers in Nutrition* 8 (January): 1–7. https://doi.org/10.3389/fnut.2021.782693.

Berbel, Julio, and Posadillo, Alejandro. 2018a. "Opportunities for the Bioeconomy of Olive Oil Byproducts." *Biomedical Journal of Scientific & Technical Research* 2 (1): 2094–96. https://doi.org/10.26717/bjstr.2017.01.000630.

Berbel, Julio, and Posadillo, Alejandro. 2018b. "Review and Analysis of Alternatives for the Valorisation of Agro-Industrial Olive Oil by-Products." *Sustainability (Switzerland)* 10 (1): 1–9. https://doi.org/10.3390/su10010237.

Bhatnagar, Amit, Fabio Kaczala, William Hogland, Marcia Marques, Christakis A. Paraskeva, Vagelis G. Papadakis, and Mika Sillanpää. 2014. Valorization of Solid Waste Products from Olive Oil Industry as Potential Adsorbents for Water Pollution Control-a Review. *Environmental Science and Pollution Research*. Vol. 21. https://doi.org/10.1007/s11356-013-2135-6.

Botsoglou, Evropi, Alexander Govaris, Ioannis Ambrosiadis, and Dimitrios Fletouris. 2012. "Lipid and Protein Oxidation of α-Linolenic Acid-Enriched Pork during Refrigerated Storage as Influenced by Diet Supplementation with Olive Leaves (Olea Europea L.) or α-Tocopheryl Acetate." *Meat Science* 92 (4): 525–32. https://doi.org/10.1016/j.meatsci.2012.05.022.

Cassano, Alfredo, Carmela Conidi, Lidietta Giorno, and Enrico Drioli. 2013. "Fractionation of Olive Mill Wastewaters by Membrane Separation Techniques." *Journal of Hazardous Materials* 248–249 (1): 185–93. https://doi.org/10.1016/j.jhazmat.2013.01.006.

Chanioti, Sofia, and Constantina Tzia. 2018. "Extraction of Phenolic Compounds from Olive Pomace by Using Natural Deep Eutectic Solvents and Innovative Extraction Techniques." *Innovative Food Science and Emerging Technologies* 48 (July): 228–39. https://doi.org/10.1016/j.ifset.2018.07.001.

Czerwińska, Monika, Anna K. Kiss, and Marek Naruszewicz. 2012. "A Comparison of Antioxidant Activities of Oleuropein and Its Dialdehydic Derivative from Olive Oil, Oleacein." *Food Chemistry* 131 (3): 940–47. https://doi.org/10.1016/j.foodchem.2011.09.082.

Darvishzadeh, Pariya, and Valerie Orsat. 2022. "Microwave-Assisted Extraction of Antioxidant Compounds from Russian Olive Leaves and Flowers: Optimization, HPLC Characterization and Comparison with Other Methods." *Journal of Applied Research on Medicinal and Aromatic Plants* 27 (August 2021): 100368. https://doi.org/10.1016/j.jarmap.2021.100368.

De Bruno, Alessandra, Rosa Romeo, Amalia Piscopo, and Marco Poiana. 2021. "Antioxidant Quantification in Different Portions Obtained during Olive Oil Extraction Process in an Olive Oil Press Mill." *Journal of the Science of Food and Agriculture* 101 (3): 1119–26. https://doi.org/10.1002/jsfa.10722.

Deng, Yong, Ting Ju, and Jun Xi. 2018. "Circulating Polyphenols Extraction System with High-Voltage Electrical Discharge: Design and Performance Evaluation." *ACS Sustainable Chemistry and Engineering* 6 (11): 15402–10. https://doi.org/10.1021/acssuschemeng.8b03827.

Dermeche, S., M. Nadour, C. Larroche, F. Moulti-Mati, and P. Michaud. 2013. "Olive Mill Wastes: Biochemical Characterizations and Valorization Strategies." *Process Biochemistry* 48 (10): 1532–52. https://doi.org/10.1016/j.procbio.2013.07.010.

Donner, Mechthild, and Ivana Radić. 2021. "Innovative Circular Business Models in the Olive Oil Sector for Sustainable Mediterranean Agrifood Systems." *Sustainability (Switzerland)* 13 (5): 1–23. https://doi.org/10.3390/su13052588.

El-Abbassi, A., H. Kiai, A. Hafidi, M. C. García-Payo, and M. Khayet. 2012. "Treatment of Olive Mill Wastewater by Membrane Distillation Using Polytetrafluoroethylene Membranes." *Separation and Purification Technology* 98: 55–61. https://doi.org/10.1016/j.seppur.2012.06.026.

Ergin M. Altuner, Cemil Işlek, Talip Çeter, and Hami Alpas. 2012. "High Hydrostatic Pressure Extraction of Phenolic Compounds from Maclura Pomifera Fruits." *African Journal of Biotechnology* 11 (4). https://doi.org/10.5897/ajb11.2506.

Galanakis, Charis M. 2021. "Functionality of Food Components and Emerging Technologies." *Foods* 10 (1): 1–26. https://doi.org/10.3390/foods10010128.

Galanakis, Charis M., Philippos Tsatalas, and Ioannis M. Galanakis. 2018. "Implementation of Phenols Recovered from Olive Mill Wastewater as UV Booster in Cosmetics." *Industrial Crops and Products* 111 (September 2017): 30–37. https://doi.org/10.1016/j.indcrop.2017.09.058.

Garcia-Castello, E., A. Cassano, A. Criscuoli, C. Conidi, and E. Drioli. 2010. "Recovery and Concentration of Polyphenols from Olive Mill Wastewaters by Integrated Membrane System." *Water Research* 44 (13): 3883–92. https://doi.org/10.1016/j.watres.2010.05.005.

Genethliou, C., M. Kornaros, and S. Dailianis. 2020. "Biodegradation of Olive Mill Wastewater Phenolic Compounds in a Thermophilic Anaerobic Upflow Packed Bed Reactor and Assessment of Their Toxicity in Digester Effluents." *Journal of Environmental Management* 255 (July 2019): 109882. https://doi.org/10.1016/j.jenvman.2019.109882.

Goldsmith, Chloe D., Quan V. Vuong, Costas E. Stathopoulos, Paul D. Roach, and Christopher J. Scarlett. 2018. "Ultrasound Increases the Aqueous Extraction of Phenolic Compounds with High Antioxidant Activity from Olive Pomace." *LWT - Food Science and Technology* 89 (June 2017): 284–90. https://doi.org/10.1016/j.lwt.2017.10.065.

Gómez-Caravaca, Ana María, Antonio Segura-Carretero, Antonio Martínez-Férez, and Javier Miguel Ochando-Pulido. 2017. Recovery of Phenolic Compounds From Olive Oil Mill Wastewaters by Physicochemical Methodologies. *Food Bioconversion*. Vol. 2. Elsevier Inc. https://doi.org/10.1016/B978-0-12-811413-1.00014-0.

Gómez-Cruz, Irene, María Del Mar Contreras, Florbela Carvalheiro, Luís C. Duarte, Luisa B. Roseiro, Inmaculada Romero, and Eulogio Castro. 2021. "Recovery of Bioactive Compounds from Industrial Exhausted Olive Pomace through Ultrasound-Assisted Extraction." *Biology* 10 (6): 1–22. https://doi.org/10.3390/biology10060514.

González-Hidalgo, Inés, Sancho Bañón, and José María Ros. 2012. "Evaluation of Table Olive By-Product as a Source of Natural Antioxidants." *International Journal of Food Science and Technology* 47: 674–81.

Guinda, Ángeles, Mirela Rada, Teresa Delgado, Pilar Gutiérrez-Adánez, and José María Castellano. 2010. "Pentacyclic Triterpenoids from Olive Fruit and Leaf." *Journal of Agricultural and Food Chemistry* 58 (17): 9685–91. https://doi.org/10.1021/jf102039t.

Hannachi, Hédia, Walid Elfalleh, Majda Laajel, Imen Ennajeh, Ridha Fethi Mechlouch, and Kamel Nagaz. 2020. "Chemical Profiles and Antioxidant Activities of Leaf, Pulp, and Stone of Cultivated and Wild Olive Trees (*Olea europaea* L.)." *International Journal of Fruit Science* 20 (3): 350–70. https://doi.org/10.1080/15538362.2019.1644574.

Hernández, Valentina, Juan M. Romero-García, Javier A. Dávila, Eulogio Castro, and Carlos A. Cardona. 2014. "Techno-Economic and Environmental Assessment of an Olive Stone Based Biorefinery." *Resources, Conservation and Recycling* 92: 145–50. https://doi.org/10.1016/j.resconrec.2014.09.008.

Jerman Klen, Tina, and Branka Mozetič Vodopivec. 2011. "Ultrasonic Extraction of Phenols from Olive Mill Wastewater: Comparison with Conventional Methods." *Journal of Agricultural and Food Chemistry* 59 (24): 12725–31. https://doi.org/10.1021/jf202800n.

Jurmanović, Sanja, Mario Jug, Toni Safner, Kristina Radić, Ana Marija Domijan, Sandra Pedisić, Sanja Šimić, Jasna Jablan, and Dubravka Vitali Čepo. 2019. "Utilization of Olive Pomace as a Source of Polyphenols:

Optimization of Microwave-Assisted Extraction and Characterization of Spray-Dried Extract." *Journal of Food and Nutrition Research* 58 (1): 51–62.

Katsinas, Nikolaos, Andreia Bento Da Silva, Amalia Enríquez-De-Salamanca, Naiara Fernández, Maria Rosário Bronze, and Soraya Rodríguez-Rojo. 2021. "Pressurized Liquid Extraction Optimization from Supercritical Defatted Olive Pomace: A Green and Selective Phenolic Extraction Process." *ACS Sustainable Chemistry and Engineering* 9 (16): 5590–5602. https://doi.org/10.1021/acssuschemeng.0c09426.

Khwaldia, Khaoula, Nouha Attour, Julia Matthes, Luisa Beck, and Markus Schmid. 2022. "Olive Byproducts and Their Bioactive Compounds as a Valuable Source for Food Packaging Applications." *Comprehensive Reviews in Food Science and Food Safety*, no. October 2021: 1–36. https://doi.org/10.1111/1541-4337.12882.

Komnitsas, Kostas, Kostas Modis, Maria Doula, Victor Kavvadias, Daphne Sideri, and Dimitra Zaharaki. 2016. "Geostatistical Estimation of Risk for Soil and Water in the Vicinity of Olive Mill Wastewater Disposal Sites." *Desalination and Water Treatment* 57 (7): 2982–95. https://doi.org/10.1080/19443994.2014.983988.

Kumar, Kshitiz, Shivmurti Srivastav, and Vijay Singh Sharanagat. 2021. "Ultrasound Assisted Extraction (UAE) of Bioactive Compounds from Fruit and Vegetable Processing by-Products: A Review." *Ultrasonics Sonochemistry* 70 (July 2020): 105325. https://doi.org/10.1016/j.ultsonch.2020.105325.

Lafka, Theodora Ioanna, Andriana E. Lazou, Vassilia J. Sinanoglou, and Evangelos S. Lazos. 2011. "Phenolic and Antioxidant Potential of Olive Oil Mill Wastes." *Food Chemistry* 125 (1): 92–98. https://doi.org/10.1016/j.foodchem.2010.08.041.

Lama-Muñoz, Antonio, María Del Mar Contreras, Francisco Espínola, Manuel Moya, Inmaculada Romero, and Eulogio Castro. 2019. "Optimization of Oleuropein and Luteolin-7-o-Glucoside Extraction from Olive Leaves by Ultrasound-Assisted Technology." *Energies* 12 (13). https://doi.org/10.3390/en12132486.

Maestri, Damián, Damián Barrionuevo, Romina Bodoira, Adoración Zafra, José Jiménez-López, and Juan de Dios Alché. 2019. "Nutritional Profile and Nutraceutical Components of Olive (*Olea europaea* L.) Seeds." *Journal of Food Science and Technology* 56 (9): 4359–70. https://doi.org/10.1007/s13197-019-03904-5.

Majbar, Z., Lahlou, K., Ben Abbou, M., Ammar, E., Triki, A., Abid, W., Nawdali, M., Bouka, H., Taleb, M., El Haji, M., & Rais, Z. 2018. "Co-Composting of Olive Mill Waste and Wine-Processing Waste: An Application of Compost as Soil Amendment." *Journal of Chemistry* 2018. https://doi.org/10.1155/2018/7918583.

Manzanares, Paloma, Encarnación Ruiz, Mercedes Ballesteros, María J. Negro, Francisco J. Gallego, Juan C. López-Linares, and Eulogio Castro. 2017. "Residual Biomass Potential in Olive Tree Cultivation and Olive Oil Industry in Spain: Valorization Proposal in a Biorefinery Context." *Spanish Journal of Agricultural Research* 15 (3): 1–12. https://doi.org/10.5424/sjar/2017153-10868.

Maroun, Richard G., Hiba N. Rajha, Nada El Darra, Sally El Kantar, Stéphanie Chacar, Espérance Debs, Eugène Vorobiev, and Nicolas Louka. 2018. *Emerging Technologies for the Extraction of Polyphenols from Natural Sources. Polyphenols: Properties, Recovery, and Applications*. Elsevier Inc. https://doi.org/10.1016/B978-0-12-813572-3.00008-7.

Martín-García, Beatriz, Soumi De Montijo-Prieto, Maria Jiménez-Valera, Alegría Carrasco-Pancorbo, Alfonso Ruiz-Bravo, Vito Verardo, and Ana María Gómez-Caravaca. 2022. "Comparative Extraction of Phenolic Compounds from Olive Leaves Using a Sonotrode and an Ultrasonic Bath and the Evaluation of Both Antioxidant and Antimicrobial Activity." *Antioxidants* 11 (558).

Martín-García, Beatriz, Federica Pasini, Vito Verardo, Elixabet Díaz-De-cerio, Urszula Tylewicz, Ana María Gómez-Caravaca, and Maria Fiorenza Caboni. 2019. "Optimization of Sonotrode Ultrasonic-Assisted Extraction of Proanthocyanidins from Brewers' Spent Grains." *Antioxidants* 8 (8). https://doi.org/10.3390/antiox8080282.

Martín-García, Beatriz, Sandra Pimentel-Moral, Ana María Gómez-Caravaca, David Arráez-Román, and Antonio Segura-Carretero. 2020. "Box-Behnken Experimental Design for a Green Extraction Method of Phenolic Compounds from Olive Leaves." *Industrial Crops and Products* 154 (February): 112741. https://doi.org/10.1016/j.indcrop.2020.112741.

Martins, Daniela, Rui C. Martins, and Mara E.M. Braga. 2021. "Biocompounds Recovery from Olive Mill Wastewater by Liquid-Liquid Extraction and Integration with Fenton's Process for Water Reuse." *Environmental Science and Pollution Research* 28 (23): 29521–34. https://doi.org/10.1007/s11356-021-12679-2.

Martiny, Thamiris Renata, Vijaya Raghavan, Caroline Costa De Moraes, Gabriela Silveira Da Rosa, and Guilherme Luiz Dotto. 2021. "Optimization of Green Extraction for the Recovery of Bioactive Compounds from Brazilian Olive Crops and Evaluation of Its Potential as a Natural Preservative." *Journal of Environmental Chemical Engineering* 9 (2): 105130. https://doi.org/10.1016/j.jece.2021.105130.

Mudimu, Ompe Aime, Martin Peters, Florian Brauner, and Gerd Braun. 2012. "Overview of Membrane Processes for the Recovery of Polyphenols from Olive Mill Wastewater Olive Mill Wastewater." *American Journal of Environmental Sciences* 8 (3): 195–201. https://doi.org/10.3844/ajessp.2012.195.201.

Mwakalukwa, Rogers, Yhiya Amen, Maki Nagata, and Kuniyoshi Shimizu. 2020. "Postprandial Hyperglycemia Lowering Effect of the Isolated Compounds from Olive Mill Wastes - An Inhibitory Activity and Kinetics Studies on α-Glucosidase and α-Amylase Enzymes." *ACS Omega* 5 (32): 20070–79. https://doi.org/10.1021/acsomega.0c01622.

Nunes, M. Antónia, Filipa B. Pimentel, Anabela S. G. Costa, Rita C. Alves, and M. Beatriz P. P. Oliveira. 2016. "Olive Byproducts for Functional and Food Applications: Challenging Opportunities to Face Environmental Constraints." *Innovative Food Science and Emerging Technologies* 35: 139–48. https://doi.org/10.1016/j.ifset.2016.04.016.

Otero, P., Garcia-Oliveira, P., Carpena, M., Barral-Martinez, M., Chamorro, F., Echave, J., Garcia-Perez, P., Cao, H., Xiao, J., Simal-Gandara, J., & Prieto, M. A. 2021. "Applications of Byproducts from the Olive Oil Processing: Revalorization Strategies Based on Target Molecules and Green Extraction Technologies." *Trends in Food Science and Technology* 116: 1084–1104. https://doi.org/10.1016/j.tifs.2021.09.007.

Otero, Paz, Somaris E. Quintana, Guillermo Reglero, Tiziana Fornari, and Mónica R. García-Risco. 2018. "Pressurized Liquid Extraction (PLE) as an Innovative Green Technology for the Effective Enrichment of Galician Algae Extracts with High Quality Fatty Acids and Antimicrobial and Antioxidant Properties." *Marine Drugs* 16 (5). https://doi.org/10.3390/md16050156.

Padilla-Rascón, C., E. Ruiz, I. Romero, E. Castro, J. M. Oliva, I. Ballesteros, and P. Manzanares. 2020. "Valorisation of Olive Stone By-Product for Sugar Production Using a Sequential Acid/Steam Explosion Pretreatment." *Industrial Crops and Products* 148 (March): 112279. https://doi.org/10.1016/j.indcrop.2020.112279.

Panzella, Lucia, Federica Moccia, Rita Nasti, Stefania Marzorati, Luisella Verotta, and Alessandra Napolitano. 2020. "Bioactive Phenolic Compounds From Agri-Food Wastes: An Update on Green and Sustainable Extraction Methodologies." *Frontiers in Nutrition* 7 (May): 1–27. https://doi.org/10.3389/fnut.2020.00060.

Papaphilippou, Petri C., Constantinos Yiannapas, Maria Politi, Vasileia M. Daskalaki, Costas Michael, Nicolas Kalogerakis, Dionissios Mantzavinos, and Despo Fatta-Kassinos. 2013. "Sequential Coagulation-Flocculation, Solvent Extraction and Photo-Fenton Oxidation for the Valorization and Treatment of Olive Mill Effluent." *Chemical Engineering Journal* 224 (1): 82–88. https://doi.org/10.1016/j.cej.2012.11.047.

Pappas, Vasileios M., Achillia Lakka, Dimitrios Palaiogiannis, Vassilis Athanasiadis, Eleni Bozinou, George Ntourtoglou, Dimitris P. Makris, Vassilis G. Dourtoglou, and Stavros I. Lalas. 2021. "Optimization of Pulsed Electric Field as Standalone 'Green' Extraction Procedure for the Recovery of High Value-Added Compounds from Fresh Olive Leaves." *Antioxidants* 10 (10). https://doi.org/10.3390/antiox10101554.

Pappas, Vasileios M., Achillia Lakka, Dimitrios Palaiogiannis, Eleni Bozinou, George Ntourtoglou, Georgia Batra, and Vassilis G. Dourtoglou 2 and Stavros I. Lalas 1, Vassilis Athanasiadis 1, Dimitris P. Makris 1. 2021. "Use of Pulsed Electric Field as a Low-Temperature and High-Performance 'Green' Extraction Technique for the Recovery of High Added Value Compounds from Olive Leaves." *Beverages* 45 (7). https://doi.org/10.3390/antiox10101554.

Pavez, Inés Cea, Jesús Lozano-Sánchez, Isabel Borrás-Linares, Hugo Nuñez, Paz Robert, and Antonio Segura-Carretero. 2019. "Obtaining an Extract Rich in Phenolic Compounds from Olive Pomace by Pressurized Liquid Extraction." *Molecules* 24 (3108): 1–17.

Pimenta Inada, Kim Ohanna, Samara Nunes, J. Alberto Martínez-Blázquez, Francisco A. Tomás-Barberán, Daniel Perrone, and Mariana Monteiro. 2020. "Effect of High Hydrostatic Pressure and Drying Methods on Phenolic Compounds Profile of Jabuticaba (*Myrciaria jaboticaba*) Peel and Seed." *Food Chemistry* 309 (November 2019): 125794. https://doi.org/10.1016/j.foodchem.2019.125794.

Ranalli, Alfonso, Luciano Pollastri, Stefania Contento, Giuseppina Di Loreto, Emilia Lannucci, Lucia Lucera, and Francesca Russi. 2002. "Sterol and Alcohol Components of Seed, Pulp and Whole Olive Fruit Oils. Their Use to Characterise Olive Fruit Variety by Multivariates." *Journal of the Science of Food and Agriculture* 82 (8): 854–59. https://doi.org/10.1002/jsfa.1116.

Rodrigues, Francisca, Filipa B. Pimentel, and M. Beatriz P.P. Oliveira. 2015. "Olive Byproducts: Challenge Application in Cosmetic Industry." *Industrial Crops and Products* 70: 116–24. https://doi.org/10.1016/j.indcrop.2015.03.027.

Romero, Concepción, Eduardo Medina, Maria Antonia Mateo, and Manuel Brenesa. 2017. "New By-products Rich in Bioactive Substances from the Olive Oil Mill Processing." *Society of Chemical Industry* 98: 225–30.

Rosa, Andréia Dalla, Alexander Junges, Ilizandra Aparecida Fernandes, Rogério Luis Cansian, Marcos L. Corazza, Elton Franceschi, Geciane Toniazzo Backes, and Eunice Valduga. 2019. "High Pressure Extraction of Olive Leaves (*Olea europaea*): Bioactive Compounds, Bioactivity and Kinetic Modelling." *Journal of Food Science and Technology* 56 (8): 3864–76. https://doi.org/10.1007/s13197-019-03856-w.

Rosa, Gabriela Silveira Da, Thamiris Renata Martiny, Guilherme Luiz Dotto, Sai Kranthi Vanga, Débora Parrine, Yvan Gariepy, Mark Lefsrud, and Vijaya Raghavan. 2021. "Eco-Friendly Extraction for the Recovery of Bioactive Compounds from Brazilian Olive Leaves." *Sustainable Materials and Technologies* 28: e00276. https://doi.org/10.1016/j.susmat.2021.e00276.

Roselló-Soto, Elena, Mohamed Koubaa, Amine Moubarik, Rita P. Lopes, Jorge A. Saraiva, Nadia Boussetta, Nabil Grimi, and Francisco J. Barba. 2015a. "Emerging Opportunities for the Effective Valorization of Wastes and Byproducts Generated during Olive Oil Production Process: Non-Conventional Methods for the Recovery of High-Added Value Compounds." *Trends in Food Science and Technology* 45: 296–310. https://doi.org/10.1016/j.tifs.2015.07.003.

———. 2015b. "Emerging Opportunities for the Effective Valorization of Wastes and Byproducts Generated during Olive Oil Production Process: Non-Conventional Methods for the Recovery of High-Added Value Compounds." *Trends in Food Science and Technology* 45 (2): 296–310. https://doi.org/10.1016/j.tifs.2015.07.003.

Şahin, Selin, Mehmet Bilgin, and Mehmet Umur Dramur. 2011. "Investigation of Oleuropein Content in Olive Leaf Extract Obtained by Supercritical Fluid Extraction and Soxhlet Methods." *Separation Science and Technology* 46 (11): 1829–37. https://doi.org/10.1080/01496395.2011.573519.

Sánchez-Arévalo, Carmen M., Álvaro Jimeno-Jiménez, Carlos Carbonell-Alcaina, María Cinta Vincent-Vela, and Silvia Álvarez-Blanco. 2021. "Effect of the Operating Conditions on a Nanofiltration Process to Separate Low-Molecular-Weight Phenolic Compounds from the Sugars Present in Olive Mill Wastewaters." *Process Safety and Environmental Protection* 148: 428–36. https://doi.org/10.1016/j.psep.2020.10.002.

Sarfarazi, Messiah, Seid Mahdi Jafari, Ghadir Rajabzadeh, and Charis M. Galanakis. 2020. "Evaluation of Microwave-Assisted Extraction Technology for Separation of Bioactive Components of Saffron (*Crocus sativus* L.)." *Industrial Crops and Products* 145 (July 2019): 111978. https://doi.org/10.1016/j.indcrop.2019.111978.

Souilem, Safa, Abdelilah El-Abbassi, Hajar Kiai, Abdellatif Hafidi, Sami Sayadi, and Charis M. Galanakis. 2017. *Olive Oil Production Sector: Environmental Effects and Sustainability Challenges. Olive Mill Waste: Recent Advances for Sustainable Management*. Elsevier Inc. https://doi.org/10.1016/B978-0-12-805314-0.00001-7.

Talhaoui, Nassima, Ana María Gómez-Caravaca, Lorenzo León, Raúl De la Rosa, Antonio Segura-Carretero, and Alberto Fernández-Gutiérrez. 2014. "Determination of Phenolic Compounds of ' Sikitita' Olive Leaves by HPLC-DAD-TOF-MS. Comparison with Its Parents 'Arbequina' and 'Picual' Olive Leaves." *LWT - Food Science and Technology* 58: 28–34. https://doi.org/10.1016/j.lwt.2014.03.014.

Talhaoui, Nassima, Ana María Gómez-Caravaca, Cristina Roldán, Lorenzo León, Raúl De la Rosa, Alberto Fernández-Gutiérrez, and Antonio Segura-Carretero. 2015. "Chemometric Analysis for the Evaluation of

Phenolic Patterns in Olive Leaves from Six Cultivars at Different Growth Stages." *Journal of Agricultural and Food Chemistry* 63: 1722–29. https://doi.org/10.1021/jf5058205.

Talhaoui, Nassima, Amani Taamalli, Ana María Gómez-Caravaca, Alberto Fernández-Gutiérrez, and Antonio Segura-Carretero. 2015a. "Phenolic Compounds in Olive Leaves: Analytical Determination, Biotic and Abiotic Influence, and Health Benefit." *Food Research International* 77: 92–108. https://doi.org/10.1016/j.foodres.2015.09.011.

———. 2015b. "Phenolic Compounds in Olive Leaves: Analytical Determination, Biotic and Abiotic Influence, and Health Benefits." *Food Research International* 77: 92–108. https://doi.org/10.1016/j.foodres.2015.09.011.

Talhaoui, Nassima, Teresa Vezza, Ana María Gómez-Caravaca, Alberto Fernández-Gutiérrez, Julio Gálvez, and Antonio Segura-Carrretero. 2016. "Phenolic Compounds and in Vitro Immunomodulatory Properties of Three Andalusian Olive Leaf Extracts." *Journal of Functional Foods* 22: 270–77. https://doi.org/10.1016/j.jff.2016.01.037.

Tapia-Quirós, Paulina, Maria Fernanda Montenegro-Landívar, Monica Reig, Xanel Vecino, Teresa Alvarino, Jose Luis Cortina, Javier Saurina, and Merce Granados. 2020. "Olive Mill and Winery Wastes as Viable Sources of Bioactive Compounds: A Study on Polyphenols Recovery." *Antioxidants* 9 (11): 1–15. https://doi.org/10.3390/antiox9111074.

Tapia-Quirós, Paulina, María Fernanda Montenegro-Landívar, Mònica Reig, Xanel Vecino, José Luis Cortina, Javier Saurina, and Mercè Granados. 2022. "Recovery of Polyphenols from Agri-Food By-Products: The Olive Oil and Winery Industries Cases." *Foods* 11 (3): 1–26. https://doi.org/10.3390/foods11030362.

Tarchoune, Imen, Cristina Sgherri, Jamel Eddouzi, Angela Zinnai, Mike Frank Quartacci, and Mokhtar Zarrouk. 2019. "Olive Leaf Addition Increases Olive Oil Nutraceutical Properties." *Molecules* 24 (3). https://doi.org/10.3390/molecules24030545.

Tundis, Rosa, Carmela Conidi, Monica R. Loizzo, Vincenzo Sicari, and Alfredo Cassano. 2020. "Olive Mill Wastewater Polyphenol-Enriched Fractions by Integrated Membrane Process: A Promising Source of Antioxidant, Hypolipidemic and Hypoglycaemic Compounds." *Antioxidants* 9 (7): 1–15. https://doi.org/10.3390/antiox9070602.

Vazquez-Roig, Pablo, and Yolanda Picó. 2015. "Pressurized Liquid Extraction of Organic Contaminants in Environmental and Food Samples." *TrAC - Trends in Analytical Chemistry* 71: 55–64. https://doi.org/10.1016/j.trac.2015.04.014.

Venturi, Francesca, Chiara Sanmartin, Isabella Taglieri, Anita Nari, Gianpaolo Andrich, Erika Terzuoli, Sandra Donnini, Cristiano Nicolella, and Angela Zinnai. 2017. "Development of Phenol-Enriched Olive Oil with Phenolic Compounds Extracted from Wastewater Produced by Physical Refining." *Nutrients* 9 (8): 1–13. https://doi.org/10.3390/nu9080916.

Vlyssides, Apostolos G., George K. Lamprou, and Anestis Vlysidis. 2017. *Industrial Case Studies on the Detoxificaton of OMWW Using Fenton Oxidation Process Followed by Biological Processes for Energy and Compost Production. Olive Mill Waste: Recent Advances for Sustainable Management*. Elsevier Inc. https://doi.org/10.1016/B978-0-12-805314-0.00006-6.

Wani, Touseef Ahmed, F. A. Masoodi, Adil Gani, Waqas Nabi Baba, Neda Rahmanian, Rehana Akhter, Idrees Ahmed Wani, and Mudasir Ahmad. 2018. "Olive Oil and Its Principal Bioactive Compound: Hydroxytyrosol – A Review of the Recent Literature." *Trends in Food Science and Technology* 77 (May): 77–90. https://doi.org/10.1016/j.tifs.2018.05.001.

Wedyan, Mohammad, Bilal Abu Hanieh, Ahmed Al-Harasheh, and Abdel Rahman Altawaha. 2017. "Chemical Characterization of Olive Pomace in the Northern Region of Jordan." *Bulgarian Journal of Agricultural Science* 23 (5): 866–72.

Xie, Pujun, Lorenzo Cecchi, Maria Bellumori, Diletta Balli, Lisa Giovannelli, Lixin Huang, and Nadia Mulinacci. 2021. "Phenolic Compounds and Triterpenes in Different Olive Tissues and Olive Oil By-Products, and Cytotoxicity on Human Colorectal Cancer Cells: The Case of Frantoio, Moraiolo and Leccino Cultivars (Olea Europaea L.)." *Foods* 10 (11). https://doi.org/10.3390/foods10112823.

Xie, Pujun, Lixin Huang, Caihong Zhang, Yejun Deng, Xiaojie Wang, and Jiang Cheng. 2019. "Enhanced Extraction of Hydroxytyrosol, Maslinic Acid and Oleanolic Acid from Olive Pomace: Process Parameters, Kinetics and Thermodynamics, and Greenness Assessment." *Food Chemistry* 276 (16): 662–74. https://doi.org/10.1016/j.foodchem.2018.10.079.

Xynos, Nikos, Georgios Papaefstathiou, Evangelos Gikas, Aikaterini Argyropoulou, Nektarios Aligiannis, and Alexios-Leandros Skaltsounis. 2014. "Design Optimization Study of the Extraction of Olive Leaves Performed with Pressurized Liquid Extraction Using Response Surface Methodology." *Separation and Purification Technology* 122: 323–30. https://doi.org/10.1016/j.seppur.2013.10.040.

Xynos, Nikos, Georgios Papaefstathiou, Marios Psychis, Aikaterini Argyropoulou, Nektarios Aligiannis, and Alexios Leandros Skaltsounis. 2012. "Development of a Green Extraction Procedure with Super/Subcritical Fluids to Produce Extracts Enriched in Oleuropein from Olive Leaves." *Journal of Supercritical Fluids* 67: 89–93. https://doi.org/10.1016/j.supflu.2012.03.014.

Yücel, Yasin. 2011. "Biodiesel Production from Pomace Oil by Using Lipase Immobilized onto Olive Pomace." *Bioresource Technology* 102 (4): 3977–80. https://doi.org/10.1016/j.biortech.2010.12.001.

Žuntar, Irena, Predrag Putnik, Danijela Bursać Kovačević, Marinela Nutrizio, Filip Šupljika, Andreja Poljanec, Igor Dubrović, Francisco J. Barba, and Anet Režek Jambrak. 2019. "Phenolic and Antioxidant Analysis of Olive Leaves Extracts (*Olea europaea* L.) Obtained by High Voltage Electrical Discharges (HVED)." *Foods* 8: 248.

Chapter 9

The Integral Use of Pitahaya and Pitaya Byproducts

María Encarnación Mena-Martínez[1]
and Miriam Fabiola Fabela-Morón[2,*]
[1]Departamento de Energía, Universidad Autónoma Metropolitana, Unidad Azcapotzalco, Ciudad de México
[2]Departamento de Ciencias de la Alimentación, Universidad Autónoma Metropolitana, Unidad Lerma, Estado de México

Abstract

Nowadays, approximately 1.3 billion tonnes of food and food byproducts are lost or wasted per year. Agro-industrial wastes, such as fruit peels, become a potential source of high value-added functional compounds, which could be used in pharmaceutical, cosmetics and food industries. In this sense, pitahaya or dragon fruit, and pitaya are fruits that belong to the family of *Cactaceae*, and have an important relevance as raw materials for their betacyanin content, nutrients, phytochemical, nutraceutical, and bioactive compounds present in the seeds, peel, and flesh, which made suitable for their integral use in diferent food applications. This chapter focuses on the principal aspects related to the integral use of pitahaya and pitaya byproducts considering the fruit processing, the chemical composition, nutrients, and bioactive compounds that can be obtained by means of different kind of extraction techniques and could be protected through out microencapsulation by spray-drying.

Keywords: integral use, Pitahaya, Pitaya, byproducts, bioactive compounds

Introduction

Agro-industrial wastes become a potential source of high value-added functional compounds, which could be used in pharmaceutical, cosmetics and food industries.

Pitahaya and pitaya are cacti fruits of the *Stenocereus* and *Hylocereus* genera, cultivated in various regions around the world, including tropical regions of America, making them very important as food (Ramírez-Rodriguez et al., 2020).

* Corresponding Author's Email: qa.miriamfabela@gmail.com.

In: Food Byproducts
Editors: S. Teresita Martín-del-Campo, A. Cardador-Martínez and J. del Pilar Ramírez-Anaya
ISBN: 979-8-88697-612-0

It is worth mentioning that both fruits have color differences in the peel and pulp depending on the region where they are found (Ramírez-Rodríguez et al., 2020, Castillo Martínez, Livera Muñoz, and Márquez Guzmán 2005, Centurión-Yah et al., 2008).

These fruits have traditionally been consumed fresh, although, due to their nutritional properties and chemical composition, it is of interest to process and preserve them to diversify their consumption (Fabela-Morón et al., 2018, Yong et al., 2018, Joshi and Prabhakar 2020, García- Cruz et al., 2016), so that they are used not only as food, but also as a source of bioactive compounds, pigments, sugars, and other macromolecules for industrial use.

By trying the fruits to various unit operations, it is possible to obtain various products such as juices, beverages, jams, dressings, ice cream and pigments, among others (Choo et al., 2018).

Its peel, pulp and seeds are processed through various extraction techniques to obtain bioactive compounds that include: betacyanins, antioxidants, polyphenols, polyunsaturated fatty acids, nutraceuticals, and others, such as probiotics obtained from pitahaya (Ferreres et al., 2017, Joshi and Prabhakar 2020, Perez-Loredo, Hernández de Jesus and Barragán-Huerta 2016).

Remarkably, its composition can vary depending on species, phenotype, cultivation, and post-harvest conditions.

Due to the large amount of macro and micronutrients present in these fruits, it is extremely important to promote their integral use and valorization; therefore, in this chapter the generalities of pitahaya and pitaya are described, as well as the chemical composition of some specific species of these fruits.

In the same way, the bioactive compounds that can be founded in the fruits and their use in the obtention and protection of bioproducts and bioactive compounds are indicated. It is worth mentioning that for the collection of information, a documentary analysis type investigation was carried out.

Pitahaya or Dragon Fruit

Pitahaya, also known as dragon fruit, is a cactus fruit from the genus *Hylocereus and Cactaceae* family, which is cultivated in America, desert environments and tropical regions around the world (Mexico, Colombia, Nicaragua, Guatemala, Costa Rica, Peru, Venezuela, China, India, Thailand, Taiwan, Philippines, Japan, Vietnam, Indonesia, Australia, New Zealand, etc.) (Le Bellec, Vaillant, and Imbert 2006, Ramírez-Rodríguez et al., 2020).

Figure 1. Pitahaya or dragon fruit.

Pitahaya fuits can be exploited in an integral use, and its cultivation is very important through family garden plantations and wild plantations, generating heterogeneity of plants and fruits, where pitahaya fruits include red skin with white pulp, with red pulp and yellow skin with white pulp varieties (Castillo Martínez, Livera Muñoz, and Márquez Guzmán 2005, Centurión-Yah et al., 2008, Damar Jaya 2018). Pitahaya fruits are shown in Figure 1.

Pitahaya Fruits Composition

Pitahaya represents an important source of nutritional macronutrients, micronutrients and phytochemicals. The fruit composition varies according to species, genotypes, and origin, which is affected by postharvest conditions. Table 1 presents the red pitahaya fruit composition as (Lezama et al., 2009i):

On the other hand, Table 2 indicates the chemical composition of some pitahaya cultivars from Mexico, according to the National Institute of Medical Sciences and Nutrition Salvador Zubirán report (INCMNSZ 2016). In addition, chemical composition for pitahaya varieties from other countries are presented:

In this sense, pitahaya fruits have important chemical and nutritional compounds of great interest to the food industry due to the oligosaccharides and pectin content present in the peel and natural pigments present in the peel and pulp of the fruit. Also, is a source of beta-carotene, lycopene, vitamin E, linoleic acid, linolenic acid and prebiotics (Ding et al., 2009, Wichienchot, Jatupornpipat, and Rastall 2010).

Bioactive phenolic compounds have been reported in *H. undatus* and *H. megalanthus* (Esquivel, Stintzing, and Carle 2007, Esquivel, Stintzing Florian, and Carle 2007, Ferreres et al., 2017), whose extracts possess antioxidant capacity due to betacyanins (betanin, isobetanin, folocactin, butyrylbetanin) (Fathordoobady et al., 2016), in addition to tocopherols and sterols (Lim et al., 2010).

Table 1. Red pitahaya fruit composition

Component	Composition per 100 g of the edible part
Water	89.40 g
Protein	0.50 g
Fat	0.10 g
Carbohydrates	9.20 g
Fiber	0.30 g
Ash	0.50 g
Ca	6.00 g
P	19.00 mg
Fe	0.40 mg
Thiamine	0.01 mg
Riboflavin	0.03 mg
Niacin	0.20 mg
Ascorbic acid	25.00 mg

Table 2. Chemical composition from Pitahaya varieties

Composition	INCMNSZ (2016)					Khalili et al., (2006)	Nurul and Asmah (2014)	
	Hylocereus ungatus	*H. monacanthus* (Lem.), *H. undatus* (Haw.)				*Hylocereus* sp. (Malaysia)	*Hylocereus polyrhizus*	
	Fruit	Yellow	Cultivars	Pink	Solferino	Red	Red (Malaysia)	Red (Australia)
Edible portion (%)	55	45	45	45	45			
Energy (kJ)	228.00	205.00	179.00	189.00	239.00		263.32*	147.91*
Water (%)	82.3	85.90	87.05	84.50	83.40	87.30	85.05	89.98
TSS [a]								
Ether extract (g)	0.11	1.20	0.94	0.82	0.70	0.23	ND	ND
Crude protein (g)	1.44	1.10	1.60	1.25	0.80	0.16	1.45	0.41
Carbohydrates (g)	13.55	9.80	8.28	9.47	13.50		12.97	8.42
Sugars (g)		1.70	5.20		7.90	1.48		
Fiber (g)	2.10	1.40	1.29	3.26	0.90	10.10	2.65	ND
Ca (mg)	5	8.26	7.50	12.00	12.80	5.70	6.72	1.55
P (mg)	15		34.00	25.00		23.00		
Fe (mg)	0.75	*	2.90	1.98		3.40	0.30	0.03
Na (mg)		137.9	52.00		62.50	50.15	35.63	14.3
K (mg)		*				56.96	158.29	437.35
Mg (mg)		253.40	272.60		170.90	28.30		
Cu (mg)		27.77	36.60		21.90	0.03		
Mn (mg)		0.20	0.20		0.30		26.4	13.7
Vit A (U.I.)		558.44	7400.79		6704.95			
Vit. B1 (mg)	0.11		0.04	0.06				
Vit B2 (mg)	0.13		0.04	0.05				
Niacin (mg)	0.37		0.22	0.26				
Ascorbic acid (mg)	25.80	7.34	13.70	9.50			0.24	0.30

[a] TSS: Total soluble solids, ° Brix.

[b] S.F.A.: Saturated Fatty Acids; M.F.A.: Monounsaturated fatty acids; P.U.F.A.: Polyunsaturated Fatty Acids.

[c] Calculated from reported values according to 100g of fruit.

Pitahaya seeds from H. undatus and H. polyrhizus present linoleic acid (48-55.63%), oleic acid (24.43-26.80%) and palmitic acid (12.78-16.53%) (Ding et al., 2009, Lim et al., 2010). Table 3 presents the bioactive compounds identified in pitahaya fruits:

Table 3. Bioactive compounds from pitahaya fruit

Bioactive compounds (Ding et al., 2009, Lim et al., 2010, Ferreres et al., 2017)	Bioactive compounds (Fathordoobady et al., 2016, Ramírez-Rodríguez et al., 2020)
Gallic acid	Isobetanin
Malonyl-glucoside	Phyllocactin I [6_-*O*- malonyl-betanin]
Tyrosine	Iso-phyllocactin I
Cyclo-Dopa malonyl- glucoside	Phyllocactin II [4_-*O*- malonyl-betanin]
2-Decarboxy-*cyclo*-dopa	Decarboxy neobetanidin
Cyclo-Dopa malonyl-glucoside	Neobetanin decarboxylated
Ellagic acid	5,5_,6,6_-Tetrahydroxy- 3,3_-biindolyl
Betalamic acid	Neobetanidin, decarboxylated, dehydrogenated
Indicaxanthin	Flavonoid-3-O-glycosyl-7-O-rhamnosides
Cyclo-Dopa malonyl-glucoside	Flavonoid-3-O-glycosides
Cyclo-Dopa malonyl-glucoside	Isohylocerenin
Acetylcoumarin	20 -Apiosyl-isophyllocactin
Betanin [betanidin- 5-O-glucoside]	20 -Apiosyl-phyllocactin

On the other hand, non-digestible oligosaccharides are considered prebiotics, which have the beneficial effect of stimulating the growth and/or activity of certain bacteria in the colon, leading to improved consumer health. Some sugars and galacturonic acid present in pitahaya (*Hylocereus* sp.) fruit mucilage have been reported (Ramírez-Truque, Esquivel, and Carle 2011), as well as the extraction of pectin from the peel to add value (Muhammad et al., 2014, Montoya-Arroyo et al., 2014).

Wichicnchot, Jatupornpipat, and Rastall (2010) reported an oligosaccharides content of 86.2 - 89.6 g/kg in pitahaya. Also, in a simulated study under gastrointestinal conditions was found that such compounds were resistant to hydrolysis of artificial human gastric juice (4.04%) and human alpha-amylase (34.88%). These oligosaccharides were able to stimulate the growth of probiotic bacteria such as lactobacilli and bifidobacteria.

The presence of vitamins, polyphenols and oligosaccharides of different molecular weights from pitahaya pulp has generated interest in taking full advantage of it as a potential source for obtaining functional ingredients from the fruit, pulp, seeds, peel as by-products (Ariffin et al., 2009, Corzo-Rios et al., 2017).

Integral Use and Applications of Pitahaya Fruit

Due to its composition and properties, pitahaya has potential agroindustrial application, considering the integral use of peel, flesh and seeds as by-products.

Pitahaya flesh can be processed by freezing, concentration, evaporation, dehydration, fermentation and thermal processing to elaborate different food products, whose conservation is by means of temperature or with the addition of preservatives that preserve its physicochemical, sensory properties and shelf life (Fabela-Morón et al., 2018, Yong et al., 2018, Joshi and Prabhakar 2020).

In addition, pitahaya is used as a principal ingredient in the formulation of different food products such as beverages, dressings, vinaigrettes, jams, jelly, ice cream, yogurt, and canned fruits (Choo et al., 2018).

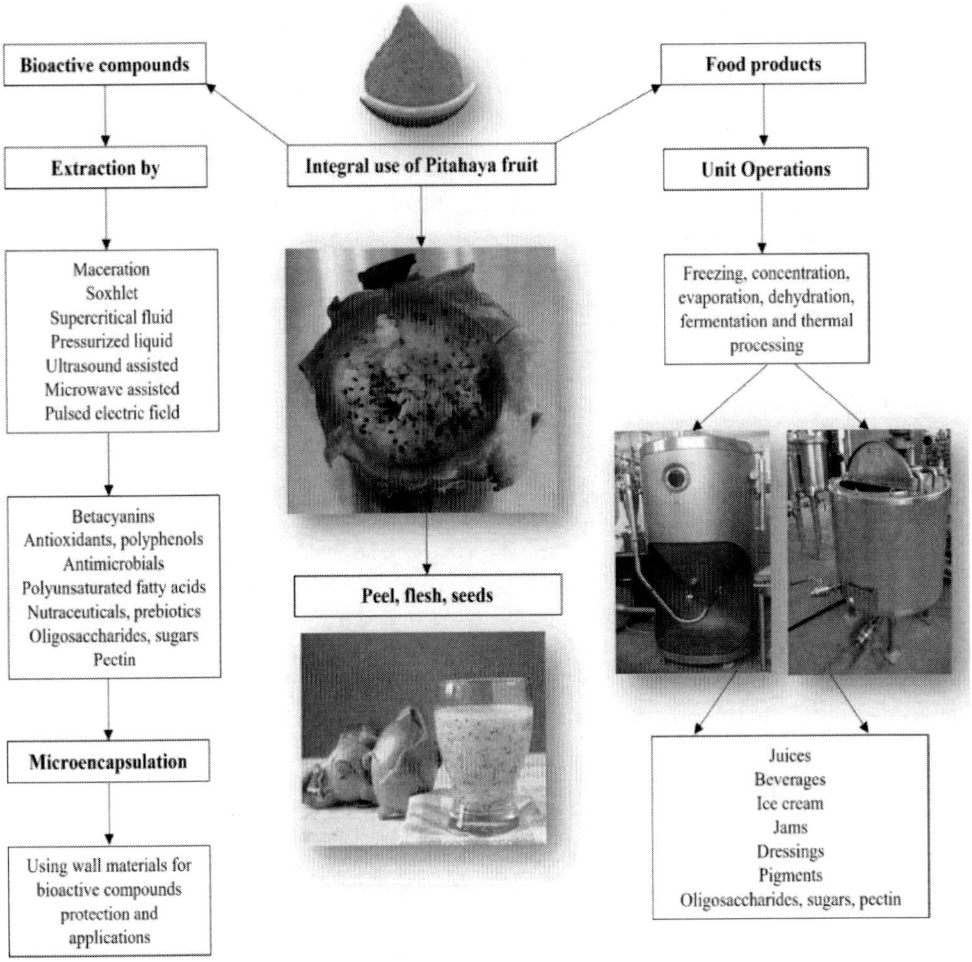

Figure 2. Integral use of pitahaya fruit.

For the obtention of natural coloring agents and bioactive compounds by different extraction techniques, pitahaya is used as a raw material considering the integral use of peel, flesh and seeds (Ferreres et al., 2017, Joshi and Prabhakar 2020).

Microencapsulation is used to preserve betacyanins obtained from pitahaya, using as wall materials or carrier agents Arabic gum, maltodextrin, and pectin matrices with 94-95% of yield and 90-95% of efficiency. This process could be done by spray-drying, freeze-drying, etc. (Rodriguez et al., 2016, Castro-Enríquez et al., 2020).

Other uses of bioactive compounds are related to betalain dyes obtained from pitahaya (*Hylocereus polyrhizus*) fruits have been used as new sensitizers for dye-sensitized solar cells, which are the third generation of photovoltaic systems capable of converting sunlight to electrical energy using sensitive devices based on natural pigments (Al-Alwani et al., 2017).

Figure 3. Pitaya fruit.

The integral use of pitahaya fruit promotes its potential use, where in addition to pigments and antioxidants as bioactive compounds, sugars, oligosaccharides and pectin can be obtained from the peel, flesh and seeds as shown in Figure 2.

Pitaya

Pitaya fruit belongs to genus *Stenocereus* (Figure 3), which are cacti growing in semiarid regions. The fruits are surrounded by large prickles (horns), and the flesh of the fruit presents colors such as white, red, orange, violet, and yellow, being distributed from the southwest U.S., Mexico, Colombia, Costa Rica, to Venezuela (Ramírez-Rodríguez et al., 2020). The pitaya species with commercial importance include *Stenocereus queretaroensis*, *Stenocereus* griseus (Haworth) Buxbaum, *Stenocereus pruinosus* (Otto) Buxbaum, and *Stenocereus stellatus* (Pfeiffer) Riccobono (García-Cruz et al., 2016).

Pitaya Chemical Composition

Table 4 indicates the chemical composition of pitaya fruits cultivated in Mexico and Colombia. Pitaya presents differences in vitamin C content from 8.49 mg/100 g to 17.04 mg/100 g of fresh pulp in 4 types of *Stenocereus stellatus* Riccobono from Puebla, Mexico (Beltrán-Orozco et al., 2009) respect to the content reported by Pérez-Loredo, García-Ochoa, and Barragán-Huerta (2016).

Table 4. Chemical composition from pitaya fruit

Composition	Cañar, Caetano, and Bonilla (2014)	Pérez-Loredo, García-Ochoa, and Barragán-Huerta (2016)
Edible portion (%)	*Selenicereus megalanthus* (Colombia)	*Stenocereus stellatus* (Mexico)
Energy (kJ)	162.12[c]	
Water (%)	84.8	85-86
TSS [a]		
Ether extract (g)	0.43	0.39-0.49
Crude protein (g)	0.65	1.08-1.30
Carbohydrates (g)	13.38	9.65-10.85
Sugars (g)		
Fiber (g)	0.77	1.39-1.63
Ascorbic acid (mg)		261-411[c]

[a] TSS: Total soluble solids, ° Brix.
[b] S.F.A.: Saturated Fatty Acids; M.F.A.: Monounsaturated fatty acids; P.U.F.A.: Polyunsaturated Fatty Acids.
[c] Calculated from reported values.

Table 5. Bioactive compounds from pitaya fruit

Bioactive compounds (Pérez-Loredo, García-Ochoa, and Barragán-Huerta 2016, García-Cruz et al., 2017, Ramírez-Rodríguez et al., 2020)
Betalains, betacyanins, betaxanthins
Proline-betaxanthin (indicaxanthin)
Isoindicaxanthin
Indicaxanthin
Betanidin-glucosides
[betanidin-6-O-β-glucoside]
Gomphrenin I
Isogomphrenin-I [isobetanidin-6-O-β-glucoside]
2-descarboxybetanin
Malonyl-2-descarboxy-betanin isomers
60-O-malonyl-2-descarboxy-betanin [2-descarboxybetanidin 5-O-(60-O-malonyl-β-D-glucoside)]
60-O-malonyl-2-descarboxy-isobetanin
Phyllocactin [betanidin 5-O-(60-O-malonyl-β-D-glucoside)]
Isophyllocactin [isobetanidin 5-O-(60-O-malonyl-β-D-glucoside)]
Etanidin-5-O-(60-O-3-hydroxy-butyryl)-bglucoside
Caffeoyl hexoside I
Caffeoyl hexoside II
Feruloyl dihexoside
p-coumaroyl quinic acid
Quercetin 3-Orutinoside
Kaempferol hexoside
Isorhamnetin hexoside
Isorhamnetin 3-Oglucoside
Eriodictyol hexoside
Eriodictyol acetylhexoside
Naringenin acetylhexoside
Tyrosol
Taxifolin acetylhexoside

Stenocereus species present betalains, betaxanthins and betacyanins, which content variates according to the pitaya varieties. *Stenocereus griseus* H. from Mexico, presents a higher concentration of betacyanins in red-fleshed fruits (betacyanins:199.6 mg/100 g dry fruit and betaxanthins: 147.61 dry fruit) and betaxanthins for orange-fleshed ones (betacyanins: 37.6 mg/100 g dry fruit and betaxanthins: 177.37 dry fruit) (García-Cruz, Salinas-Moreno, and Valle-Guadarrama 2012).

Higher betaxanthins content with respect to betacyanins in *S. pruinosus* (red-fleshed and orange-fleshed) and *S. stellatus* (red-fleshed) was reported from Mexico varieties. Also, was identified total betaxanthin concentration of 17706-21681 μg/g of dry fruit, where the most abundant was Indicaxanthin, while betacyanins concentrations were from 162-5423 μg/g of dry fruit (García-Cruz et al., 2017).

Similar results were reported by Pérez-Loredo, García-Ochoa, and Barragán-Huerta (2016) for yellow, purple, and red *S. stellatus* fruits; betalains were not detected in white fruits. Table 5 presents the bioactive compounds identified in pitaya fruits:

Integral Use and Applications of Pitaya Fruit

Pitaya varieties produced are mostly consumed as fresh fruit as they are highly perishable, which makes their preservation very important.

As a non-climacteric fruit, pitaya is sensitive to abrasion damages, the elimination of thorns makes the fruit more sensitive to cracking during processing (Quiroz-González et al., 2018).

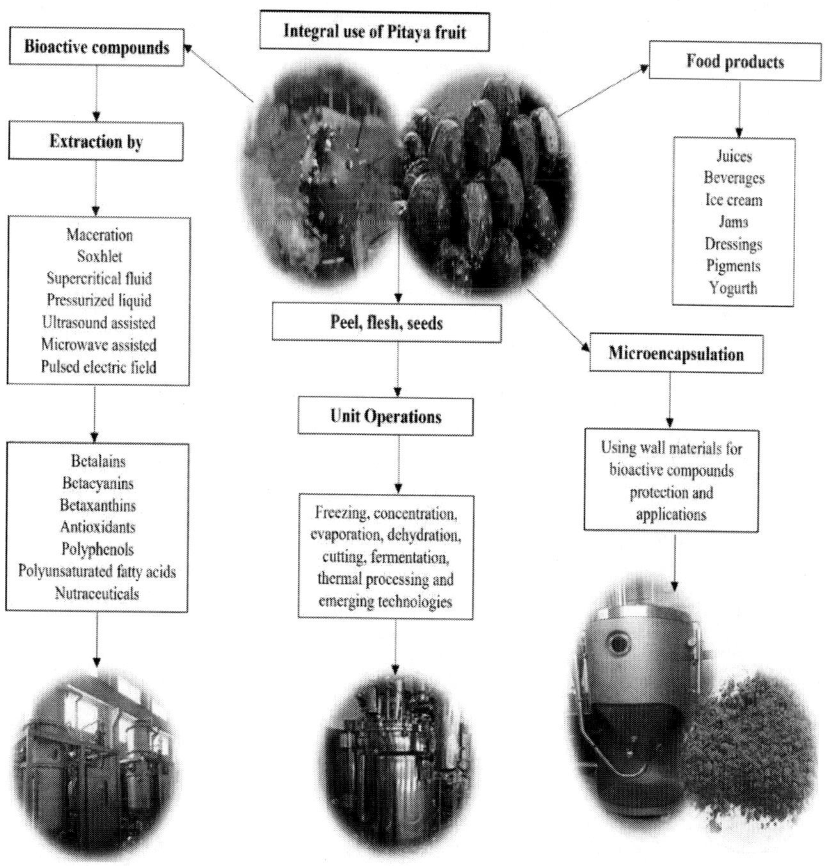

Figure 4. Integral use of pitaya fruit.

Pitayas from the genus *S. pruinosus* (red and orange flesh) and *S. stellatus* (red and white flesh) from Mexico could be preserved 6 days at 24ºC and 90% of relative humidity after harvestingnot only with good organoleptic characteristics but also preserving the biological activities (García-Cruz et al., 2016).

Fresh pitaya fruit is peeled and cut for consumers since the size of some varieties could be relatively large. Li et al., (2017) describe the effect of cutting styles on fruit quality throughout 4 d storage at 15ºC in polypropylene containers.

The cutting style did not produce an important effect on the organoleptic quality of the fruit but induces a little deterioration in total soluble solids, total acidity, vitamin C content, as well as an increase in b* color value due to oxidation attributed to total phenolics making the fruit unacceptable after 4 days of storage. On the other hand, these authors mentioned cutting style could induce the production of reactive oxygen species, increase the antioxidant activity, and the accumulation of phenolics.

For this reason, it is important to preserve the flesh of this fruit to produce juices, beverages, jams, ice cream, and dressings with properly quality, using emergent technologies. In this sense, emerging technologies have been applied to pitaya juices to decrease the microbial charge while keeping the biological activities instead of the traditional pasteurization methods.

Ochoa-Velasco and Guerrero-Beltrán (2012) applied UV-C light *to Stenocereus grises* juice from Mexico during 25 min at 30.33 mL/s flow and 4 ºC (juice temperature), with a reduction of 2.11 log reduction of aerobic mesophyll bacteria and 1.14 log reduction of molds plus yeasts maintaining the juice color; nevertheless, some parameters suffered a reduction such as phenolic compounds (13.27% reduction), antioxidant activity (3.55% reduction), and betalains (26.02% reduction).

The integral use of pitaya fruit includes the obtention of pigments, antioxidants and bioactive compounds, sugars, food products from the peel, flesh and seeds by-products as shown in Figure 4.

Final Remarks

Actually, is very relevant finding new applications to the food wastes to contribute for a sustainable valorization of food byproducts.

Pitahaya and pitaya fruits represent an important source of nutrients, betacyanin, and phytochemical compounds, which made suitable as an alternative raw material for obtaining bioactive compounds for their use in food industry.

The obtention of bioactive compounds from pitahaya and pitaya fruits and their peel, seeds, and flesh by different extraction techniques, contributes to the integral use of byproducts that in other way would be discarded after primary use.

Unit operations applied to process the pitahaya, pitaya fruits, contribute to the integral use to elaborate food products: juice, ice-cream, jams, dressings.

Microencapsulation by spray-drying allows protect the bioactive compounds obtained from pitahaya and pitahaya (microcapsules). Also, enabling to diversify their use to process the flesh into a powder form and increase its shelf-life.

References

Al-Alwani, M. A. M., N. A. Ludin, A. B. Mohamad, A. A. H. Kadhum, and K. Sopian. 2017. "Extraction, preparation and application of pigments from *Cordyline fruticosa* and *Hylocereus polyrhizus* as sensitizers for dye-sensitized solar cells." *Spectrochimica Acta Part A: Molecular and Biomolecular Spectroscopy* 179:23-31. doi: 10.1016/j.saa.2017.02.026.

Ariffin, A. A., J. Bakar, C. P. Tan, R. A. Rahman, R. Karim, and C. C. Loi. 2009. "Essential fatty acids of pitaya (dragon fruit) seed oil." *Food Chemistry* 114 (2):561-564. doi: 10.1016/j.foodchem.2008.09.108.

Beltrán-Orozco, M. C., T. G. Oliva-Coba, T. Gallardo-Velázquez, and G. Osorio-Revilla. 2009. "Ascorbic acid, phenolic content, and antioxidant capacity of red, cherry, yellow and white types of pitaya cactus fruit (*Stenocereus stellatus* Riccobono)." *Agrociencia* 43 (2):153-161.

Cañar, D., C. Caetano, and M. Bonilla. 2014. "Caracterización fisicoquímica y proximal del fruto de pitahaya amarilla (*Selenicereus megalanthus* (k. schum. ex vaupel) moran) cultivada en Colombia [Physicochemical and proximal characterization of the fruit of yellow pitahaya (*Selenicereus megalanthus* (k. schum. ex vaupel) moran) cultivated in Colombia]." *Revista Agronomía* 22 (1):77-87.

Castillo Martínez, R., M. Livera Muñoz, and G. J. Márquez Guzmán. 2005. "Caracterización morfológica y compatibilidad sexual de cinco genotipos de pitahaya (*Hylocereus undatus*) [Morphological characterization and sexual compatibility of five pitahaya (*Hylocereus undatus*) genotypes]." *Agrociencia* 39 (2):183-194.

Castro-Enríquez, D. D., B. Montaño-Leyva, C. L. Del Toro-Sánchez, J. E. Juaréz-Onofre, E. Carvajal-Millan, S. E. Burruel-Ibarra, J. A. Tapia-Hernández, C. G. Barreras Urbina, and F. Rodríguez-Félix. 2020. "Stabilization of betalains by encapsulation—a review." *Journal of Food Science and Technology* 57 (5):1587-1600. doi: 10.1007/s13197-019-04120-x.

Centurión-Yah, A. R., S. Solís-Pereira, C. Saucedo-Veloz, R. Báez-Sañudo, and E. Sauri-Duch. 2008. "Cambios físicos, químicos y sensoriales en frutos de pitahaya (*Hylocereus undatus*) durante su desarrollo [Physical, chemical and sensory changes in pitahaya (*Hylocereus undatus*) fruits during their development]." *Revista Fitotecnia Mexicana* 31 (1):1-5fe. doi: 10.35196/rfm.2008.1.1.

Choo, K. Y., C. Kho, Y. Y. Ong, Y. Y. Thoo, R. L. H. Lim, C. P. Tan, and C. W. Ho. 2018. "Studies on the storage stability of fermented red dragon fruit (*Hylocereus polyrhizus*) drink." *Food Science and Biotechnology* 27 (5):1411-1417. doi: 10.1007/s10068-018-0367-4.

Corzo-Rios, L. J., M. E. Bautista-Ramírez, Y. d. l. M. Gómez y Gómez, and L. G. Torres-Bustillos. 2017. "Frutas de cactáceas: Compuestos bioactivos y sus propiedades nutracéuticas [Cactus fruits: bioactive compounds and their nutraceutical properties]." In *Propiedades funcionales de hoy*, edited by M.E. Ramírez-Ortiz, 57-60. OmniaScience.

Damar Jaya, I. K. 2018. "Morphology and physiology of pitaya and its future prospects in Indonesia." *CROP AGRO, Journal Ilmiah Budidaya* (1):44-50.

Ding, P., M. K. Chew, S. Abdul Aziz, O. M. Lai, and J. O. Abdullah. 2009. "Red-fleshed pitaya (*Hylocereus polyrhizus)* fruit colour and betacyanin content depend on maturity." *International Food Research Journal* 16:233-242.

Esquivel, P., F. C. Stintzing, and R. Carle. 2007. "Pigment pattern and expression of colour in fruits from different *Hylocereus* sp. genotypes." *Innovative Food Science & Emerging Technologies* 8 (3):451-457. doi: 10.1016/j.ifset.2007.03.022.

Esquivel, P., C. Stintzing Florian, and R. Carle. 2007. "Phenolic Compound Profiles and their Corresponding Antioxidant Capacity of Purple Pitaya (*Hylocereus* sp.) Genotypes." *Zeitschrift für Naturforschung C* 62 (9-10):636. doi: 10.1515/znc-2007-9-1003.

Fabela-Morón, M. F., J. C. Cuevas-Bernardino, N. A. Pacheco-López, T. Ayora-Talavera, T. González-Flores, M. Canul-Chan, and M. Barba-Bellettini. 2018. "Potencial del chile habanero (*Capsicum chinense* Jacq.) y de la pitahaya (*Hylocereus* spp.) de Yucatán, y de sus subproductos para la obtención de productos de alto valor agregado [Potential of habanero pepper (*Capsicum chinense* Jacq.) and pitahaya (*Hylocereus* spp.) from Yucatan, and their by-products to obtain high value-added products]." In *Avances y perspectivas de la biotecnologia en la peninsula de Yucatán*, edited by J. J. Sandoval-Gío, 17-32. Mexico: Tecnológico Nacional de México.

Fathordoobady, F., H. Mirhosseini, J. Selamat, and M. Y. A. Manap. 2016. "Effect of solvent type and ratio on betacyanins and antioxidant activity of extracts from Hylocereus polyrhizus flesh and peel by supercritical fluid extraction and solvent extraction." *Food Chemistry* 202:70-80. doi: 10.1016/j.foodchem.2016.01.121.

Ferreres, F., C. Grosso, A. Gil-Izquierdo, P. Valentão, A. T. Mota, and P. B. Andrade. 2017. "Optimization of the recovery of high-value compounds from pitaya fruit by-products using microwave-assisted extraction." *Food Chemistry* 230:463-474. doi: 10.1016/j.foodchem.2017.03.061.

García-Cruz, L., M. Dueñas, C. Santos-Buelgas, S. Valle-Guadarrama, and Y. Salinas-Moreno. 2017. "Betalains and phenolic compounds profiling and antioxidant capacity of pitaya (*Stenocereus* spp.) fruit from two species (*S. Pruinosus* and *S. stellatus*)." *Food Chemistry* 234:111-118. doi: 10.1016/j.foodchem.2017.04.174.

García-Cruz, L., Y. Salinas-Moreno, and S. Valle-Guadarrama. 2012. "Betalaínas, compuestos fenólicos y actividad antioxidante en pitaya de mayo (*Stenocereus griseus* H.) [Betalains, phenolic compounds and antioxidant activity in May dragon fruit (*Stenocereus griseus* H.)]." *Revista fitotecnia mexicana* 35:01-05. doi: 10.35196/rfm.2012.Especial_5.1.

García-Cruz, L., S. Valle-Guadarrama, Y. Salinas-Moreno, and C. d. C. Luna-Morales. 2016. "Postharvest quality, soluble phenols, betalains content, and antioxidant activity of *Stenocereus pruinosus* and *Stenocereus stellatus* fruit." *Postharvest Biology and Technology* 111:69-76. doi: 10.1016/j.postharvbio.2015.07.004.

INCMNSZ. 2016. *Tablas de Composición de Alimentos y Productos Alimenticios Mexicanos* (Versión Condesada 2015) [*Tables of Composition of Mexican Food and Food Products* (Condensed Version 2015)]: Instituto Nacional de Ciencias Médicas y Nutrición Salvador Zubirán: México City, México.

Joshi, M., and B. Prabhakar. 2020. "Phytoconstituents and pharmaco-therapeutic benefits of pitaya: A wonder fruit." *Journal of Food Biochemistry* 44 (7):e13260. doi: 10.1111/jfbc.13260.

Khalili, R. M. A., A. Norhayati, M. Rokiah, R. Asmah, M. M. Nasir, and M. S. Muskinah. 2006. "Proximate composition and selected mineral determination in organically grown red pitaya (*Hylocereus* sp.)." *Journal of Tropical Agriculture and Food Science* 34 (2):269.

Le Bellec, F., F. Vaillant, and E. Imbert. 2006. "Pitahaya (*Hylocereus* spp.): a new fruit crop, a market with a future." *Fruits* 61 (04):237-250. doi: 10.1051/fruits:2006021.

Lezama, A., A. Tapia, G. Muñoz, and V. Zepeda. 2009i. "El cultivo de pitahaya [The cultivation of pitahaya]." *SAGARPA (Secretaría de Agricultura, Ganadería, Desarrollo Rural, Pesca y Alimentación). Puebla, México.*

Li, X., Q. Long, F. Gao, C. Han, P. Jin, and Y. Zheng. 2017. "Effect of cutting styles on quality and antioxidant activity in fresh-cut pitaya fruit." *Postharvest Biology and Technology* 124:1-7. doi: 10.1016/j.postharvbio.2016.09.009.

Lim, H. K., C. P. Tan, R. Karim, A. A. Ariffin, and J. Bakar. 2010. "Chemical composition and DSC thermal properties of two species of *Hylocereus cacti* seed oil: *Hylocereus undatus* and *Hylocereus polyrhizus*." *Food Chemistry* 119 (4):1326-1331. doi: 10.1016/j.foodchem.2009.09.002.

Montoya-Arroyo, A., R. M. Schweiggert, M.-L. Pineda-Castro, M. Sramek, R. Kohlus, R. Carle, and P. Esquivel. 2014. "Characterization of cell wall polysaccharides of purple pitaya (*Hylocereus* sp.) pericarp." *Food Hydrocolloids* 35:557-564. doi: 10.1016/j.foodhyd.2013.07.010.

Muhammad, K., N. I. Mohd. Zahari, S. P. Gannasin, N. Mohd. Adzahan, and J. Bakar. 2014. "High methoxyl pectin from dragon fruit (*Hylocereus polyrhizus*) peel." *Food Hydrocolloids* 42:289-297. doi: 10.1016/j.foodhyd.2014.03.021.

Nurul, S., and R. Asmah. 2014. "Variability in nutritional composition and phytochemical properties of red pitaya (*a*) from Malaysia and Australia." *International Food Research Journal* 21 (4):1689-1697.

Ochoa-Velasco, C. E., and J. Á. Guerrero-Beltrán. 2012. "Ultraviolet-C light effect on pitaya (*Stenocereus griseus*) juice." *Journal of Food Research* 1 (2):60. doi: 10.5539/jfr.v1n2p60.

Pérez-Loredo, M. G., F. García-Ochoa, and B. E. Barragán-Huerta. 2016. "Comparative Analysis of Betalain Content in *Stenocereus stellatus* Fruits and Other Cactus Fruits Using Principal Component Analysis." *International Journal of Food Properties* 19 (2):326-338. doi: 10.1080/10942912.2015.1022259.

Quiroz-González, B., M. García-Mateos, J. Corrales-García, and M. Colinas-León. 2018. "Pitaya (*Stenocereus* spp.): An under-utilized fruit." *Journal of the Professional Association for Cactus Development.*

Ramírez-Rodríguez, Y., M. Martínez-Huélamo, J. Pedraza-Chaverri, V. Ramírez, N. Martínez-Tagüeña, and J. Trujillo. 2020. "Ethnobotanical, nutritional and medicinal properties of Mexican drylands Cactaceae Fruits: Recent findings and research opportunities." *Food Chemistry* 312:126073. doi: 10.1016/j.foodchem.2019.126073.

Ramírez-Truque, C., P. Esquivel, and R. Carle. 2011. "Neutral sugar profile of cell wall polysaccharides of pitaya (*Hylocereus* sp.) fruits." *Carbohydrate Polymers* 83 (3):1134-1138. doi: 10.1016/j.carbpol.2010.09.042.

Rodriguez, E. B., M. L. P. Vidallon, D. J. R. Mendoza, and C. T. Reyes. 2016. "Health-promoting bioactivities of betalains from red dragon fruit (*Hylocereus polyrhizus* (Weber) Britton and Rose) peels as affected by carbohydrate encapsulation." *Journal of the Science of Food and Agriculture* 96 (14):4679-4689. doi: 10.1002/jsfa.7681.

Wichienchot, S., M. Jatupornpipat, and R. A. Rastall. 2010. "Oligosaccharides of pitaya (dragon fruit) flesh and their prebiotic properties." *Food Chemistry* 120 (3):850-857. doi: 10.1016/j.foodchem.2009.11.026.

Yong, Y. Y., G. Dykes, S. M. Lee, and W. S. Choo. 2018. "Effect of refrigerated storage on betacyanin composition, antibacterial activity of red pitahaya (*Hylocereus polyrhizus*) and cytotoxicity evaluation of betacyanin rich extract on normal human cell lines." *LWT* 91:491-497. doi: 10.1016/j.lwt.2018.01.078.

Chapter 10

Opuntia Fruits Byproducts: Important Sources of Phytochemicals

Rosa Vázquez-García[1]
Jessica del Pilar Ramírez-Anaya[2]
and Sandra T. Martín-del-Campo[1,*]
[1]Tecnologico de Monterrey, School of Engineering and Sciences, Querétaro, Qro, Mexico
[2]Department of Computer Science and Technological Innovation, CUSUR-University of Guadalajara, Ciudad Guzmán, Mexico

Abstract

Opuntia genus comprises more than 1500 species of cactus. Many produce edible fruits with a well-accepted flavor in well-established cultures. But others, as wild species, are used for different purposes, such as fences and animal feeding. Thie genus adapts easily to arid and semiarid climates, becoming an interesting crop due to the low amount of water required and their huge potential for biomass production and CO_2 fixation. There is an increasing interest in the obtention of natural additives for food, cosmetic and pharmaceutical applications, such as colorants, antioxidants, and functional ingredients.

On the other hand, avoiding food waste has become an important development subject. It has been demonstrated that *Opuntia* fruits are good sources of different phytochemicals in all the fruit parts, such as pulp, peels, flowers, and seeds. The valorization of the *Opuntia* byproducts of their industrialization or distribution to produce colorants and other functional ingredients has aroused great interest. Different extraction and stabilization methods have been developed. Compounds extraction with water and ethanol has been described for greener processes, while another organic solvent has been used for non-polar compounds or specific fractionations. Stabilization of extracts, a key step, has been done with spray-drying, lyophilization, and gelation, among others. This document explains the most important compounds that this genus have been shown.

Keywords: prickly pear, phytochemicals, betalains, polyphenols, natural pigments

[*] Corresponding Author's Email: teresitamartin@gmail.com.

In: Food Byproducts
Editors: S. Teresita Martín-del-Campo, A. Cardador-Martínez and J. del Pilar Ramírez-Anaya
ISBN: 979-8-88697-612-0

Introduction

Prickly pear or cactus fruit is a berry produced by the cactus of the genus *Opuntia*. They originated in the semiarid and arid Mexican regions, and nowadays, their culture has been established on other continents. Mexico is the biggest producer, followed by Italy, Chile, South Africa, and Israel; they produce important amounts of this fruit since the cultures have been established well.

The adaptation of *Opuntia* to arid and semiarid climates turns prickly pear into an interesting crop. Their low requirements for culture and the low amount of water required for their growth promote a huge potential for biomass production and CO_2 fixation. Besides, they grow where other crops could not grow. Additionally, this is a multipurpose culture that grows in different lands that could become important due to the water shortage and population increment.

Even though there is a huge number of known *Opuntia* species, few of them are used to establish cultures destined for human consumption. Although wild species are potential food sources, they arrive only in regional markets due to the low knowledge about their nutritional value and nutraceutical properties.

These times, a huge increment in the selection of natural antioxidants has been observed instead of synthetic antioxidants. These increment has originated from the restriction on certain synthetic antioxidants due to their carcinogenic or toxic effects, but also consumers seem to be interested in more natural food products.

Cactus fruits contain an important concentration of phenolic compounds with antioxidant properties that helps to prevent different diseases such as diabetes, hypercholesterolemia, cancer, atherosclerosis, etc.

Prickly Pear Phytochemicals

Prickly pear fruit is a berry with a thick pericarp with clefs of small prickles, a taste, and a sweet pulp containing small seeds (Abdel-Hameed et al. 2014). These fruits present an important commercial value since they are nutritious fruits with a well-accepted and appreciated flavor and present a variety of colors such as white, red, yellow, and purple, among others (Figure 1). This variety of colors is present not only in the pulp but also in the peel.

(A) (B)

Figure 1. Different *Opuntia ficus-indica* varieties from Mexico. (A) Yellow and white pulp, (B) purple and white pulp.

(A) (B)

Figure 2. Open fruits of *Opuntia ficus-indica* varieties from Mexico showing peel and pulp. (A) White prickly pear, (B) purple prickly pear.

(A) (B)

Figure 3. Peel damages in *Opuntia ficus-indica* varieties from Mexico. (A) White pulp with fungus damage, (B) purple pulp with mechanical damage.

The eatable portion of the most commercialized and domesticated species of prickly pear fruits is similar, representing between 55% and 57% of the fruit (INCMNSZ 2016). Nevertheless, in the case of wild species, the eatable portion presents more variability, going from 29.9% to 63.7% (Aparicio et al. 2017, Chougui et al. 2013). On the other hand, seeds represent 2-10% of the fruit (Barba et al. 2017, Chougui et al. 2013) and peel 37-67% of the fruit (Barba et al. 2017).

Most prickly pears are consumed as fresh fruits, but they have been traditionally used to elaborate a wide variety of food products such as jams, juices, alcoholic beverages, etc. For the elaboration of most of these traditional products, only the pulp is used, so peels and seeds are discarded. On the other hand, peels are discarded when fresh fruit is consumed; nevertheless, they often have the same color as the pulp (Figure 2).

On the other hand, depending on the prickly pear species or cultivars, shelf life could be short due to fruit dehydration, loss of physicochemical characteristics, and/or microbial contamination (Figure 3). Even though some treatments have been used to increase shelf life, such as natural waxes, fungicides, hot water treatments, thermoultrasonication, and cold storage, in most of the production zones, these conservation methods are unaffordable and do not apply to all the varieties and cultivars.

The prickly pear fruits distribution issues and the transformation processes generate a large number of byproducts that could be used as a source of bioactive compounds and nutrients with a high added value. Different groups of phytochemicals have been identified and quantified in pulp, peel, seeds, and flowers.

Pulp

The prickly pear pulp has an important nutritive value. The macro- and micro- nutrients proportion variates depending on the species, cultivars, and production regions. Globally, prickly pear fruits have a low energy apport, but they are a good fiber and microelements source (Hernández Galán and Martín del Campo 2021, INCMNSZ 2016, Barba et al. 2017).

On the other hand, the prickly pear pulp is a good source of phytochemicals presenting different biological activities. Among the phytochemicals, the presence of polyphenols, betalains, carotenoids, flavonoids, and flavonols has been reported. The phytochemicals profile, concentration, and biological activities variates among the different species and production regions.

Tables 1A-B show antioxidant activity and the concentration of the most important phytochemicals associated with the activity. The concentration and presence of polyphenols, flavonoids, and flavonols are not constant, so depending on the origin of the fruit, one group predominates over the others (Tables 1A-B). This variability is also observed for individual polyphenols, as shown in Table 2. The predominant polyphenols reported in most prickly pear varieties are gallic acid, catechin, epicatechin, trans-ferulic acid, and kaempferol (Table 2).

Depending on the variety or cultivar, different pigments have been identified in prickly pear pulp belonging to both betalains and carotenoids, being the betalains the most abundant and described pigments (Allegra et al. 2015, Cano et al. 2017, Castellanos-Santiago and Yahia 2008, El Kharrassi et al. 2016).

For betalains, the variety and origin also influence the proportion between betacyanins and betaxanthins (Tables 3A-B). Cano et al. (2017) reported a higher concentration of betaxanthins for the Spanish Sanguinos (red) in Verdal (orange) *Opuntia* varieties. Verdal variety presented a higher proportion of betaxanthins than the Sanguinos (82% and 57%, respectively). Castellanos-Santiago and Yahia (2008) reported that betacyanins represented between 54% and 74% of total betalains for the Mexican cultivars Camuesa (*O. robusta* Wendl), Roja Pelota (*O. robusta*), Cardona (*O. streptacantha* Lemaire), 2142(*O. ficus-indica*), Liria (*O. ficus-indica*) and Roja Lisa (*O. ficus-indica* (L.) Mill). On the other hand, the betaxanthins represented between 67% and 85% for cultivars Naranjona (*O. megacantha*), and Reyna (*O. albi-carpa*), lines 2651 (*O. ficus-indica*), 21441 (*O. ficus-indica*). The most abundant betacyanin was indicaxanthin, followed by muscaaurin, while the most abundant betaxanthins were betanin, followed by betanidin, iso-betanin, neo-betanin in relative abundance order.

Carotenes are the less studied or reported pigments in prickly pear; nevertheless, they could present interesting concentrations depending on the *Opuntia* species, varieties, or cultivars. Allegra et al. (2015) reported 0.005-0.01 g of β-carotene/100 g of the pulp of *O. ficus-indica*, cv. Gialla fuits from Italy. El Kharrassi et al. (2016) evaluated the total carotenoid content of 27 *O. ficus-indica* accessions and seven *O. megacantha* accessions from Morocco. The carotenoid concentration variates depending on the *Opuntia* species but also the accession. For *O. ficus-indica*, carotenoid concentration was between 5.4 μg/l of juice to 20.8 μg/l of juice, while for *O. megacantha* ranged from 4.8 μg/l of juice to 18.6 μg/l of juice. Cano et al. (2017) reported total carotenoid concentrations of 255.93 mg/100 g of FW for Spanish Sanguinos (red) *Opuntia* and 379.45 mg/100 g of FW for Verdal (orange) *Opuntia*. For both varieties, the most abundant carotenoid was (all-E)-lutein, representing 78.7% and 53.7% of total carotenoids, respectively.

Prickly pear pulp has not a significant concentration of fat. Depending on the variety, cultivar, and origin, the ether extract represents 0.04-0.90 g/100 g of fresh fruit (Hernández Galán and Martín del Campo 2021, Ramadan and Mörsel 2003a). However, important fatty acids have been quantified in the prickly pear fruit.

Ramadan and Mörsel (2003a) evaluated the pulp fat and fatty acids profiles in the total fat and different fat fractions of red to purple *Opuntia ficus-indica* fruits bought in local markets in Germany. They reported that the major fatty acids of the pulp's total fat were linolenic (37% of the fatty acids), palmitic (34.4% of the fatty acids), and oleic (10.8% of the fatty acids); linoleic acid was present only in very low concentration (0.68% of the fatty acids). They also reported that the most abundant sterols quantified in pulp were the β-sitosterol (11.2 g/kg of total lipids), campesterol (8.74 g/kg of total lipids), and Δ5-avenasterol (1.43 g/kg of total lipids).

Melgar et al. (2017) evaluated the fatty acids profile of *Opuntia ficus-indica* var. Gialla, *Opuntia ficus-indica* var. sanguigna, and *Opuntia engelmannii*. For the three varieties, the most abundant fatty acid was linoleic acid (59%, 68.3%, and 75.73% of the fatty acids, respectively), followed by palmitic (16%, 13.9%, and 7.92% of the fatty acids, respectively), and oleic (11.7%, 9.82%, and 10.23% of the fatty acids, respectively) acids. Additionally, they found that total tocopherols concentration ranged from 43-158 μg/100 g of fresh weight, which has been α-tocopherol the most abundant in the three varieties. Similar results were reported by Andreu-Coll et al. (2019) for 6 cultivars of *Opuntia* from Spain. They also reported that linoleic acid (20.2-53.9% of the fatty acids) was the most abundant one, followed by palmitic acid (16.4-29% of the fatty acids) and oleic acid (16.9-40.2% of the fatty acids).

Peels

Even though most studies are focused on pulp, it has been found that prickly pear peels are a rich source of functional and bioactive compounds that could be used as functional compounds in different industries. Peels often present the same or similar color as the pulp (Figures 2 and 4). Habibi, Mahrouz, and Vignon (2009) reported that the main components of *Opuntia ficus-indica* from Maroc were other polysaccharides (35% dry weight), cellulose (27% dry weight), fat, and waxes (11% dry weight), proteins (8.6% dry weight), mucilage (4.1% dry weight), among other components.

(A) (B)

Figure 4. Peel and pulp colors of *Opuntia ficus-indica* varieties from Mexico. (A) Purple pulp and peel, (B) purple and green peels.

Table 1A-B. Antioxidant activities, total phenolics, flavonoids, and flavonols in the pulp of different varieties and cultivars of prickly pear

Table 1A

Parameter	Abdel-Hameed et al. (2014)		Albano et al. (2015)		Butera et al. (2002)			Cano et al. (2017)		Cayupán, Ochoa, and Nazareno (2011)		
	Opuntia ficus indica Mill. (Saudi Arabia) Pulp+seeds		*Opuntia ficus-indica* (L.) Mill. (Italy) Pulp		*O. ficus indica*, (Italy) Pulp			*Opuntia ficus-indica*, spp.) (Spain)		*Opuntia* (Argentina)		
	Red	Yellow	Purple	9 varieties	Verdal (orange)	Red	White	Sanguinos (red)	Verdal (orange)	*O. megacanta* (Argentina)	*O.* spp. (Argentina)	*O. ficus-indica* (Argentina)
Total Phenols (mg GAE)	1065.15	667.82	89.2	69.8				457.25	432.50	0.36-0.75	0.51-1.03	0.41-0.93
ABTS [#]			0.61[a]	0.37[a]	5.31[b]	4.20[b]	4.36[b]	108.85[c]	122.47[c]			
TEAC (mM)												
TAC [$]	735.35[b]	204.31[b]								0.25-0.51[b]	0.46-0.50[b]	0.48-0.57[b]
ORAC (mmol)			1.28	0.98				30.28	28.53			
DPPH [%]	81.90[a]	391.90[a]										
Reducing power activity (mg equivalent to ascorbic acid)	123.23	81.38										
Total Flavonoids (mg rutin equivalent)	159.49	80.35										
Total flavonols (mg rutin equivalent)	341.81	160.02										

[#] ABTS concentration units: a) mmol; b) µmol of Trolox equivalent/g eatable pulp; c) µmol of Trolox equivalent/100 g eatable pulp.
[$] TAC concentration units: a) mmol/Trolox equivalent; b) mg ascorbic acid equivalent.
[%] DPPH concentration units: a) mg ascorbic acid equivalent; b) mmol/Trolox equivalent; c) % discoloration; d) IC50 mg/ml.

Table 1B

Parameter	Chang, Hsieh, and Yen (2008)	Chavez-Santoscoy, Gutierrez-Uribe, and Serna-Saldívar (2009)	Cruz-Bravo et al. (2019)		de Wit et al. (2019)	Castañeda-Yañez et al. (2018)		Juhaimi et al. (2020)
	Opuntia dillenii Haw (Taiwan)	*Opuntia* (Mexico)	*Opuntia* (Mexico)		*Opuntia ficus.indica* and *Opuntia robusta*	*Opuntia* (Mexico) Juices		*Opuntia ficus-barbarica* A. Berger (Turkey)
	Red	9 types	Roja Lisa	Amarilla Olorosa	8 varieties	Red	Purple	5 harvesting times
Total Phenols (mg GAE)	91.5	22.3-226.3	97	94	1.58-22.8			90.97-156.77
ABTS #								
TEAC (mM)	1.01							
TAC $		8.82^{a}	5.15^{a}					
ORAC (mmol)	0.10	17.4-25.8	7.03	5.79				
DPPH %			0.15b	0.06^{b}		7.50^{c}	12.62^{c}	2.55-9.81^{c}
Reducing power activity (mg equivalent to ascorbic acid)								
Total Flavonoids (mg rutin equivalent)	29.5	95.8-374.3	31	47				
Total flavonols (mg rutin equivalent)								

ABTS concentration units: a) mmol; b) µmol of Trolox equivalent/g eatable pulp; c) µmol of Trolox equivalent/100 g eatable pulp.

$ TAC concentration units: a) mmol/Trolox equivalent; b) mg ascorbic acid equivalent.

% DPPH concentration units: a) mg ascorbic acid equivalent; b) mmol/Trolox equivalent; c) % discoloration; c) % discoloration; d) IC50 mg/ml.

Table 2. Phenolic compounds in the pulp of prickly pear from different species and origins

Parameter*	Abdel-Hameed et al. (2014)		Chang, Hsieh, and Yen (2008)	Juhaimi et al. (2020)
	Opuntia ficus-indica Mill. Pulp+seeds (Saudi Arabia)		*Opuntia dillenii* Haw (Taiwan)	*Opuntia ficus-barbarica* A. Berger (Turkey)
	Red pulps	Yellow pulps	Red	5 harvesting times
Gallic acid	452.22[a]	748.22[a]	4.0b	7.16-33.43[c]
Catechin	305.22[a]	491.94[a]	22.7b	4.55-10.41[c]
Epicatechin			10.9b	
Quercetin-3-glucose-(1→6)-gallic acid	Tr	Tr		
Rutin	2234.89	516.11[a]	ND	1.35-27.22[c]
Caffeic				0.27-8.99[c]
trans-Ferulic acid				4.06-13.48[c]
Quercetin 3-O-b-D-glucoside	Tr	Tr		
Kaempferol 3-O-b-D-glucoside	152.09[a]	481.42[a]		
Apigenin 7 glucoside				1.01-2.08[c]
Quercetin	Tr	Tr	ND	1.97-5.67[c]
Apigenin	ND	ND		
Syringic acid				1.10-7.74[c]
trans-Cinnamic				0.26-0.74[c]
p-Coumaric				0.28-1.32[c]
Resveratrol				1.49-2.15[c]
Kaempferol				5.92-9.17[c]
Isorhamnetin				1.20-5.46[c]
Naringenin				ND-0.55[c]
3,4-Dihydroxybenzoic acid				2.10-16.49[c]
1,2-Dihydroxybenzene				2.78-12.13[c]

[*] Concentration units: a) g/100 ml juice; b) mg/100 g fresh sample, c) mg/100 g Dry Weight (DW)), Tr: traces, ND: nondetected.

Table 3A-B. Betalains of prickly pear pulp from different varieties and origins

Table 3A

Parameter*	Albano et al. (2015)			Butera et al. (2002)			Cano et al. (2017)		Castellanos-Santiago and Yahia (2008)	Cayupán, Ochoa, and Nazareno (2011)		
	Opuntia ficus-indica (L.) Mill. (Italy) Pulp		*Opuntia* (Mexico)	*O. ficus-indica*, (Italy) Pulp			*Opuntia ficus-indica*, spp.) (Spain)		*Opuntia* (Mexico)	*Opuntia* (Argentina)		
	Purple	Orange	18 wild varieties	Yellow	Red	White	Sanguinos (red)	Verdal (orange)	10 cultivars	*O. megacanta* (Argentina)	Roja Lisa	Amarilla Olorosa
Betacyanin	39.3[a]	3.6[a]					1.98[a]	0.37[a]	0.05-5.29[b]			
Indicaxanthin				8.42[a]	2.61[a]	5.86[a]						
Betanin				1.04[a]	5.12[a]	0.10[a]						
Isobetanin												
Betaxanthins			1.08-11.49[a]				2.61[a]	1.70[a]	0-12-2.86[b]	16-52[c]	47-150[c]	3.1-140[c]
Betacyanins			0.50-20.95[a]							11-18[c]	58-322[c]	0.8-344[c]
Total betalains			1.61-32.44[a]				4.59[a]	2.07[a]	0.17-8.15[b]	27-70[c]	105-472[c]	3.9-484

*Concentration units: a) mg/100 g Fresh Weight, b) mg/g dry weight, c) µg/g Fresh Weight; d) mg/kg fruit.

Table 3B

Parameter*	Chang, Hsieh, and Yen (2008)	Aparicio et al. (2017)	Cruz-Bravo et al. (2019)		de Wit et al. (2019)	Castañeda-Yañez et al. (2018)		Chavez-Santoscoy, Gutierrez-Uribe, and Serna-Saldívar (2009)
	Opuntia dillenii Haw (Taiwan)	*Opuntia* (Mexico)	*Opuntia* (Mexico)		*Opuntia ficus-indica* and *Opuntia robusta*	*Opuntia* (Mexico) juices		*Opuntia* (Mexico)
	Red	18 wild varieties	Roja lista	Amarilla olorosa	8 varieties	Red	Purple	9 types
Betacyanin								
Indicaxanthin								
Betanin	18.2[a]							
Isobetanin	19.1[a]							
Betaxanthins		1.08-11.49[a]	10[d]	11[d]	1.22-112.62[d]	39	157	3.1-189.9[c]
Betacyanins		0.50-20.95[a]	26[d]	2[d]	0.72-74.47[d]	56	257	1.6-300.5[c]
Total betalains		1.61-32.44[a]						

* Concentration units: a) mg/100 g Fresh Weight, b) mg/g dry weight, c) µg/g Fresh Weight; d) mg/kg fruit.

Table 4. Antioxidant activities, total phenolics, flavonoids, and flavonols in the peels of different varieties and cultivars of prickly pear

Parameter	Abdel-Hameed et al. (2014)		Cano et al. (2017)		Cardador-Martínez, Jimenez Martinez, and Sandoval (2011)				Chang, Hsieh, and Yen (2008)	de Wit et al. (2019)
	Opuntia ficus-indica Mill. (Saudi Arabia) Peels		*Opuntia ficus-indica*, spp.) (Spain)		*Opuntia* (Mexico)				*Opuntia dillenii* Haw (Taiwan)	*Opuntia ficus-indica* and *Opuntia robusta*
	Red	Yellow	Sanguinos (red)	Verdal (orange)	Esmeralda (green-yellow)	Pelón-liso (red-purple)	Cristalina (clear-green)	Montesa (yellow-brown)	Red	8 varieties
Total phenols (mg GAE)	1152.97	786.01	698.37	630.30	362	194	141	229	133.4	7.44-58.88
ABTS µmol of Trolox equivalent			141.80	141.60						
TEAC (Trolox equivalent)					322	291	430	484	1.24	
TAC (mg ascorbic acid equivalent)	760.61	228.57								
ORAC (mmol)			88.31	48.35					0.14	
DPPH %	70.17[a]	234.96[a]			21[c]	14[c]	33[c]	14[c]		
Reducing power activity (mg equivalent to ascorbic acid)	149.49	83.12								
Total flavonoids (mg rutin equivalent)	113.32	69.56			66	21	36	35	32.5	
Total flavonols (mg rutin equivalent)	195.15	98.71								
Tannins (mg equivalent of catechin)					53	144	30	144		

% DPPH concentration units: a) DPPH (mg ascorbic acid equivalent); b) mmol/Trolox equivalent; c) % discoloration.

Table 5. Pigments identified and quantified in prickly pear peels

Pigment*	Cano et al. (2017)		Chang, Hsieh, and Yen (2008)	de Wit et al. (2019)
	Opuntia ficus-indica, spp.) (Spain)		*Opuntia dillenii* Haw (Taiwan)	*Opuntia ficus-indica* and *Opuntia robusta*
	Sanguinos (red)	Verdal (orange)	Red	Eight varieties
Betalains				
Betacyanin (mg betanin eq/100 g FW)	2.52	1.17		
Betaxanthins (mg Indicaxanthin eq/100 g FW)	2.00	1.73		
Betanin (mg/100 g FW)			15.7	
Isobetanin (mg/100 g FW)			19.2	
Betaxanthin (mg/100 g)				0.62-29.84
Betacyanins (mg/100 g)				0.89-42.62
Total Betalain (mg/100 g FW)	4.54	2.90		
Carotenoids (mg/100 g fresh weight)				
(all-E)-violaxanthin	93.64	87.67		
(all-E)-neoxanthin	34.51	59.05		
(9Z)-violaxanthin	29.55	29.49		
(all-E)-anteraxanthin	39.94	40.7		
(9Z)-neoxanthin	Tr	Tr		
(all-E)-lutein	1132.51	767.98		
(all-E)-zeaxanthin	63.26	52.98		
Lutein-5,6- epoxide	63.26	11.59		
(all-E)-β-criptoxanthin	8.82	5.76		
(all-E)-α-carotene	10.34	8.11		
(all-E)-β-carotene	200.4	173.5		
(9Z)-β-carotene	26.01	21.44		
Lycopene	45.61	NS.		
Total Carotenoids	1693.38	1257.74	1.79-6.06	

* FW: Fresh weight, DM: Dry matter, Tr: Traces; ND: Non-detected.

Table 6. Fatty acids profile in prickly pear seeds from different varieties and cultivars

Fatty acids (g/100 g of oil)*	Amador-Rodríguez et al. (2019)		Brahmi et al. (2020)	Chahdoura et al. (2015)		Chougui et al. (2013)	Ennouri et al. (2005)		Juhaimi et al. (2020)	Ramadan and Mörsel (2003a)
	Opuntia ficus-indica (11 Opuntia genotypes)		*Opuntia ficus-indica* mil. (Algeria)	*O. macrorhiza* (Engelm.)	*O. microdasys* (Lhem.)	*Opuntia ficus-indica* mil. (Algeria)	*O. ficus-indica* (Tunisia)	*O. stricta* (Tunisia)	*Opuntia ficus-barbarica* A. Berger (Turkey)	*O. ficus-indica* (Germany)
	Prickly pear	Xoconoxtle	Cold press oil	Red	Red	4 cultivars	2 seasons	2 seasons	5 harvesting times	red to purple
Lauric			0.7				ND	0.19		
Myristic			0.15	0.13	0.15		ND	0.32	0.09-1.21	ND
Palmitic	11.1-14.28	10.97	10.08	13.9	13.0	12.7-13.4	9.32	7.21	10.65-22.61	20.1
Palmitoleic			0.53	0.51	0.59		1.42	0.38		1.80
Stearic	2.61-3.79	2.63	2.97	2.8	2.9	3.3-4.2	3.11	3.83	3.92-9.20	2.72
Oleic	9.75-16.75	9.98	12.92	10.0	7.9	15.2-24.3	16.8	12.8	1.97-28.51	18.3
Vaccenic			4.3			0.0-5.3				
Linoleic	63.24-70.32	73.09	55.81	71	74	58.7-63.1	70.3	74.8	0.76-57.50	53.5
Linolenic	3.70-5.59	3.34	0.24	0.27	0.67				ND-0.69	2.58
Arachidic			0.24	0.27	0.27				0.39-12.36	
Gondoic			0.08	0.11	0.11					
Paullinic			0.2							
Eicosadienoic			0.04							
Heneicosylic			0.02							
Behenic			0.05	0.18	0.17				ND-0.84	
Erucic			0.04						ND-0.34	
15-docosenoic			0.17							
Nervonic			0.01							
Lignoceric			0.05							
Arachidonic									0.17-0.93	
Total SFA						16.6-16.9				22.8
Total UFA						83.1-83.5				76.1
UFA/SFA						4.9-5	7.11	7.61		3.38

* SFA: Saturated fatty acids; UFA: Unsaturated fatty acids; ND: Non detected.

Table 7. Antioxidant activities, total phenolics, flavonoids, tannins, and pigments in the seeds of different varieties and cultivars of prickly pear

Parameter	Amador-Rodríguez et al. (2019)		Brahmi et al. (2020)	Caldas-Cueva et al. (2016)	Cardador-Martínez, Jimenez Martinez, and Sandoval (2011)			Chang, Hsieh, and Yen (2008)	Chougui et al. (2013)	El-Guezzane et al. (2021)
	Opuntia ficus-indica (11 Opuntia genotypes)		*Opuntia ficus-indica* mil. (Algeria)	*Opuntia soehrensii* Britton & Rose (Peru)	Opuntia (Mexico)			*Opuntia dillenii* Haw (Taiwan)	*Opuntia ficus-indica* mil. (Algeria)	*Opuntia ficus-indica* and *Opuntia megacantha*
	Prickly pear	Xoconoxtle	Cold press oil	Dry seeds	Pelón-liso (red-purple)	Cristalina (clear-green)	Montesa (yellow-brown)	Red	4 cultivars Soxhlet extracted	Methanolic extract
Total phenols (mg GAE/100 g)	60.73-109.64	141.81	55.82	101.43	345	337	373	212.8	48-89	13.51-22.08
ABTS µmol of Trolox equivalent	13.18-43.62	5.53		4.74						
TEAC (Trolox equivalent)					47	24	19			
DPPH %					38[c]	32[c]	38[c]			0.78-0.96[d]
Total flavonoids (mg QE/100 g)			3.08		46	46	49	144.1	1.5-2.6	0.18-0.71
Tannins (mg rutin equivalent)					171	174	172		4.1-6.6	0.16-0.34
Carotenoids (mg/kg)			65.5						40.15-69.29	
Betacyanins (mg/kg)									17.48-160.88	
Betaxanthins (mg/kg)									12.24-12-62	

% DPPH concentration units: a) mg ascorbic acid equivalent; b) mmol/Trolox equivalent; c) % discoloration; d) IC5? mg/ml.

Table 8. Phenolic compounds in the seeds of different varieties and cultivars of prickly pear

Parameter*	Chahdoura et al. (2015)		Chang, Hsieh, and Yen (2008)	Chougui et al. (2013)
	O. macrorhiza (Engelm.)	*O. microdasys* (Lhem.)	*Opuntia dillenii* Haw (Taiwan)	*Opuntia ficus-indica* mil. (Algeria)
			Red	Four cultivars Soxhlet extracted
Gallic acid			2.6[b]	
Catechin			35.6[b]	
Epicatechin			31.8[b]	
p-Coumaric acid			2.2[b]	
Ferulic acid			10.2[b]	
Sinapinic acid			26.8[b]	
Rutin			0.3[b]	
Quercetin			33.5[b]	
Apigenin				
Caffeoyl hexoside	46[a]	33[a]		
Feruloyl di-hexoside	45[a]	37[a]		
Feruloyl hexoside derivative	25[a]	77[a]		
Feruloyl-sucrose isomer 1				6.7-14.2[c]
Feruloyl-sucrose isomer 2				5.6-33.0[c]
Feruloyl-sucrose isomer 3				105-717[c]
Feruloyl derivative				34-233[c]
Isorhamnetin-O-(rhamnosyl)rutinoside	24[a]	ND		
Quercetin 3-O-b-D-glucoside				
Kaempferol 3-O-b-D-glucoside				

* Concentration units: a) µg/g extract; b) mg/100 g fresh sample; c) µmol/100 g dry matter; ND: Non detected.

Peels present an important antioxidant activity, and phytochemicals concentrations, in some varieties, could be superior to the same parameters in the pulp. Table 4 shows the antioxidant activity and the polyphenols, flavonoids, flavonols, and tannins quantified in the peel of different species and varieties of *Opuntia* fruits from different origins. Polyphenols are the most abundant compounds reported for most varieties, followed by flavonols or tannins, depending on the fruit (Table 4).

Regarding individual polyphenols, Chang, Hsieh, and Yen (2008) found for the peel of red *Opuntia dillenii* Haw (Taiwan) that catechin, epicatechin, gallic acid were the most abundant (18.0 mg/100 g fresh sample, 17.1 mg/100 g fresh sample, and 2.7 mg/100 g fresh sample, respectively). On the other hand, Abdel-Hameed et al. (2014) reported different compounds and proportions by analyzing the red and yellow peels of *Opuntia ficus-indica* Mill. peels (Saudi Arabia). For both varieties, the most abundant polyphenol was rutin (4374.13 µg/100 ml juice and 5735.92 µg/100 ml juice, respectively), followed by gallic acid (802.75 µg/100 ml juice and 721.04 µg/100 ml juice, respectively), and catechin (411.33 µg/100 ml juice and 581.98 µg/100 ml juice, respectively).

Another important group of bioactive compounds found in prickly pear peels is pigments (Table 5). Betalains are considered the most abundant group of pigments in prickly pear; however, carotenoids could be an important group for some varieties of prickly pear.

As in the pulp, prickly pear peels not have an important concentration of fat. Peel fat content is correlated with variety, cultivar, and origin (Andreu-Coll et al. 2019, Ramadan and Mörsel 2003b) since it is related to the pulp. The concentration of lipids could arrive until 2.6-3.6% of the peel (Ramadan and Mörsel 2003b).

Ramadan and Mörsel (2003b) evaluated the peel fat and fatty acids profiles in the total fat and different fat fractions of red to purple *Opuntia ficus-indica* fruits bought in local markets in Germany. For the peels, they reported that the major fatty acids of the peel's total fat were linolenic (32.3% of the fatty acids), oleic (24.1% of the fatty acids), palmitic (23.1% of the fatty acids), and linoleic acid (8.60% of the fatty acids). They also reported a total sterols content of 37.0 g/kg of total lipids, the most abundant the β-sitosterol (21.1 g/kg of total lipids), campesterol (8.76 g/kg of total lipids), Δ5-avenasterol (2.71 g/kg of total lipids) and stigmasterol (2.12 g/kg of total lipids).

Andreu-Coll et al. (2019) evaluated the fatty acids profile from nine *Opuntia* fruits varieties, finding that the fatty acids profile depended strongly on the variety. Three fatty acids were found in all the varieties and presented the higher concentrations. The most abundant was linoleic acid (37-52% of the fatty acids), followed by palmitic (21.5-32.1% of the fatty acids), and oleic (6.88-31% of the fatty acids). Linolenic acid was not detected in all the varieties.

Seeds

Seeds represent an important proportion of prickly pear fruits. Their proportion depends on the variety. For *Opuntia ficus*-indica Mill. Cultivars 'Gialla' and 'Rossa' from Italy, seeds represented 2.9-3.8% dry weight of the fruit (Barbera, Inglese, and Mantia 1994). The nutritional composition of seed has been slightly studied nevertheless depending on the *Opuntia* species; the most abundant compounds are the fiber (40-67%) and the fat (9-17%), followed by carbohydrates (5-10%) and crude protein (5-16%) (Amador-Rodríguez et al. 2019, Chahdoura et al. 2015, Núñez-Gastélum 2018, Sawaya, Khalil, and Al-Mohammad 1983).

Additionally, seeds contain minerals such as Zn (1.45-10.35 mg/100 g dry weight), Fe (2.8-152 mg/100 g dry weight), Ca (0.11-16 mg/100 g dry weight), and K (0.16-163 mg/100 g dry weight) (Amador-Rodríguez et al. 2019, Sawaya, Khalil, and Al-Mohammad 1983).

Prickly pear seeds oil or fat composition has been studied in different works (Table 6). Since fat represents an important proportion of the seed, the oil has gained interest. The fatty acids profile differs depending on the species, variety, and origin. Nevertheless, according to different works (Table 6), linoleic acid is the most abundant fatty acid; its concentrations could represent 55-73% of the oil for most of the species, and only in a few cases has not been found or only in traces. Next in concentration are the palmitic and oleic acids, with 10-22% and 1.9-28.5%, respectively. Other fatty acids are reported in lower amounts.

Like other parts of the fruit, Prickly pear seeds present antioxidant activity as well as polyphenols, flavonoids, and tannins (Table 7). The most abundant compounds reported are polyphenols, followed by tannins and flavonoids. On the other hand, some pigments have also been reported (Table 7). As expected, in seeds, carotenoids have shown higher concentration than betalains due to their hydrophobic nature. Individual polyphenols reported in prickly pear seeds are shown in Table 8.

Flowers

One of the less studied byproducts of prickly pear production is the flowers. They are discarded after their separation from the fruit. Cactus flower infusions have traditionally been used for medicinal purposes in different countries.

Polyphenols are among the main compounds identified in flowers, followed by flavonoids and tannins (Table 9); as for other compounds, their concentration and predominance depend on the species, varieties, and origin. The same phenomenon is observed for the individual polyphenols, as shown in Table 10. *Opuntia* flower extracts have shown dose-dependent anti-inflammatory activities *in vitro* tests without cytotoxic effects (Benayad et al. 2014).

A less studied *Opuntia* flowers compounds group corresponds to volatile compounds (De Leo et al. 2010, Ammar et al. 2012). De Leo et al. (2010) evaluated volatiles emitted in vivo from Italian *Opuntia* flowers. They reported 18 volatile compounds that represented 84.2% of whole volatiles, were the most abundant were germacrene D (12.6%), 1-hexanol (12.3%), n-tetradecane (9.1%), and decanal (8.2%). Ammar et al. (2012) identified 50 volatile compounds in hexane extracts of *O. ficus-indica* (L.) Mill. and *O. stricta* (Haw.) Haworth flowers from Tunisia. The compounds present, their concentration, and their biological activity depended on the flowering stage and the specie. These flower hexane extracts showed similar or higher antioxidant activities than ascorbic acid, BHT, and BHA.

Alternatively, some pigments have been reported in prickly pear flowers. Brahmi et al. (2022) reported a total betalains content of 1.79 mg/g of the dry weight of orange flowers of *Opuntia ficus-indica* from Algeria. The most abundant were betaxanthins (0.93 mg/g of dry weight), followed by betacyanins (8.86 mg/g of dry weight).

Table 9. Antioxidant activities, total phenolics, flavonoids, and flavonols in the flowers of different varieties and cultivars of prickly pear

Parameter[a]	Benayad et al. (2014)	Brahmi et al. (2022)	El-Guezzane et al. (2021)
	Opuntia ficus-indica (Morocco)	*Opuntia ficus-indica* (Algeria)	*Opuntia ficus-indica* and *Opuntia megacantha*
	Yellow	Orange	Methanolic extract
Total phenols (mg GAE/100 mg)	3.18	24.38	44.25-50.01
Procyanidins (mg CCE/g DM)	0.07		
TAA (μmol Trolox)		3091	
FRAP (μmol Trolox Equivalent/g DW)		4078	
ORAC (μmol TE/g DM)	272.84		
DPPH %		13.76[b]	0.20-0.24[d]
Total flavonoids (mg CE/g DM)	1.48*		11.22-13.33
Total flavonoids (mg QE/g DM)		9.70	
Tannins (mg equivalent of catechin)			0.70-1.18

% DPPH concentration units: a) mg ascorbic acid equivalent; b) mmol/Trolox equivalent; c) % discoloration.

Table 10. Phenolic compounds in flowers of prickly pear from different species and origins

Phenolic compound content* ([a] % of total phenols, [b] mg/g DW; [c] mg/g FW)	Benayad et al. (2014)	Brahmi et al. (2022)	De Leo et al. (2010)
	Opuntia ficus-indica (Morocco)	*Opuntia ficus-indica* (Algeria)	*Opuntia ficus-indica* (Italy)
	Yellow	Orange	Deffated flower extracts
5-Hydroxyferulic acid-rhamnoside-hexoside	3.50		
Caffeoyl methoxycinnamoyl quinic acid	3.65		
1,4-Diferuloyl syringic acid	1.02		
4-p-Coumaroyl caffeic acid	0.63		
1,5-Dicaffeoyl ferulic acid	1.51		
1,4-Syringicferuloyl 4-coumaroyl caffeic acid	2.54		
Quercetin-derivative	3.40		
Isorhamnetin 3-O-rhamnosyl 7-O-rutinoside	32.84		
Isorhamnetin glucoside		1.26	
Quercetin 3-O-rutinoside			7.09[c]
Quercetin 7-O-rutinoside	3.19		
Quercetin hexoside		0.22	
Quercetin 3-O-glucoside			4.47c
Isorhamnetin 3-O-robinobioside			42.69[c]
Isorhamnetin 3-O-galactoside			9.79[c]
Isorhamnetin 7-O-rutinoside	47.72	3.34	
Isorhamnetin 3-O-glucoside			7.24[c]
Piscidic acid		6.65	
Eucomic acid		13.4	
Rutin		0.40	
Kaempferol 3-O-rutinoside			4.00[c]
Kaempferol 3-O-arabinoside			3.24[c]

* Concentration units: a) % of total phenols; b) mg/g DW; c) mg/g FW.

Phytochemicals Recovery and Applications

The methods applied for recuperating the phytochemicals from prickly pear byproducts should be focused on the compound's nature and future applications, considering not only extraction yields but also the purity, stability, and required qualifications for the applications.

Some studies have been carried out on the direct use of pulp juices (Chauhan, Sheth, and Suhagia 2015, Chavez-Santoscoy, Gutierrez-Uribe, and Serna-Saldívar 2009) and seed oil without fractionation (Ghazi et al. 2015). The obtained products have shown important health protective effects.

Chavez-Santoscoy, Gutierrez-Uribe, and Serna-Saldívar (2009) applied the juices obtained from 9 types of Mexican prickly pears to different cancer cell lines to evaluate their viability after 12 h. The observed effect depended on the prickly pear type since, in all the cases, the same juice concentration was used. They found that Moradillo fruits (*Opuntia violaceae*) affected prostate and colon cancer cells and the normal fibroblast cells. Rastrero fruits (*Opuntia rastrera*) affected the four cancer cells evaluated without affecting normal fibroblast cells. Gavia (*Opuntia robusta*) affected colon cancer cells, while Cardon (*Opuntia streptacantha*) affected prostate and colon cancer cells. They emphasize that the viability of hepatic cancer cells was affected only by the juices of Gavia and Rastrero prickly pears.

Chauhan, Sheth, and Suhagia (2015) evaluated the antinociceptive and anti-inflammatory activities of *Opuntia elatior* Mill. pulp juices from India. They reported that the oral administration of 15 ml of juice/kg showed similar antinociceptive action to sodium diclofenac. Additionally, they reported that significant anti-inflammatory effect with 10 ml/kg and 15 ml/kg that was similar to sodium diclofenac during the same period.

Opuntia seeds oil has shown an interesting composition to be used for human consumption. Ghazi et al. (2015) evaluated seed oil composition and antioxidant capacity obtained from *O. ficus-indica* and *O. dillenii* (Morocco). They reported that *O. dillenii* seeds oil was the most unsaturated (79.83%) than *O. ficus-indica* seeds oil (58.79%); in both cases, linoleic acid was the most abundant. Additionally, both seed oils showed higher antioxidant activities (19.79-27.21 IC50 μg/ml) than Vitamin C (16.56 IC50 μg/ml). These results suggest that *Opuntia* prickly pear seeds could be used as a good and cheap source of products that could be industrially exploited.

On the other hand, fractionation of prickly pear phytochemicals made it possible to obtain high-value products with more specific applications. In all the cases, the profile of recovered compounds, the extraction yield, and the biological activities depend on the extractive method and its conditions, as well as the combination of the solvent used for phytochemicals fractionation (Benayad et al. 2014, Brahmi et al. 2022, Castellar, Obón, and Fernández-López 2006, Coavoy Sánchez 2015).

Polyphenols

Prickly pear pulp is the part of the fruit that has been widely studied for polyphenols extraction, mainly because their colors are directly correlated with the polyphenols concentration, suggesting the highest extraction yields. The chemical structures of the most important compounds are shown in Figure 5. These compounds correlate with the fruits' antioxidant activity and their extracts. Coavoy Sánchez (2015) reported an antioxidant capacity of 40.18% DPPH discoloration and 1,002.5 mg of total phenols/l of juice (as gallic acid) of purple prickly pears from Peru using 80% Ethanol solution and 60ºC.

de Souza et al. (2015) evaluated the inhibitory effect of Brazilian *Opuntia ficus-indica* ethanolic extracts over the α-amylase and α-glucosidase enzymes in in-vitro studies. The highest inhibition of both evaluated enzymes (50.65% and 70.75% inhibition, respectively) was obtained with 70% ethanol solution extracts. This inhibition could be a good strategy to manage the early stages of type 2 diabetes by reducing the starch and saccharose digestion and, consequently, glycemia. The authors highlight that it is necessary to have an appropriate purification step before considering its implementation.

In the case of Prickly pear flower extracts, Benayad et al. (2014) obtained better results with a mixture of acetone/water (80:20) by using Assisted Solvent Extraction (ASE) than using the traditional maceration. On the other hand, greener extraction methods with GRASS solvents offer better results when the recovered compounds are destined for human consumption. An important increase in bioactive compounds extraction from *Opuntia* flowers was reported by Brahmi et al. (2022) when the extraction was performed with a hydroalcoholic mixture (35.70% ethanol in water). These authors found that ultrasound-assisted extraction (UAE) made it possible to obtain higher polyphenols, flavonoids, and betalain pigments recovery yields in a shorter time.

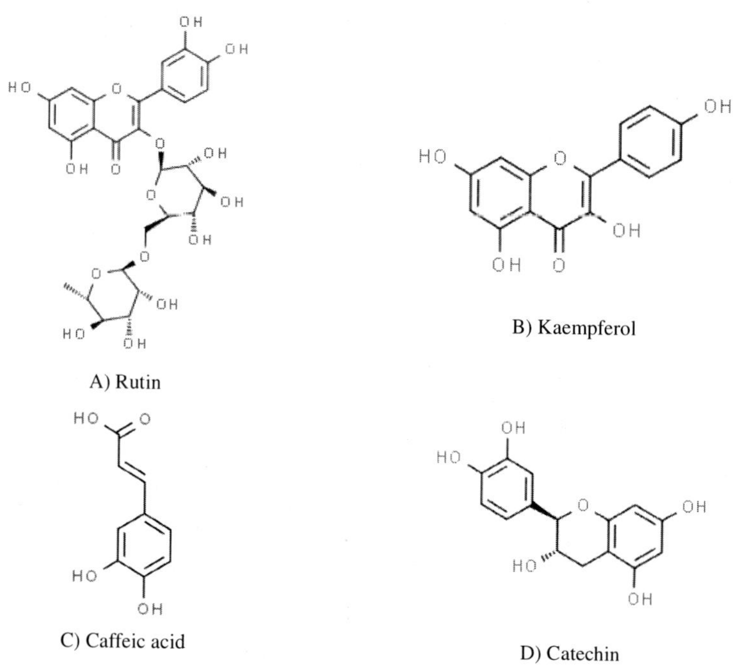

Figure 5. Molecular structures of four important polyphenols identified in prickly pear.

Betalains

Betalains are the most important pigments in *Opuntia* fruits, been the betanin and indicaxanthin the most abundant free betalains (Figure 6). Different protocols have been developed for prickly pears betalains extraction and conservation with different composition, yield and stability results depending on the cultivars (Otálora et al. 2015, Castellar, Obón, and Fernández-López 2006, Castellar et al. 2003, Castañeda-Yañez et al. 2018, Gandia-Herrero et al. 2010, Otálora et al. 2016).

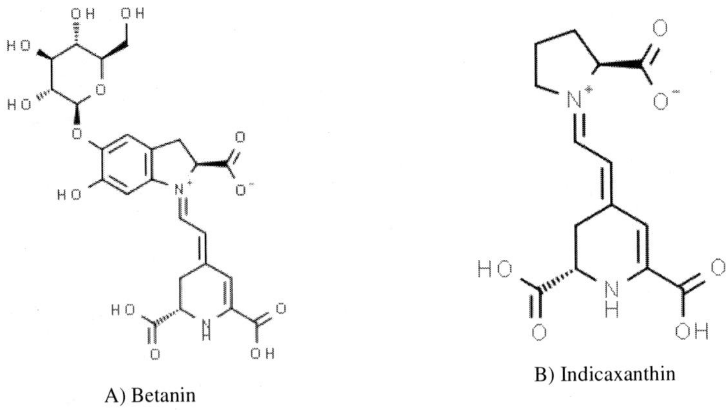

Figure 6. Molecular structures of two important betalains identified in prickly pear.

Castellar et al. (2003) extracted betacyanins from three red-violet cultivars: *Opuntia stricta* Haw., *Opuntia undulata* Griff., and *O. ficus-indica* L. Mill. whole fruit homogenates. The highest recovery was obtained using water for the extraction and variated depending on the variety for *Opuntia stricta* Haw. (80.1 mg of betanin/100 g fresh fruit), followed by *Opuntia undulata* (19.6 mg of Betanin/100 g fresh fruit) and *O. ficus-indica* (15.2 mg of Betanin/100 g fresh fruit). The extracts showed very high storage stability (t1/2 = 392 d at 4ºC and t1/2 = 19 d at 25ºC). In more recent works, Castellar, Obón, and Fernández-López (2006) describe an optimized method for betacyanins extraction from the *Opuntia stricta* (red-violet) fruits. The pigments of the homogenized whole prickly pear fruit were extracted using a mixture of ethanol/water (60:40 v/v). This method made it possible to obtain a 40-fold concentrated extract with a 4.7 g/l betanin concentration and to reduce the viscosity due to the fruit mucilage and pectins. The obtained extract showed high stability (t1/2 = 236.6 d at 4ºC) and color equivalent to commercial liquid colorants of red beet, red carrot, and red grape skin.

Different studies have been focused on the stabilization and protection of betalains after the extraction procedure such as spray-drying (Castañeda-Yañez et al. 2018, Díaz Sánchez et al. 2006, Gandia-Herrero et al. 2010, Otálora et al. 2015) or encapsulation (Otálora et al. 2016).

Díaz Sánchez et al. (2006) extracted the betalains of the Mexican *Opuntia lasiacantha* Pfeiffer LS 6813 juices with a thermal treatment consisting of heating (80 ºC, 5 min) and fast cooling (8-10ºC) followed to centrifugation for solids separation. The obtained extracts contained only betacyanins. Extracts were then spray-dried using a maltodextrin 10 ED as a protector. The powders showed 86.20% of pigments retention after 24 weeks of storage at 24ºC and light absence.

In another study, Gandia-Herrero et al. (2010) studied the stabilization of betalains of the different fractions (peel, flesh, and epidermis) of edible yellow and violet *Opuntia* from Spain. For both varieties, the most abundant betalains was indicaxanthin showing similar concentrations on flesh (145.3 μg/g of fresh weight and 54.1 μg/g of fresh weight for yellow and violet, respectively) and peel (146.2 μg/g of fresh weight and 40.2 μg/g of fresh weight for yellow and violet, respectively). After spray-drying the purified indicaxanthin with maltodextrin as a protective agent, the powder did not show significant color changes after 6 months in darkness, even at 20ºC. Authors conclude this is a good method to obtain a natural yellow pigment due to the demonstrated stability.

On the other hand, Otálora et al. (2015) evaluated the pertinence to preserve betalains of *O. ficus-indica* (purple pulp) from Argentina using maltodextrins and the mucilage extracted from the cladodes of the same *Opuntia* species. Using a phosphate buffer, they obtained a concentrated betalains extract from the freeze-dried pulp. Next, they evaluated using maltodextrin or a mixture of maltodextrin and the extracted mucilage as drying protectants. Both treatments gave the same encapsulation yields and non-significant different betalain contents. Additionally, both powders presented pigments retention >70% when stored at 18ºC and 57% RH. Authors highlight that using cladodes mucilage increases the total dietary fiber, adding value to the obtained colorant. These add value to another prickly pear byproduct, the mucilage.

Castañeda-Yañez et al. (2018) describe a procedure to preserve betalains of prickly pear juices (red and purple *Opuntia* sp.) by spray-drying microencapsulation of concentrated ethanolic extracts. The obtained extracts contained both betacyanins and betaxanthins. These authors evaluated the protective effect of three different maltodextrins to preserve the betalains and their antioxidant capacity. Even if they found the best combination of the prickly pear color, maltodextrin, and spry-drying conditions to protect betalains, they mention the necessity of further evaluating the obtained powders' stability and shelf life.

The betalains encapsulation with alginate gels was described by Otálora et al. (2016) for *Opuntia ficus-indica* fruits from Argentina. A betalains-rich lyophilized extract was obtained from the pulp with phosphate and citric acid buffer. Then, for encapsulation, the authors evaluated two methods for bead making; the first used alginate (CA), and in the second one, bovine serum albumin was combined with the alginate (CAB) to enhance matrix structure. Even though both encapsulation methods showed good results, alginate beds performed better than alginate + bovine serum albumin beds when they were stored at 25ºC and 34.6% RH for 25 days. They reported a betalains retention of 48.5% for CA and 37.2% for CAB, with a betalains degradation $t_{1/2}$ of 6.3 d and 5.1 d, respectively. The antiradical activity retention, evaluated by the DPPH method, was 88.5% for CA and 86.6% for CAB. Authors conclude that ionic gelation could be a potential method to obtain a red-violet colorant to be used in the food industry as a functional ingredient.

Flavonoids

Dok-Go et al. (2003) isolated specific flavonoids from the Korean *Opuntia ficus-indica* var. Saboten and evaluated specific biological activities. The isolation of quercetin, (+)-dihydroquercetin, and quercetin 3-methyl ether, was done by successive fractionation with ethyl acetate and n-butanol of the fruit methanolic extracts. The three isolated compounds showed antioxidant activity and inhibited neuronal injuries in primary cultured rat cortical cells. Even though quercetin is the most studied compound, and its properties are well studied, these authors found that quercetin 3-methyl ether was 6-9 times more potent than quercetin.

Non-fractioned Extracts

Extracts obtained with different organic solvents or water containing a mixture of compounds have been interesting alternatives for phytochemicals recuperation and use without compound fractionation and purification. The most described method used for their stabilization is spray-drying.

Obón et al. (2009) evaluated the suitability of spray-drying to obtain a natural red-purple colorant from *O. stricta* fruit juices to be used as a food colorant. Whole fruit juices (without peel and seeds separation) were dried, then prepared using Glucose syrup (DE 29) as a drying aid. Authors succeeded in the obtention of non-sticky powders that retained >98% color during the drying process and gave drying yields of 58%. When the powders were tested in food model systems (yogurt and soft drink), the red tonality remained vivid and attractive for consumers after one month at 4°C.

Saénz et al. (2009) evaluated the effect of encapsulation of the pulp and ethanolic extracts of *Opuntia ficus-indica* from Chile and the protective agent (maltodextrin or inulin). They observed that the highest phytochemical recuperation after spray-drying was obtained with maltodextrin, 98% encapsulation yield for betacyanins, and 92% for indicaxanthin. They also observed that microencapsulation of pulp extracts required less encapsulating agent than ethanolic extracts. They attribute this to mucilage in fiber in this kind of sample.

Fernández-López et al. (2010) aim to find an economic and easy method to obtain betacyanin-pigment extracts to be used as food colorants, obtained water extracts of the whole fruits (without peel separation) of *O. stricta* Haw., *O. undulata* Griff. and *O. ficus-indica* (L.) Mill. Extracts composition and antioxidant activity were different for the three species evaluated. *O. stricta* extracts showed the highest concentration of ascorbic acid (23.3 mg/100 g fresh fruit), betacyanins (80.1 mg betanin/100 g fresh fruit), total betalains (80.1 mg/100 g fresh fruit), and total flavonoids (161.1 µg/g fresh fruit); betaxanthins were not detected. *O. undulata* presented the highest concentration of total carotenoids (6.68 µg β-carotene equivalents/100 g fresh fruit) and taurine (11.22 mg/100 g fresh fruit), and the lowest levels of antioxidant activity by ABTS (1.58 µmol Trolox equivalent/g fresh fruit) and DPPH (1.08 µmol Trolox equivalent/g fresh fruit). Finally, *O. ficus-indica* extracts presented the highest concentrations of total phenolics (218.8 mg gallic acid equivalents/100 g fresh fruit) and betaxanthins (25.4 mg indicaxanthin/100 g fresh fruit), as well as the highest levels of antioxidant activity by ABTS (6.70 µmol Trolox equivalent/g fresh fruit) and DPPH (5.22 µmol Trolox equivalent/g fresh fruit). These authors conclude that obtained extracts have the potential to be used as a source of bioactive compounds but also as a coloring product due to their betalains content.

Ammar et al. (2012) evaluated hexane extracts of *O. ficus-indica* (L.) Mill. and *O. stricta* (Haw.) Haworth flowers from Tunisia. They describe the polyphenols and volatile compounds present in the extracts depending on the variety and flowering stage. Their antioxidant activity was similar to or higher than synthetic antioxidants such as BHT and BHA. Additionally, these extracts showed antibacterial activity against *Pseudomonas aeruginosa,* about 50%-90% of the activity of neomycin.

El-Hawary et al. (2020) obtained purified polysaccharides and freeze-dried powders from methanolic extracts from Egyptian *Opuntia ficus-indica* (L.) Mill fruit's pulp and peel. The powdered methanolic extracts were composed of various compounds (37 secondary metabolites); the most abundant compounds were O-methylated flavonoids. Additionally, phenolic acids, organic and fatty acids were identified and quantified. Due to composition, peels and methanolic pulp extracts showed DPPH antioxidant activity, 10.65 IC50, mg/ml and 16.57 IC50, mg/ml, respectively. On the other hand, purified polysaccharides showed DPPH antioxidant activity higher than methanolic extracts (74.13 IC50, mg/ml). The evaluation of the neuroprotective effects of the extracts (methanolic and polysaccharides) showed that a daily dose of 100 mg of the extract/kg of weight supplied orally for 6 weeks made it possible to revert the neuroinflammation and the oxidative stress induced by $AlCl_3$. This was due to restoring the brain monoamine neurotransmitter and the acetylcholinesterase in the brain of the studied animals.

Looking to develop a greener procedure to recover phytochemicals from the freeze-dried pulp of *Opuntia stricta* var. Dillenii's, a wild prickly pear from Spain, Gómez-López et al. (2022), improved pressurized liquid extraction (PLE), a greener process, and compared the compounds' recovery as well as the antioxidant and anti-inflammatory activities of the extracts with the conventional ethanol/water (50/50) liquid extraction. PLE extraction with a mixture of ethanol/water (50% v/v) and 25°C improved betalains and phenolic compounds extraction. The compounds profile obtained was similar between the PLE and the conventional method. PLE extraction made it possible to improve antioxidant and anti-inflammatory activities, 160% and 11% higher than the conventional method, respectively.

Recovery of phytochemicals from prickly pear peels has a special interest since it is the principal byproduct discarded during fresh fruit consumption or transformation. Gómez-Salazar et al. (2022), using mathematical modeling of real data, improved an extraction method to recover phytochemicals from *Opuntia streptacantha* peels (red prickly pear) from Mexico. The developed method consisted of two steps: peels drying and phytochemicals extraction. In the drying step, the authors found that the highest dehydration was obtained at 75-85°C, but the highest retention of total flavonoids, betaxanthins, and anthocyanins was at 65-75°C, so they concluded that the best drying temperature was 72°C. Next, conventional and microwave-assisted (MAE) extraction methods were evaluated using water and 70% ethanol in both cases. Results were similar for both methods, with water being the best solvent to recover phenolic compounds, flavonoids, and betacyanins. MAE had the advantage of shorter extraction times, 5 min instead of 50 min.

The methanolic extracts (methanol/water 80:20) obtained by El-Guezzane et al. (2021) from *Opuntia ficus-indica* and *Opuntia megacantha* flowers and seeds presented various of compounds. They reported extraction yields from 10.22-16.42% for seeds and 14.37-15.84% in the case of flowers. Phenols, flavonoids, and tannins were quantified in seeds and flower extracts. Additionally, in the case of the saponins presence test, they reported a low reaction for flowers and a negative for seeds. The test for alkaloids was negative for both seeds and flowers. The authors did not evaluate extracts' stability; nevertheless, they suggest that these extracts could be valorized in food, biomedical or cosmetic industries.

Conclusion

Most research about phytochemicals recuperation from *Opuntia* fruits byproducts has focused on prickly pear since it has established cultures and arrives in bigger markets, while wild varieties arrive only to local markets in lower quantities. On the other hand, a variety of *Opuntia* fruits are not destined for human consumption. In all the cases, the lower valued fruits and the byproducts of prickly pear's industrialization are sources of different phytochemical compounds with interesting applications in different industries. More research is needed to characterize undervalued *Opuntia* species to be used as phytochemical sources.

Concerning extraction and stabilization methods, new technologies have been developed. Applying electric pulses, high pressure or instant controlled pressure drop technology may increase extraction yields while phytochemical loss decreases.

References

Abdel-Hameed, E. S. S., M. A. Nagaty, M. S. Salman, and S. A. Bazaid. 2014. "Phytochemicals, nutritionals and antioxidant properties of two prickly pear cactus cultivars (*Opuntia ficus indica* Mill.) growing in Taif, KSA." *Food Chemistry* 160:31-38. doi: 10.1016/j.foodchem.2014.03.060.

Albano, C., C. Negro, N. Tommasi, C. Gerardi, G. Mita, A. Miceli, L. De Bellis, and F. Blando. 2015. "Betalains, Phenols and Antioxidant Capacity in Cactus Pear [*Opuntia ficus-indica* (L.) Mill.] Fruits from Apulia (South Italy) Genotypes." *Antioxidants* 4 (2):269-280. doi: 10.3390/antiox4020269.

Allegra, A., G. Sortino, G. Miciletta, M. Riotto, T. Fasciana, and P. Inglese. 2015. "The influence of harvest period and fruit ripeness at harvest on minimally processed cactus pears (*Opuntia ficus-indica* L. Mill.) stored under passive atmosphere." *Postharvest Biology and Technology* 104:57-62. doi: 10.1016/j.postharvbio.2015.03.001.

Amador-Rodríguez, K. Y., R. Serralta-Macías, S. Flores-Benítez, L. L. Valera-Montero, S. H. Guzmán-Maldonado, C. Gallegos-Vázquez, and H. Silos-Espino. 2019. "Chemical and potentially functional compounds of selected prickly pear fruit seeds (*Opuntia* spp.)." *Mitteilungen Klosterneuburg, Rebe und Wein, Obstbau und Früchteverwertung* 69 (3):141-153.

Ammar, I., M. Ennouri, B. Khemakhem, T. Yangui, and H. Attia. 2012. "Variation in chemical composition and biological activities of two species of *Opuntia* flowers at four stages of flowering." *Industrial Crops and Products* 37 (1):34-40. doi: 10.1016/j.indcrop.2011.11.027.

Andreu-Coll, L., M. Cano-Lamadrid, E. Sendra, Á. Carbonell-Barrachina, P. Legua, and F. Hernández. 2019. "Fatty acid profile of fruits (pulp and peel) and cladodes (young and old) of prickly pear [*Opuntia ficus-indica* (L.) Mill.] from six Spanish cultivars." *Journal of Food Composition and Analysis* 84:103294. doi: 10.1016/j.jfca.2019.103294.

Aparicio, X., S. Loza-Cornejo, M. G. Torres-Bernal, N. J. Velázquez-Placencia, and H. J. Arreola-Nava. 2017. "Características físico-químicas de frutos de variedades silvestres de *Opuntia* de dos regiones semiáridas de Jalisco, México [Physicochemical characteristics of fruits from wild *Opuntia* species from two semiarid regions of Jalisco, Mexico]." *Polibotanica* 43:219-244. doi: 10.18387/polibotanica.43.10.

Barba, F. J., P. Putnik, D. Bursać Kovačević, M. M. Poojary, S. Roohinejad, J. M. Lorenzo, and M. Koubaa. 2017. "Impact of conventional and non-conventional processing on prickly pear (*Opuntia* spp.) and their derived products: From preservation of beverages to valorization of by-products." *Trends in Food Science & Technology* 67:260-270. doi: 10.1016/j.tifs.2017.07.012.

Barbera, G., P. Inglese, and T. L. Mantia. 1994. "Seed content and fruit characteristics in cactus pear (*Opuntia ficus-indica* Mill.)." *Scientia Horticulturae* 58 (1):161-165. doi: 10.1016/0304-4238(94)90136-8.

Benayad, Z., C. Martinez-Villaluenga, J. Frias, C. Gomez-Cordoves, and N. E. Es-Safi. 2014. "Phenolic composition, antioxidant and anti-inflammatory activities of extracts from Moroccan *Opuntia ficus-indica* flowers obtained by different extraction methods." *Industrial Crops and Products* 62:412-420. doi: 10.1016/j.indcrop.2014.08.046.

Brahmi, F., F. Blando, R. Sellami, S. Mehdi, L. De Bellis, C. Negro, H. Haddadi-Guemghar, K. Madani, and L. Makhlouf-Boulekbache. 2022. "Optimization of the conditions for ultrasound-assisted extraction of phenolic compounds from *Opuntia ficus-indica* [L.] Mill. flowers and comparison with conventional procedures." *Industrial Crops and Products* 184:114977. doi: 10.1016/j.indcrop.2022.114977.

Brahmi, F., S. Haddad, K. Bouamara, D. Yalaoui-Guellal, E. Prost-Camus, J. P. P. de Barros, M. Prost, A. G. Atanasov, K. Madani, L. Boulekbache-Makhlouf, and G. Lizard. 2020. "Comparison of chemical composition and biological activities of Algerian seed oils of *Pistacia lentiscus* L., *Opuntia ficus indica* (L.) mill. and *Argania spinosa* L. Skeels." *Industrial Crops and Products* 151:112456. doi: 10.1016/j.indcrop.2020.112456.

Butera, D., L. Tesoriere, F. Di Gaudio, A. Bongiorno, M. Allegra, A. M. Pintaudi, R. Kohen, and M. A. Livrea. 2002. "Antioxidant activities of Sicilian prickly pear (*Opuntia ficus indica*) fruit extracts and reducing properties of its betalains: betanin and indicaxanthin." *Journal of Agricultural and Food Chemistry* 50 (23):6895-901. doi: 10.1021/jf025696p.

Caldas-Cueva, J. P., P. Morales, F. Ludeña, I. Betalleluz-Pallardel, R. Chirinos, G. Noratto, and D. Campos. 2016. "Stability of Betacyanin Pigments and Antioxidants in Ayrampo (*Opuntia soehrensii* Britton and Rose) Seed Extracts and as a Yogurt Natural Colorant." *Journal of Food Processing and Preservation* 40 (3):541-549. doi: 10.1111/jfpp.12633.

Cano, M. P., A. Gómez-Maqueo, T. García-Cayuela, and J. Welti-Chanes. 2017. "Characterization of carotenoid profile of Spanish Sanguinos and Verdal prickly pear (*Opuntia ficus-indica*, spp.) tissues." *Food Chemistry* 237:612-622. doi: 10.1016/j.foodchem.2017.05.135.

Cardador-Martínez, A., C. Jimenez Martinez, and G. Sandoval. 2011. "Revalorization of cactus pear (*Opuntia* spp.) wastes as a source of antioxidants." *Ciência e Tecnologia de Alimentos, Campinas* 31:782-788. doi: 10.1590/S0101-20612011000300036.

Castañeda-Yañez, A., S. T. Martín-del-Campo, A. San-Martin, and A. Cardador-Martínez. 2018. "Color and Antioxidant Capacity Preservation of *Opuntia* spp. Juices by Spray-drying Microencapsulation." *Journal of Food Research* 7 (3):16-26. doi: 10.5539/jfr.v7n3p16.

Castellanos-Santiago, E., and E. M. Yahia. 2008. "Identification and quantification of betalains from the fruits of 10 Mexican prickly pear cultivars by high-performance liquid chromatography and electrospray ionization mass spectrometry." *Journal of Agricultural and Food Chemistry* 56 (14):5758-64. doi: 10.1021/jf800362t.

Castellar, M. R., J. M. Obón, and J. A. Fernández-López. 2006. "The isolation and properties of a concentrated red-purple betacyanin food colourant from *Opuntia stricta* fruits." *Journal of the Science of Food and Agriculture* 86 (1):122-128. doi: 10.1002/jsfa.2285.

Castellar, R., J. M. Obón, M. Alacid, and J. A. Fernández-López. 2003. "Color properties and stability of betacyanins from *Opuntia fruits*." *Journal of Agricultural and Food Chemistry* 51 (9):2772-2776. doi: 10.1021/jf021045h.

Cayupán, Y. S. C., M. J. Ochoa, and M. A. Nazareno. 2011. "Health-promoting substances and antioxidant properties of *Opuntia* sp. fruits. Changes in bioactive-compound contents during ripening process." *Food Chemistry* 126 (2):514-519. doi: 10.1016/j.foodchem.2010.11.033.

Chahdoura, H., J. C. M. Barreira, L. Barros, C. Santos-Buelga, I. C. F. R. Ferreira, and L. Achour. 2015. "Seeds of *Opuntia* spp. as a novel high potential by-product: Phytochemical characterization and antioxidant activity." *Industrial Crops and Products* 65:383-389. doi: 10.1016/j.indcrop.2014.11.011.

Chang, S. F., C. L. Hsieh, and G. C. Yen. 2008. "The protective effect of *Opuntia dillenii* Haw fruit against low-density lipoprotein peroxidation and its active compounds." *Food chemistry* 106 (2):569-575. doi: 10.1016/j.foodchem.2007.06.017.

Chauhan, S. P., N. R. Sheth, and B. N. Suhagia. 2015. "Analgesic and Anti-inflammatory action of *Opuntia elatior* Mill fruits." *Journal of Ayurveda & Integrative Medicine* 6 (2):75-81. doi: 10.4103/0975-9476.159025.

Chavez-Santoscoy, R. A., J. A. Gutierrez-Uribe, and S. O. Serna-Saldívar. 2009. "Phenolic Composition, Antioxidant Capacity and *In Vitro* Cancer Cell Cytotoxicity of Nine Prickly Pear (*Opuntia* spp.) Juices." *Plant Foods for Human Nutrition* 64 (2):146-152. doi: 10.1007/s11130-009-0117-0.

Chougui, N., A. Tamendjari, W. Hamidj, S. Hallal, A. Barras, T. Richard, and R. Larbat. 2013. "Oil composition and characterisation of phenolic compounds of *Opuntia ficus-indica* seeds." *Food Chemistry* 139 (1):796-803. doi: 10.1016/j.foodchem.2013.01.054.

Coavoy Sánchez, I. A. 2015. "*Evaluación de la capacidad Opuntia ficus-indica antioxidante de los compuestos fenólicos de la tuna morada (Opuntia ficus-indica) del Distrito de San Bartolomé, Huarochirí, Lima [Evaluation of the antioxidant capacity of Opuntia ficus-indica of the phenolic compounds of the purple tuna (Opuntia ficus-indica) from the District of San Bartolomé, Huarochirí, Lima*]." Licence, Universidad Peruana Unión.

Cruz-Bravo, R. K., S. H. Guzmán-Maldonado, H. A. Araiza-Herrera, and J. A. Zegbe. 2019. "Storage alters physicochemical characteristics, bioactive compounds and antioxidant capacity of cactus pear fruit." *Postharvest Biology and Technology* 150:105-111. doi: 10.1016/j.postharvbio.2019.01.001.

De Leo, M., M. B. D. Abreu, A. M. Pawlowska, P. L. Cioni, and A. Braca. 2010. "Profiling the chemical content of *Opuntia ficus-indica* flowers by HPLC–PDA-ESI-MS and GC/EIMS analyses." *Phytochemistry Letters* 3 (1):48-52. doi: 10.1016/j.phytol.2009.11.004.

de Souza, R., M. S. Santana, E. S. de Macedo, E. de Brito, and R. P. Correia. 2015. "Physicochemical, bioactive and functional evaluation of the exotic fruits *Opuntia ficus-indica* and *Pilosocereus pachycladus* Ritter from the Brazilian caatinga." *Journal of Food Science and Technology*:1-8. doi: 10.1007/s13197-015-1821-4.

de Wit, M., A. du Toit, G. Osthoff, and A. Hugo. 2019. "Cactus pear antioxidants: a comparison between fruit pulp, fruit peel, fruit seeds and cladodes of eight different cactus pear cultivars (*Opuntia ficus-indica* and *Opuntia robusta*)." *Journal of Food Measurement and Characterization* 13 (3):2347-2356. doi: 10.1007/s11694-019-00154-z.

Díaz Sánchez, F., E. M. Santos López, S. Filardo Kerstupp, R. Villagómez Ibarra, and L. Scheinvar. 2006. "Colorant extraction from red prickly pear (*Opuntia lasiacantha*) for food application." *Electronic Journal of Environmental, Agricultural and Food Chemistry* 5 (2):1330-1337.

Dok-Go, H., K. H. Lee, H. J. Kim, E. H. Lee, J. Lee, Y. S. Song, Y. H. Lee, C. Jin, Y. S. Lee, and J. Cho. 2003. "Neuroprotective effects of antioxidative flavonoids, quercetin, (+)-dihydroquercetin and quercetin 3-methyl ether, isolated from *Opuntia ficus-indica* var. saboten." *Brain Research* 965 (1-2):130-6. doi: 10.1016/S0006-8993(02)04150-1.

El-Guezzane, C., H. El-Moudden, H. Harhar, N. Chahboun, M. Tabyaoui, and A. Zarrouk. 2021. "A comparative study of the antioxidant activity of two Moroccan prickly pear cultivars collected in different regions." *Chemical Data Collections* 31:100637. doi: 10.1016/j.cdc.2020.100637.

El-Hawary, S. S., M. Sobeh, W. K. Badr, M. A. O. Abdelfattah, Z. Y. Ali, M. E. El-Tantawy, M. A. Rabeh, and M. Wink. 2020. "HPLC-PDA-MS/MS profiling of secondary metabolites from *Opuntia ficus-indica* cladode, peel and fruit pulp extracts and their antioxidant, neuroprotective effect in rats with aluminum chloride induced neurotoxicity." *Saudi Journal of Biological Sciences* 27 (10):2829-2838. doi: 10.1016/j.sjbs.2020.07.003.

El Kharrassi, Y., M. A. Mazri, H. Benyahia, H. Benaouda, B. Nasser, and E. H. El Mzouri. 2016. "Fruit and juice characteristics of 30 accessions of two cactus pear species (*Opuntia ficus indica* and *Opuntia megacantha*) from different regions of Morocco." *LWT - Food Science and Technology* 65:610-617. doi: 10.1016/j.lwt.2015.08.044.

Ennouri, M., B. Evelyne, M. Laurence, and A. Hamadi. 2005. "Fatty acid composition and rheological behaviour of prickly pear seed oils." *Food Chemistry* 93 (3):431-437. doi: 10.1016/j.foodchem.2004.10.020.

Fernández-López, J. A., L. Almela, J. M. Obón, and R. Castellar. 2010. "Determination of Antioxidant Constituents in Cactus Pear Fruits." *Plant Foods for Human Nutrition* 65 (3):253-259. doi: 10.1007/s11130-010-0189-x.

Gandia-Herrero, F., M. Jimenez-Atienzar, J. Cabanes, F. Garcia-Carmona, and J. Escribano. 2010. "Stabilization of the bioactive pigment of *Opuntia* fruits through maltodextrin encapsulation." *Journal of Agricultural and Food Chemistry* 58 (19):10646-52. doi: 10.1021/jf101695f.

Ghazi, Z., M. Ramdani, M. Tahri, R. Rmili, H. Elmsellem, B. El Mahi, and M. L. Fauconnier. 2015. "Chemical Composition and Antioxidant Activity of seeds oils and fruit juice of *Opuntia ficus indica* and *Opuntia dillenii* from Morocco." *Journal of Materials and Environmental Science* 6 (8):2338-2345.

Gómez-López, I., J. A. Mendiola, M. P. Portillo, and M. P. Cano. 2022. "Pressurized green liquid extraction of betalains and phenolic compounds from *Opuntia stricta* var. Dillenii whole fruit: Process optimization and biological activities of green extracts." *Innovative Food Science & Emerging Technologies* 80:103066. doi: 10.1016/j.ifset.2022.103066.

Gómez-Salazar, J. A., J. Patlán-González, M. E. Sosa-Morales, J. G. Segovia-Hernandez, E. Sánchez-Ramírez, and C. Ramírez-Márquez. 2022. "Multi-objective optimization of sustainable red prickly pear (*Opuntia streptacantha*) peel drying and biocompounds extraction using a hybrid stochastic algorithm." *Food and Bioproducts Processing* 132:155-166. doi: 10.1016/j.fbp.2022.01.003.

Habibi, Y., M. Mahrouz, and M. R. Vignon. 2009. "Microfibrillated cellulose from the peel of prickly pear fruits." *Food Chemistry* 115 (2):423-429. doi: 10.1016/j.foodchem.2008.12.034.

Hernández Galán, L., and S. T. Martín del Campo. 2021. "Cactus fruits: An ancient and Emerging Crop." In *Super and Nutraceutical Foods: Composition and Technology*, edited by S.T. Martín del Campo, J. D. Ramírez Anaya and A. Cardador Martínez, 47-84. New York: Nova Science Publisher.

INCMNSZ. 2016. *Tablas de Composición de Alimentos y Productos Alimenticios Mexicanos* (Versión Condesada 2015) [*Tables of Composition of Mexican Foods and Food Products* (Condensed Version 2015)]: Instituto Nacional de Ciencias Médicas y Nutrición Salvador Zubirán: México City, México.

Juhaimi, F. A., K. Ghafoor, N. Uslu, I. A. Mohamed Ahmed, E. E. Babiker, M. M. Özcan, and G. J. Fadimu. 2020. "The effect of harvest times on bioactive properties and fatty acid compositions of prickly pear (*Opuntia ficus-barbarica* A. Berger) fruits." *Food Chemistry* 303:125387. doi: 10.1016/j.foodchem. 2019.125387.

Melgar, B., E. Pereira, M. B. P. P. Oliveira, E. M. Garcia-Castello, A. D. Rodriguez-Lopez, M. Sokovic, L. Barros, and I. C. F. R. Ferreira. 2017. "Extensive profiling of three varieties of *Opuntia* spp. fruit for innovative food ingredients." *Food Research International* 101:259-265. doi: 10.1016/j.foodres. 2017.09.024.

Núñez-Gastélum, J. A. 2018. "Morphological characteristics, chemical composition and antioxidant activity of seeds by four wild *Opuntia* species from North of Mexico." *Journal of the Professional Association for Cactus Development* 20:23-33.

Obón, J. M., M. R. Castellar, M. Alacid, and J. A. Fernández-López. 2009. "Production of a red–purple food colorant from *Opuntia stricta* fruits by spray drying and its application in food model systems." *Journal of Food Engineering* 90 (4):471-479. doi: 10.1016/j.jfoodeng.2008.07.013.

Otálora, M. C., J. G. Carriazo, L. Iturriaga, M. A. Nazareno, and C. Osorio. 2015. "Microencapsulation of betalains obtained from cactus fruit (*Opuntia ficus-indica*) by spray drying using cactus cladode mucilage and maltodextrin as encapsulating agents." *Food chemistry* 187:174-81. doi: 10.1016/j.foodchem.2015. 04.090.

Otálora, M. C., J. G. Carriazo, L. Iturriaga, C. Osorio, and M. A. Nazareno. 2016. "Encapsulating betalains from *Opuntia ficus-indica* fruits by ionic gelation: Pigment chemical stability during storage of beads." *Food Chemistry* 202 (Supplement C):373-382. doi: 10.1016/j.foodchem.2016.01.115.

Ramadan, M. F., and J. T. Mörsel. 2003a. "Oil cactus pear (*Opuntia ficus-indica* L.)." *Food Chemistry* 82 (3):339-345. doi: 10.1016/S0308-8146(02)00550-2.

Ramadan, M. F., and J. T. Mörsel. 2003b. "Recovered lipids from prickly pear [*Opuntia ficus-indica* (L.) Mill] peel: a good source of polyunsaturated fatty acids, natural antioxidant vitamins and sterols." *Food Chemistry* 83 (3):447-456. doi: 10.1016/S0308-8146(03)00128-6.

Saénz, C., S. Tapia, J. Chávez, and P. Robert. 2009. "Microencapsulation by spray drying of bioactive compounds from cactus pear (*Opuntia ficus-indica*)." *Food Chemistry* 114 (2):616-622. doi: 10.1016/j.foodchem.2008.09.095.

Sawaya, W. N., J. K. Khalil, and M. M. Al-Mohammad. 1983. "Nutritive value of prickly pear seeds, *Opuntia ficus-indica*." *Plant Foods for Human Nutrition* 33 (1):91-97. doi: 10.1007/BF01093742.

Section 2. Cereals, Pseudo Cereals and Legumes

Chapter 11

Bean (*Phaseolus vulgaris*) Processing and Generation of Bioactive Compounds

M. S. Cid-Gallegos[1,2]
C. Jiménez-Martínez[1]
L. J. Corzo-Ríos[2]
X. M. Sanchez-Chino[3]
M. F. Juárez-Chairez[1]
D. del R. Moguel-Concha[1]
J. E. Borges-Martínez[1]
and Y. M. Gómez y Gómez[2,*]
[1]Escuela Nacional de Ciencias Biológicas, Instituto Politécnico Nacional.
Unidad Profesional Adolfo López Mateos, Delegación Gustavo A. Madero,
Ciudad de México, México
[2]Unidad Profesional Interdisciplinaria de Biotecnología, Instituto Politécnico Nacional,
Delegación Gustavo A. Madero, Ciudad de México, México
[3]Cátedra-CONACyT, Departamento de Salud, El Colegio de la Frontera Sur-Villahermosa,
Tabasco, México

Abstract

Beans (*Phaseolus vulgaris*) are the most consumed legume worldwide, followed by soybeans and chickpeas. They are part of a plant-based diet, which promotes a high consumption of nutrient-dense plant foods while minimizing the consumption of ultra-processed foods, added sugars, and animal-based foods. In Mexico, it is part of the traditional milpa system, which is a reservoir of biological and nutritional resources, of which scientific evidence has been reported health benefits when consuming its products, as well as the generation of bioactive compounds. Among the main components of beans are carbohydrates (60-70%) and proteins (15-25%), which is why they are considered one of the main sources of protein in developing countries, considering their low cost; additionally, it has a content of micronutrients and bioactive compounds. Bean seeds need to be processed for consumption, so soaking and cooking are usually used; but there are

* Corresponding Author's Email: ygomezipn@hotmail.com.

In: Food Byproducts
Editors: S. Teresita Martín-del-Campo, A. Cardador-Martínez and J. del Pilar Ramírez-Anaya
ISBN: 979-8-88697-612-0

also other methods, such as microwaves and germination, among others, that will allow the generation of various bioactive compounds, which can be applied in the food or pharmaceutical industry.

Keywords: beans, consumption, processing methods, bioactive compounds

Introduction

Ultra-processed foods have been described as unhealthy foods within the dietary patterns of the population, where large amounts of rolls, soft drinks, and fried foods are included in the diet, which do not provide adequate nutrients for a quality diet (Ferreira-Tavares et al., 2012; Lawrence and Baker 2019; Fardet and Rock 2019; Poti, Braga and Qin, 2017). Therefore, they are considered a factor in the appearance of chronic non-communicable diseases and eating-related disorders (Monteiro and Cannon 2019). Likewise, the evidence suggests a higher risk of mortality from various causes to increased consumption of ultra-processed foods, and no health benefits have been reported (Elizabeth et al., 2020). Therefore, reducing the consumption of ultra-processed foods and opting for a plant-based diet that includes whole grains, fruits, vegetables, nuts, and legumes has been associated with health benefits such as the reduced risk of cardiovascular diseases, metabolic syndrome, and type 2 diabetes in up to 50% (Kahleova, Levin, and Barnard 2017; Satija and Hu 2018). Plant-based diets are characterized by reducing or eliminating the consumption of animal products. However, they are created no equal as they depend on the food quality (Satija and Hu 2018).

In Mexico, the milpa system is being promoted since it is a reservoir of biological and nutritional resources that is mainly made up of plants such as corn (*Zea mays*), beans (*Phaseolus* spp.), and pumpkin (*Cucurbita* spp.), which is called the Mesoamerican triad. Likewise, this system includes a great variety of chili peppers, quelites, tubers, and tomatoes that grow in the milpa. The combination of products will depend on the management of each family and the plants that grow in their locality (Méndez-Flores et al., 2021). On the other hand, the bioactive compounds in these dietary patterns where legumes are included have shown beneficial biological activities for health, such as antioxidant, anticancer, anti-inflammatory, antihypertensive, and antihyperglycemic, among others (Rajaram, Jones and Lee 2019).

In addition, legumes are seeds with excellent nutrient content and are one of the main sources of protein in developing countries (Tharanathan and Mahadevamma, 2003; Shimelis and Rakshit, 2005). The genus *Phaseolus* comprises several wild and cultivated species, including *P. lunatus* (lima bean), *P. acutifolius* (tepary bean), *P. polyanthus* (year bean), *P. coccineus* (Ayocote bean), and *P. vulgaris* (common bean) (Alvarado-López et al., 2019). As the common bean, it predominates among the legumes with the highest production and consumption in Africa, India, Latin America, and Mexico, where 47 of the 52 classified species are found (Chávez-Mendoza and Sánchez 2017). Likewise, *P. vulgaris* is an annual, climbing, or erect herbaceous plant (Figure 1) and is one of the most studied legumes due to its main components such as carbohydrates, proteins, lipids, and micronutrients (vitamins and minerals); as well as its content of various phytochemicals, among which are: total phenols, anthocyanins, tannins, flavonoids, lectins, phytic acid, among others (Abdulrahman, Bala and Bello 2021; Alcázar-Valle et al., 2020; Méndez-Flores et al., 2021). On the other hand, the main components and phytochemicals of the common bean have been attributed to a series of biological activities such as anticancer, anti-inflammatory, and antioxidant, among others; for use in the pharmaceutical and food industry (Alcázar-Valle et al., 2020; Méndez-Flores et al., 2021, Moreno-García et al., 2022).

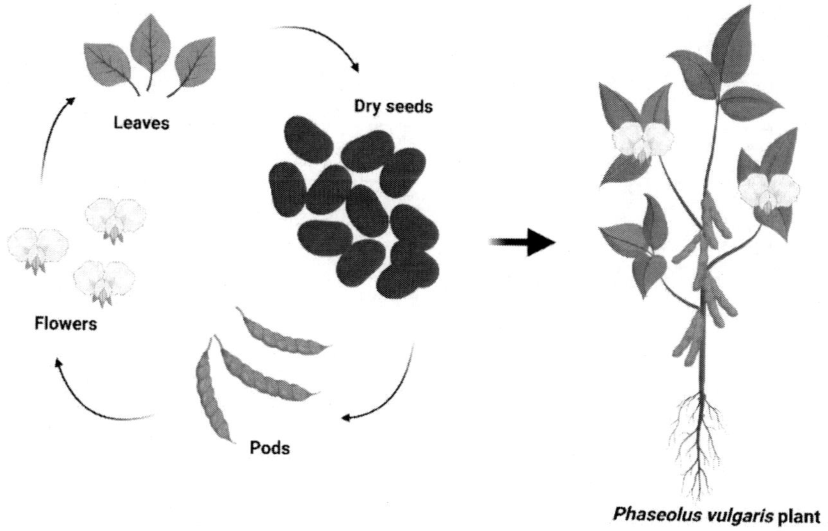

Figure 1. *Phaseolus vulgaris* plant.

Phaseolus vulgaris Chemical Composition and Processing

The common bean is composed of three main components, as shown in Table 1, which are carbohydrates (60-70%), proteins (15-25%), and lipids (1.1-2.7%); they have a lower proportion of micronutrients such as vitamins (thiamin, riboflavin, niacin, B6 and folic acid) and minerals (iron, zinc, copper, phosphorus and aluminum) (Hayat et al., 2014; Dueñas et al., 2016; Sánchez-Arteaga et al., 2015; Corzo-Ríos et al., 2020; Chávez-Mendoza and Sánchez 2017). Regarding carbohydrates, starch (amylose and amylopectin) occupies approximately 45%, and the rest is dietary fiber (1.7-27.2%) or non-starch polysaccharides (Dona et al., 2010). In addition, the bean has four protein fractions (globulins, albumins, glutelins, and prolamins) and the globulins are the ones found in the highest proportion (50-70%) (Tang and Sun 2011). Finally, the main lipid components in beans are phospholipids and triacylglycerides and unsaturated fatty acids such as linolenic acid (Hayat et al., 2014).

On the other hand, the processing is given to the seed, whether thermal (drying, cooking, or boiling) or non-thermal (soaking, germination, fermentation), can contribute to the inactivation, reduction, or elimination of certain anti-nutritional compounds or increase the concentration of certain phenolic compounds, as well as contributing to the hydrolysis of proteins that will produce peptides with biological activity (Chigwedere et al., 2018; López-Martínez et al., 2017; Nicolás-García et al., 2021; Gan et al., 2017).

Table 1. Chemical composition (%, on dry basis) of raw common beans (*Phaseolus vulgaris* L.)

Phaseolus vulgaris	Ash	Lipids	Protein	TCH[a]	fiber	Reference
Dark bean seeds	3.8 ± 0.0	1.1 ± 0.1	17.7 ± 0.5	77.4	27.2[c]	Dueñas et al., 2016
Common black beans	5.7 ± 0.0	1.8 ± 0.3	24.2 ± 1.1	66.6	1.7 ± 0.3[b]	Sánchez-Arteaga et al., 2015
National peanut	4.2 ± 0.1	1.7± 0.1	19.3 ± 0.1	-	25.5±0.4[c]	Corzo-Ríos et al., 2020
Common beans	3.2 ± 0.06	2.7±0.09	24.1 ± 0.97	-	3.9±0.04[b]	Liu et al., 2013
	4.6	1.6	24.0	61.3	2.2[b]	Chávez-Mendoza and Sánchez 2017

[a] TCH= Total carbohydrates, calculated by difference
[b] Crude fiber
[c] Dietary total fiber

Bioactive Compounds

Bioactive compounds are found in plants; their presence depends on environmental conditions since they are generated through secondary metabolism. Their main functions are reserve, defense, or interacting with other plants, microorganisms, and insects. (Méndez-Flores et al., 2021). Furthermore, the synthesis of these compounds will depend on geographic location, climate, genetic factors, interaction with other species, and their distribution and concentration. They are generally compounds that differ in their chemical nature, molecular weight, polarity, solubility, bioavailability, metabolic pathways, and excretion; therefore, it is essential to maintain biodiversity in plants to ensure access to these compounds (Phan et al., 2018). They can be proteins (protease inhibitors, α-amylases, lectins), glycosides (α-galactosides, vicine, and convicine), tannins, saponins, or alkaloids (Cid-Gallegos et al., 2022; Muzquiz et al., 2012).

On the other hand, seed processing usually improves the nutrient composition of legumes by increasing the in vitro digestibility of proteins and carbohydrates and reducing some antinutritional compounds. Since, in general, most of the antinutritional factors are thermolabile, the thermal treatment would diminish any possible adverse effect when consuming it. Although tannins, saponins, and phytic acid are heat stable, they can be reduced by dehulling, soaking, germination, and/or fermentation (Muzquiz et al., 2012).

Additionally, phenolic compounds are one of the most studied bean bioactive compounds along with bioactive peptides since these compounds usually exert beneficial properties such as modulation of metabolism, homeostasis, and cell proliferation; in addition to having antioxidant properties for the prevention of conditions such as oxidative stress and diseases (Conti et al., 2021). In recent years, bioactive peptides have been studied in vitro and in vivo to determine their therapeutic effect in various chronic-degenerative diseases. Some of their bioactive properties are based on the ability to induce only the arrest of cell division of cancer cells, but not from healthy cells. In addition, it has been shown that processes such as germination can increase the concentration of phenolic compounds in the plant or promote protein hydrolysis to obtain bioactive peptides (Gallegos-Tintoré et al., 2013; Conti et al., 2021).

Phenolic Compounds during Processing of *P. vulgaris*

Phenolic compounds are secondary metabolites that can be found distributed in various plant foods and are classified as phenolic acids, flavonoids, stilbenes, lignans, and tannins (Los et al., 2018; Nicolás-García et al., 2021). Likewise, phenolic acids and flavonoids in beans represent 50% of the total phenolic compounds (Los et al., 2018; Singh et al., 2017). Furthermore, various studies have reported biological activities of bean phenolic compounds (Table 2) as antioxidant, antiproliferative, anticancer, and anti-inflammatory after processes such as germination, and cooking, among others (Mamilla and Mishra 2017; Xue et al., 2016; López et al., 2013; Mendoza-Sánchez et al., 2019; Xu and Chang 2012; Ranilla, Genovese and Lajolo 2009).

Mamilla and Mishra (2017) reported that sprouted kidney beans had an antioxidant activity of 25.4 to 32.2%, which was higher than that reported for chickpea, lentils, and mung bean seeds. In addition, they attributed it to the activity of hydrolases and polyphenol oxidases during the germination process and the differences in seed matrices. Similarly, Xue et al., (2016) found that germination could cause an accumulation of polyphenols in germinated black beans. As a result, they could present 1.54 times more total phenolic compounds and higher antioxidant activity than raw beans after six days of germination.

López et al., (2013) studied common bean sprout extracts. They found anticancer activity in breast adenocarcinoma (MCF-7) and melanoma (UACC-62) cell lines, and said activity was attributed to their antioxidant capacity. Additionally, Mendoza-Sánchez et al., (2019) analyzed the effect of elicitors such as chitosan and salicylic acid on common bean sprouts. They observed that they improved the hypotriglyceridemic product and exerted antiobesogenic and hypolipidemic activity in rats. In addition, another study with two selected Brazilian bean cultivars (black and yellow-brown seed coat color) showed that regardless of the cooking temperature, if it is not subjected to soaking and the cooking water is conserved, it can favor the increase of certain phenolic compounds—antioxidants such as flavonols and free phenolic acids.

Table 2. Phenolic compounds bioactivities from *Phaseolus vulgaris*

Common bean (*Phaseolus vulgaris*)	Biological effect	Reference
Germinated	Antioxidant activity	Mamilla and Mishra 2017; Xue et al., 2016; López et al., 2013
	Anticancer Neuro-protective Anti-inflammatory Cardiovascular protective	López et al., 2013
	Antiobesogenic, Hypolipidemic	Mendoza-Sánchez et al., 2019
Cooked	Antioxidant	Ranilla, Genovese and Lajolo 2009

Peptides Produced through Various Methods to Process Beans

Peptides are small sequences of between 2-20 amino acids that usually show beneficial health effects when consumed; these are inactive in the protein and are released during protein hydrolysis by proteases. Some of the proteases that have been used to produce bioactive peptides from common beans are alcalase, flavourzyme, pepsin, pancreatin, chymotrypsin, papain, trypsin, and thermolysin (Luna-Vital et al., 2015). On the other hand, studies have described those processes such as germination promote protein hydrolysis and, in turn, the generation of peptides with biological activity such as antihypertensive, anti-inflammatory, anticancer, and antibacterial (Table 3) (Mamilla and Mishra 2017; de Souza Rocha et al., 2015; López-Barrios, Antunes-Ricardo and Gutiérrez-Uribe 2016; Luna-Vital, González de Mejía, and Loarca-Piña 2016; Luna-Vital, González de Mejía, and Loarca-Piña 2017; Ariza-Ortega et al., 2014; Galdino-Alves et al., 2016).

Table 3. Biological activities from *Phaseolus vulgaris* peptides

Common bean (*Phaseolus vulgaris*)	Biological effect	Reference
Germinated	Antihypertensive	Mamilla and Mishra 2017
	Reduce risk of type-2 diabetes	de Souza Rocha et al., 2015
	Anti-inflammatory	López-Barrios, Antunes-Ricardo and Gutiérrez-Uribe, 2016
Cooked	Anticancer Antihypertensive	Luna-Vital, González de Mejía, and Loarca-Piña 2016
	Anticancer	Luna-Vital, González de Mejía, and Loarca-Piña 2017

A study with bean sprouts showed that the formation of low molecular weight bioactive peptides during germination increased the antihypertensive activity of beans due to the activity of proteinases (Mamilla and Mishra, 2017). Likewise, de Souza Rocha et al., (2015) observed that the germination of Phaseolus vulgaris promoted protein hydrolysis, generating bioactive peptides that inhibit α-amylase in such a way that it blocks the digestion of starch and decreases the intestinal absorption of glucose which can reduce the risk of type 2 diabetes. Furthermore, López-Barrios et al., (2016) reported that during the 1-5 days of germination of black beans, the content of amino acids such as threonine, proline, and methionine increased, as well as the activity of absorbing oxygen radicals. They also observed anti-inflammatory activity by inhibiting nitric oxide production in macrophages, which was attributed to peptides derived from germination (López-Barrios et al., 2016).

Additionally, Luna-Vital, González de Mejía, and Loarca-Piña (2016) analyzed dietary peptides from the non-digestible fraction of common bean (GLSTK and GEGSGA); they demonstrated the inhibition of the proliferation of the cell line HCT-116 and reduced the angiotensin-II-dependent STAT3 activation. Likewise, Luna-Vital, González de Mejía, and Loarca-Piña (2017) studied peptides from the non-digestible fraction of *Phaseolus vulgaris* cv 'Azufrado Higuera' in a model of azoxymethane/dextran sodium sulfate-induced colitis-associated colon carcinogenesis in Balb/c mice and found that the peptide fraction reduced the activity index of the disease and the formation of neoplasms.

Conclusion

Phaseolus vulgaris is a legume with privileged chemical composition as it has an important content of carbohydrates, proteins, lipids, vitamins, and minerals that provide health benefits. Likewise, when subjected to specific processes, the secondary metabolites that have no direct relationship with the development of the plant will be present and will be concentrated depending on the treatment applied, such as germination, soaking, and cooking, among others. In such a way phenolic compounds and bioactive peptides may exert biological activities such as anticancer, antihypertensive, antioxidant, anti-inflammatory, and lipid-lowering to prevent and treat chronic non-communicable diseases. Therefore, it can be concluded that in vitro and in vivo evaluations and clinical trials could give more accurate answers about how each processing supports the seed to increase secondary metabolites and the mechanism of action with which these compounds exert those biological activities, to promote greater consumption.

References

Abdulrahman, Balkisu O., Bala, Muntari, and Bello Oluwasesan M. (2021). Bioactive Compounds of Black Bean (*Phaseolus vulgaris* L.). In: *Bioactive Compounds in Underutilized Vegetables and Legumes*. H.N. Murthy and K. Y. Paek (Eds.). Springer International Publishing. pp.623-641. DOI:10.1007/978-3-030-57415-4_38.

Alcázar-Valle Monserrat, Lugo-Cervantes Eugenia, Mojica Luis, Morales-Hernández Norma, Reyes-Ramírez Heydi, Enríquez-Vara Jhony Navat, and García-Morales Soledad. (2020). Bioactive Compounds, Antioxidant Activity, and Antinutritional Content of Legumes: A Comparison between four *Phaseolus* Species. *Molecules* (Basel, Switzerland), 25(15), 3528. DOI:10.3390/molecules25153528.

Alvarado-López Alejandra Nabil, Gómez-Oliván Leonardo Manuel, Heredia J. Basilio., Baeza-Jiménez Ramiro, Garcia-Galindo H ugo Sergio, and Lopez-Martinez Leticia Xochitl. (2019). Nutritional and bioactive characteristics of Ayocote bean (*Phaseolus coccienus* L.): An underutilized legume harvested in Mexico. *CyTA – Journal of Food*, 17(1), 199-206. DOI:10.1080/19476337.2019.1571530.

Ariza-Ortega Teresita de Jesús., Zenón-Briones Elia Yadet, Castrejón-Flores, José Luis, Yáñez-Fernández Jorge., Gómez-Gómez Yolanda de las Mercedes, and Oliver-Salvador María del Carmen (2014). Angiotensin-I-converting enzyme inhibitory, antimicrobial, and antioxidant effect of bioactive peptides obtained from different varieties of common beans (*Phaseolus vulgaris* L.) with in vivo antihypertensive activity in spontaneously hypertensive rats. *European Food Research and Technology*, 239(5), 785-794. DOI:10.1007/s00217-014-2271-3.

Chávez-Mendoza Celia and Sánchez Esteban. (2017). Bioactive Compounds from Mexican Varieties of the Common Bean (*Phaseolus vulgaris*): Implications for Health. *Molecules*, 22(8). DOI:10.3390 /molecules22081360.

Chigwedere Claire Maria, Olaoye Taye Foyeke, Kyomugasho Clare, Jamsazzadeh Kermani, Zahara., Pallares Pallares Andrea,Van Loey Ann M., Grauwet Tara and Hendrickx Mark. E. (2018). Mechanistic insight into softening of Canadian wonder common beans (*Phaseolus vulgaris*) during cooking. *Food Research International*, 106,522-531. DOI:10.1016/j.foodres.2018.01.016.

Cid-Gallegos María Stephanie, Corzo-Ríos Luis Jorge, Jiménez-Martínez Cristian, Sánchez-Chino Xariss Miryam. (2022). Protease Inhibitors from Plants as Therapeutic Agents- A Review. *Plant Foods for Human Nutrition* 77, 20–29. DOI:10.1007/s11130-022-00949-4.

Conti María Victoria, Guzzetti Lorenzo, Panzeri Davide, De Giuseppe Rachele, Coccetti Paola, Labra Massimo, and Cena Hellas. (2021). Bioactive compounds in legumes: Implications for sustainable nutrition and health in the elderly population. *Trends in Food Science & Technology*, 117,139-147. DOI:10.1016/j.tifs.2021.02.072.

Corzo-Ríos Luis Jorge, Sánchez-Chino Xariss Miryam, Cardador-Martínez Anaberta, Martínez-Herrera Jorge, Jiménez-Martínez Cristian.(2020). Effect of cooking on nutritional and non-nutritional compounds in two species of Phaseolus (*P. vulgaris* and *P. coccineus*) cultivated in Mexico. *International Journal of Gastronomy and Food Science*, 20. 100206. DOI:10.1016/j.ijgfs.2020.100206.

de Souza Rocha Thaís., Real Hernandez Luis Manuel, Mojica Luis, Johnson Michelle H, Chang Yoon Kil, and González de Mejía Elvira. (2015). Germination of *Phaseolus vulgaris* and alcalase hydrolysis of its proteins produced bioactive peptides capable of improving markers related to type-2 diabetes invitro. *Food Research International*, 76, 150-159. DOI:10.1016/j.foodres.2015.04.041.

Dona Anthony C., Pages Guilhem, Gilbert Robert G. and Kuchel Philip W. (2010). Digestion of starch: In vivo and in vitro kinetic models used to characterise oligosaccharide or glucose release. *Carbohydrate Polymers*, 80(3), 599-617. DOI:10.1016/j.carbpol.2010.01.002.

Dueñas Monserrat, Sarmento Thaise, Aguilera Yolanda, Benitez Vanesa., Mollá Esperanza, Esteban Rosa María, and Martín-Cabrejas María A. (2016). Impact of cooking and germination on phenolic composition and dietary fibre fractions in dark beans (*Phaseolus vulgaris* L.) and lentils (*Lens culinaris* L.). *LWT-Food Science and Technology*, 66, 72-78. DOI:10.1016/j.lwt.2015.10.025.

Elizabeth Leonie, Machado Priscila, Zinöcker Marit, Baker Phillip, and Lawrence Mark. (2020). Ultra-Processed Foods and Health Outcomes: A Narrative Review. *Nutrients*, 12(7). DOI:10.3390/nu12071955.

Fardet Anthony and Rock Edmond. (2019). Ultra-processed foods: A new holistic paradigm?. *Trends in Food Science & Technology*, 93, 174-184. https://doi.org/10.1016/j.tifs.2019.09.016.

Ferreira Tavares Leticia, Costa-Fonseca Sandra, Garcia Rosa María Luisa, and Massae Yokoo Edna (2012). Relationship between ultra-processed foods and metabolic syndrome in adolescents from a Brazilian Family Doctor Program. *Public Health Nutrition*, 15(1), 82-87. DOI:10.1017/s1368980011001571.

Galdino Alves Natália Elizabeth, Gonzalez de Mejía Elvira, Mileib Vasconcelos Christiane, Zaczuk Bassinello Priscila, and Duarte Martino Hércia Stampini S. (2016). Postharvest storage of Carioca bean (*Phaseolus vulgaris* L.) did not impair inhibition of inflammation in lipopolysaccharide-induced human THP-1macrophage-like cells. *Journal of Functional Foods*, 23, 154-166. DOI:10.1016/j.jff.2016.02.029.

Gallegos Tintoré Santiago, Chel Guerrero Luis, Corzo Ríos Luis Jorge, Martínez Ayala Alma Leticia. (2013). Péptidos con actividad antioxidante de proteínas vegetales [Peptides with antioxidant activity from plant proteins]. En *Bioactividad de péptidos derivados de proteínas alimentarias.* M. Segura Campos, L. Chel Guerrero and D. Betancur Ancona (Eds.), Barcelona: OmniaScience, pp. 111-122). DOI:10.3926/oms.94.

Gan Ren-You, Lui Wing-Yee, Wu Kao, Chan Chak-Lun, Dai Shu-Hong, Sui Zhong Qian and Corke Harold. (2017).Bioactive compounds and bioactivities of germinated edible seeds and sprouts: An updated review. *Trends in Food Science & Technology*, 59, 1-14. https://doi.org/10.1016/j.tifs.2016.11.010.

Hayat Imran, Ahmad Asif, Masud Tariq, Ahmed Anwaar, and Bashir Shaukat. (2014). Nutritional and health perspectives of beans (*Phaseolus vulgaris* L.): an overview. *Critical Reviews in Food Science and Nutrition*, 54(5), 580-592. DOI:10.1080/10408398.2011.596639.

Kahleova Hana, Levin Susan and Barnard Neal. (2017). Cardio-Metabolic Benefits of Plant-Based Diets. *Nutrients*, 9(8). DOI:10.3390/nu9080848.

Lawrence Mark A., and Baker Phillip I. (2019). Ultra-processed food and adverse health outcomes. *BMJ*, 365, l2289. DOI:10.1136/bmj.l2289.

Liu Beiyu, Peng Jiaying, Zhang Shurong, Zou Boyu and Zhong Geng. (2013). Chemical composition, in vitro starch digestibility and amino acid analysis of a underexplored kidney bean (*Phaseolus vulgaris* L.) grown in Chongqing, China. *International Journal of Food Science & Technology*, 48(3), 527-532. DOI:10.1111/j.1365-2621.2012.03214.x.

López Ana, El-Naggar Tarek, Dueñas Monserrat, Ortega Teresa, Estrella Isabel, Hernández Teresa, Gómez-Serranillos María Pilar, Palomino Olga M., and Carretero María Emilia. (2013). Effect of cooking and germination on phenolic composition and biological properties of dark beans (*Phaseolus vulgaris* L.). *Food Chemistry*, 138(1), 547-555. DOI:10.1016/j.foodchem.2012.10.107.

López-Barrios Lidia, Antunes-Ricardo Marilena, and Gutiérrez-Uribe Janeth A. (2016). Changes in antioxidant and antiinflammatory activity of black bean (*Phaseolus vulgaris* L.) protein isolates due to germination and enzymatic digestion. *Food Chemistry*, 203, 417-424. DOI:10.1016/j.foodchem.2016.02.048.

López-Martínez Leticia X., Leyva-López Nayely., Gutiérrez-Grijalva Erick P., and Heredia J. Basilio. (2017). Effect of cooking and germination on bioactive compounds in pulses and their health benefits. *Journal of Functional Foods*, 38, 624-634. DOI:10.1016/j.jff.2017.03.002.

Los Francine Gomes Basso, Zielinski Ferreira Acácio Antonio, Wojeicchowski José Pedro, Nogueira Alessandro, Demiate Ivo Mottin. (2018). Beans (*Phaseolus vulgaris* L.): whole seeds with complex chemical composition. *Current Opinion in Food Science*, 19,63-71. DOI:10.1016/j.cofs.2018.01.010.

Luna-Vital Diego A., González de Mejía Elvira, and Loarca-Piña Guadalupe. (2016). Selective mechanism of action of dietary peptides from common bean on HCT116 human colorectal cancer cells through loss of mitochondrial membrane potential and DNA damage. *Journal of Functional Foods*, 23, 24-39. DOI:10.1016/j.jff.2016.02.021.

Luna-Vital Diego A., González de Mejía Elvira, and Loarca-Piña Guadalupe. (2017). Dietary Peptides from *Phaseolus vulgaris* L. Reduced AOM/DSS-Induced Colitis-Associated Colon Carcinogenesis in Balb/c Mice. *Plant Foods for Human Nutrition*, 72(4), 445-447. DOI:10.1007/s11130-017-0633-2.

Luna-Vital Diego A., Mojica Luis, González de Mejía Elvira, Mendoza Sandra, and Loarca-Piña Guadalupe. (2015). Biological potential of protein hydrolysates and peptides from common bean (*Phaseolus vulgaris* L.): A review. *Food Research International*, 76, 39-50. DOI:10.1016/j.foodres.2014.11.024.

Mamilla Ravi Kumar and Mishra Vijay Kumar. (2017). Effect of germination on antioxidant and ACE inhibitory activities of legumes. *LWT-Food Science and Technology*, 75, 51-58. DOI:10.1016/j.lwt.2016.08.036.

Méndez-Flores, O.G., Ochoa-Díaz López, H., Castro-Quezada, I., Olivo-Vidal, Z. E., García-Miranda, R., Rodríguez-Robles, U., Irecta-Nájera, C. A., López-Ramírez, G., and Sánchez-Chino, X. M. (2021). The Milpa as A Supplier of Bioactive Compounds: A Review. *Food Reviews International*, 1-18. DOI:10.1080/87559129.2021.1934001.

Mendoza-Sánchez Magdalena, Pérez-Ramírez Iza F., Wall-Medrano Abraham, Martinez-Gonzalez Alejandra I., Gallegos-Corona Marco A., and Reynoso-Camacho Rosalía. (2019). Chemically induced common bean (*Phaseolus vulgaris* L.) sprouts ameliorate dyslipidemia by lipid intestinal absorption inhibition. *Journal of Functional Foods*, 52, 54-62. DOI:10.1016/j.jff.2018.10.032.

Monteiro Carlos A., and Cannon Geoffrey J. (2019). The role of the transnational ultra-processed food industry in the pandemic of obesity and its associated diseases: problems and solutions. *World Nutrition*, 10(1), 89 99. DOI:10.26596/wn.201910189-99.

Moreno-García Kathia L., Antunes-Ricardo Marilena, Martínez-Ávila Mariana, Milán-Carrillo, Jorge, and Guajardo-Flores Daniel. (2022). Evaluation of the antioxidant, anti-inflammatory and antihyperglycemic activities of black bean (*Phaseolus vulgaris* L.) by-product extracts obtained by supercritical CO2. *The Journal of Supercritical Fluids*, 183, 105560. DOI:10.1016/j.supflu.2022.105560.

Muzquiz Mercedes, Varela Alejandro, Burbano Carmen, Cuadrado Carmen, Guillamón Eva, and Pedrosa Martín Mercedes. (2012). Bioactive compounds in legumes: pronutritive and antinutritive actions. Implications for nutrition and health. *Phytochemistry Reviews*, 11(2), 227-244. DOI:10.1007/ s11101-012-9233-9.

Nicolás-García Mayra, Perucini-Avendaño Madeleine, Jiménez-Martínez Cristian, Perea-Flores María Jesús, Gómez-Patiño Mayra B., Arrieta-Báez Daniel and Dávila-Ortiz Gloria. (2021). Bean phenolic compound changes during processing: Chemical interactions and identification. *Journal of Food Science*, 86(3), 643-655. DOI:10.1111/1750-3841.15632.

Phan Minh Amh Thu., Paterson Janet, Bucknall Martín, and Arcot Jayashree. (2018). Interactions between phytochemicals from fruits and vegetables: Effects on bioactivities and bioavailability. *Critical Reviews in Food Science and Nutrition*, 58(8), 1310-1329. DOI:10.1080/10408398.2016.1254595.

Poti Jennifer M., Braga Bianca, and Qin Bo. (2017). Ultra-processed Food Intake and Obesity: What Really Matters for Health-Processing or Nutrient Content? *Current obesity reports*, 6(4),420-431. DOI:10.1007/ s13679-017-0285-4.

Rajaram Sujatha, Jones Julie and Lee Grace J. (2019). Plant-Based Dietary Patterns, Plant Foods, and Age-Related Cognitive Decline. *Advances in Nutrition* (Bethesda, Md.), 10(4), S422-s436. DOI:10.1093/ advances/nmz081.

Ranilla Lena G., Genovese María Inés and Lajolo Franco M. (2009). Effect of different cooking conditions on phenolic compounds and antioxidant capacity of some selected Brazilian bean (*Phaseolus vulgaris* L.) cultivars. *Journal of Agricultural and Food Chemistry*, 57(13), 5734-5742. DOI:10.1021/jf900527v.

Sánchez-Arteaga, H.M., Urías-Silvas, J. E., Espinosa-Andrews, H., and García-Márquez, E. (2015). Effect of chemical composition and thermal properties on the cooking quality of common beans (*Phaseolus vulgaris*). *CyTA - Journal of Food*, 13(3), 385-391. DOI:10.1080/19476337.2014.988182.

Satija Ambika and Hu Frank B. (2018). Plant-based diets and cardiovascular health. *Trends in Cardiovascular Medicine*, 28(7), 437-441. DOI:10.1016/j.tcm.2018.02.004.

Shimelis Emire Admassu and Rakshit Sudip Kumar. (2005). Proximate composition and physico-chemical properties of improved dry bean (*Phaseolus vulgaris* L.) varieties grown in Ethiopia. *LWT-Food Science and Technology*, 38(4),331-338. DOI:10.1016/j.lwt.2004.07.002.

Singh Balwinder, Singh Jatinder Pal, Kaur Amritpal and Singh Narpinder. (2017). Phenolic composition and antioxidant potential of grain legume seeds: A review. *Food Research International*, 101, 1-16. DOI:10.1016/j.foodres.2017.09.026.

Tang Chuang He, and Sun Xin. (2011). A comparative study of physicochemical and conformational properties in three vicilins from *Phaseolus* legumes: Implications for the structure–function relationship. *Food Hydrocolloids*, 25(3), 315-324. DOI:10.1016/j.foodhyd.2010.06.009.

Tharanathan, R.N., and Mahadevamma, S. (2003). Grain legumes—a boon to human nutrition. *Trends in Food Science & Technology*, 14(12), 507-518. DOI:10.1016/j.tifs.2003.07.002.

Xu Baojun and Chang Sam K. C. (2012). Comparative study on antiproliferation properties and cellular antioxidant activities of commonly consumed food legumes against nine human cancer cell lines. *Food Chemistry*, 134(3), 1287-1296. https://doi.org/10.1016/j.foodchem.2012.02.212.

Xue Zhaohui, Wang Cen, Zhai Lijuan, Yu Wancong, Chang Huiru, Kou Xiaohong, and Zhou Fengjuan. (2016). Bioactive compounds and antioxidant activity of mung bean (*Vigna radiata* L.), soybean (*Glycine max* L.) and black bean (*Phaseolus vulgaris* L.) during the germination process. Czech *Journal of Food Science*, 34(1), 68-78. DOI:10.17221/434/2015-CJFS.

Chapter 12

Quinoa Grain Byproducts: Their Valuable Components and Applications

Anayansi Escalante-Aburto[1]
Nestor Ponce-García[2,*]
and Rosa María Mariscal-Moreno[3,†]
[1]Tecnologico de Monterrey, The Institute for Obesity Research, Monterrey, Nuevo Leon, Mexico
[2]Faculty of Agricultural Sciences, UAEMex, Toluca, Estado de Mexico
[3]Health Department, Universidad Iberoamericana, Mexico City, Mexico

Abstract

Food industries produced many waste products each year, and it causes serious environmental problems that result in economic losses and social impact. However, some studies have reported that many wastes have potential and valuable bioactive compounds. These biomaterials have a high potential for food additives, cosmetic products, pharmaceutical applications, and biomass that can be beneficial in terms of ensuring global food sustainability.

The quinoa industry is not the exception to waste production; however, in comparison with the studies from other pseudo-cereals, the quinoa industry needs more attention from the scientific community because the information from byproducts from quinoa is not widely studied, and many of the processes that have been proposed for recovering bioactive compounds from quinoa are not industrially scalable. Additionally, the compounds obtained from the quinoa process are generally discarded as waste or used in animal feed. However, as described in this chapter, compounds obtained from quinoa processing can be potentially valuable bioactive compounds. The main uses of quinoa byproducts are presented and discussed, and the mainstream sectors of applications for these byproducts, such as in the pharmaceutical, health, food, and cosmetic industries, are highlighted.

Keywords: quinoa byproducts, quinoa saponins, bioactive compounds, biomass, applications

---

* Corresponding Author's Email: nponceg@uaemex.mx.
† Corresponding Author's Email: rosa.mariscal@ibero.mx.

In: Food Byproducts
Editors: S. Teresita Martín-del-Campo, A. Cardador-Martínez and J. del Pilar Ramírez-Anaya
ISBN: 979-8-88697-612-0

Introduction

Quinoa (*Chenopodium quinoa* Willd.) is the Amaranthaceae family's summer annual dicotyledonous herbaceous crop. It was domesticated first in the Andean countries of South America about 7000 years ago (Pathan and Siddiqui 2022). The Andean people began the cultivation of quinoa maybe for its nutritional values, drought tolerance, and ability to grow in high salt conditions. The Incas considered quinoa a sacred crop and called it the "mother grain". Although overlooked for thousands of years, this crop's agronomic and nutritional importance was rediscovered during the last 50 years, leading to a resurgence in its production (Pathan and Siddiqui 2022).

The quinoa seed is as small as a millet seed, flat and oval-shaped, with colors ranging from dark red to pale yellow. Although quinoa is called a complete food for several reasons, Wu (2015) reported that protein and lipid contents in quinoa are higher than in other cereals; protein accounts for about 16.5% and lipids 6.3%. Carbohydrates are the primary component in quinoa and comprise around 69%, and the fiber and ash content are around 3.8%. Additionally, quinoa contains many essential amino acids, especially lysine, tryptophan, cysteine, vitamins (vitamin E, B, C), and minerals such as calcium, iron, manganese, magnesium, and copper, and potassium compared to most cereal grains (Dakhili et al. 2019). It is also rich in antioxidant compounds such as polyphenols (Repo-Carrasco-Valencia and Serna 2011).

Usually, quinoa is consumed as whole grain, and another portion is submitted to different processes, mainly to increase palatability and sensory acceptance. Methods applied to quinoa are usually: milling, cooking, extrusion, toasted, and microwave heating.

Quinoa processing generates many byproducts, causing environmental disposal problems and other additional outgoings. However, many of those byproducts have promising potential to generate food additives, cosmetic products, and pharmaceutical applications that can be beneficial in terms of ensuring global food sustainability.

Additionally, processing also provides good nutritional alternatives byproducts for animal feeding and applications in the pharmaceutical industry and biomass production. Quinoa has been evaluated as a source of dietary fiber to be included in food products for celiacs. Considering the concerns about improving the sustainability of the food industry and environmental problems involved in managing the residues, a growing interest in valorizing these agri-food byproducts to obtain valuable compounds has increased significantly.

This chapter reports the outcomes of current research regarding the byproducts obtained during quinoa processing based on identifying their most valuable constituents.

Chemical Compounds in Quinoa Wastes and Their Exploitation

Proteins

Increasing the utilization of plant-based proteins in the human diet has been touted as one of the most promising approaches to support the consumer demands for protein-rich. Quinoa is a pseudo-cereal that has been recognized for its high protein nutritional value and has more protein content than the majority of cereals, varying from around 12.8 to 16.5%, and it has well-balanced essential amino acids content, similar to casein and milk (Kuktaite et al. 2021), of which the major protein components are globulins (37%) and albumins (35%) (Dakhili et al. 2019). Moreover, the *in vitro* digestibility of quinoa protein is reported to be between 75.9% and 84.8%, comparable to other plant proteins such as rice, corn, and beans, though lower than animal protein digestibility (Shi et al. 2019). Quinoa proteins contain more histidine than wheat and rice proteins but are comparable to that of corn. While the sulfur and aromatic amino acids in quinoa are adequate to meet the needs of young children and adults, they are present in quantities lower than those of wheat, corn, and rice. According to the WHO/FAO/UNU scoring patterns for young children and adults, quinoa proteins contain sufficient amounts of tryptophan, isoleucine, cysteine, methionine, phenylalanine, tyrosine, and histidine (Scanlin and Lewis 2017). Vega-Gálvez et al. (2010) reported that quinoa protein could supply over 180% of the requirement of histidine estimated by WHO/FAO/UNU, 274% for isoleucine, 338% for lysine, 212% for methionine + cysteine, 320% for phenylalanine + tyrosine, 331% for threonine, 228% for tryptophan and 323% for valine. Quinoa protein also has good performance in the functionality characteristics, including foam stability, emulsion stability, water/oil holding capacity, gelation property, and film-forming ability (Shen, Tang, and Li 2021).

Waste generated by agro-industries can be a rich source of valuable compounds, among which are proteins, and therefore could become a sustainable alternative to reduce malnutrition and hunger in developing countries (Montesano et al. 2020). Depending on the sequence of amino acids, these peptides can exhibit immunomodulatory, antimicrobial, antioxidant, anti-thrombotic, hypocholesterolemic, and anti-hypertensive actions (Montesano et al. 2020).

Most quinoa is usually consumed as whole seeds (Vega-Gálvez et al. 2010). However, quinoa protein concentrate or isolated from industrial waste could have applications as food ingredients (Scanlin and Lewis 2017); even though they are not yet commercially produced because there are some limitations in extraction methods proposed at the laboratory, they are not economically feasible. Also, research techniques focused on protein may degree other ingredients such as starch, fiber, and oil, which would not be practical for commercial purposes. Some of the starting materials for protein extraction include (1) whole seed flour; (2) defatted flour; (3) protein-rich bran; and (4) defatted, protein-rich bran.

On the other hand, a commercially available byproduct of popped quinoa, toasted protein-rich bran, is not optimal since the EAA (essential amino acids) lysine is degraded with Maillard browning (Scanlin and Lewis 2017). Accordingly, with Scanlin and Lewis (2017), the protein-rich brand can be obtained 84% of protein recovery.

In addition, bioactive quinoa protein hydrolysate may be a potential plant-based alternative to commercially available animal peptides.

Protein as a byproduct from quinoa has not been widely studied, more research is needed, and new laboratory processes to isolate protein from quinoa are required to provide new opportunities for quinoa waste.

Peptides

An important factor determining the nutritional quality of food proteins is the potential regulatory activity of the bioactive peptides contained in the amino acid sequence. Recent studies have shown that some peptides contained within the primary sequence of proteins, released by *in vitro* enzymatic hydrolysis during food production processes or during *in vivo* digestion, can exert critical biological functions (Sharma, Singh and Rana 2011), as they can bind to receptors belonging to cells implicated in specific metabolic processes.

Quinoa protein presents a great potential to obtain bioactive peptides with multifunctional properties, such as anti-diabetic, anti-hypertensive, anti-inflammatory, and antioxidant activity, making it suitable for applications in functional foods (Guo et al. 2021). In addition, solubility and bioactive properties of protein hydrolysate have been shown to increase relative to QPC (quinoa protein concentrate) and may be a potential plant-based alternative to commercially available animal peptides (Aluko and Monu 2003).

The functional properties of proteins are connected to the physicochemical properties, which govern protein behavior in foods. For example, emulsifying capacity (EC) and emulsion stability (ES) are two important functional characteristics of proteins that affect the behavior of various industrial products, including adhesives, cosmetics, and packing material (Httiarachy and Kalapathy 1998).

Quinoa seeds germination facilitates the natural release of peptides and the fermentation process during quinoa drought and yogurt beverages (Guo et al. 2021). However, overall, the peptide composition has not been profiled for most of the bioactive hydrolysates from quinoa protein, and extensive studies are required further to identify peptide sequences responsible for the observed biological effect.

Food protein hydrolysates are a good candidate to produce anti-diabetic peptides, which may affect the regulation of blood glucose levels by multiple mechanisms (Guo et al. 2021).

In the study by Zheng et al. (2019), a novel ACE (angiotensin-I converting enzyme) inhibitory peptide RGQVIYVL derived from quinoa bran albumin has been identified, and oral administration of the peptide (100-150 mg❿kg^{-1} body weight) to spontaneously hypertensive rats resulted in significant decreases in both systolic and diastolic blood pressure.

Like protein hydrolysate, the industrial obtention of those products has been not feasible until nowadays because discovery and characterization of bioactive peptides are conventionally realized through a series of operations, including protein extraction and hydrolysis peptides purification, identification, and biological test (Guo et al. 2021).

Saponins

The name 'saponin' is derived from the Latin word *sapo*, which means 'soap' because saponin molecules form soap-like foams when shaken with water. They are structurally diverse molecules that are chemically referred to as triterpene and steroid glycosides. They consist of non-polar aglycones coupled with one or more monosaccharide moieties (Oleszek 2002). This combination of polar and non-polar structural elements in their molecules explains their soap-like behavior in aqueous solutions. Saponins are minor components in quinoa, broadly studied due to their biological properties. One of the essential steps is the removal of bitter saponins, which are located primarily in the pericarp, the outermost layer that is polished to reveal a smooth seed coat layer. The traditional method to remove those components involves manually washing with water on an abrasive stone until the outer layer or hull is removed (Quiroga et al. 2017). Saponins confer a bitter taste and are primarily found in the outer seed coat. Therefore, some procedures like cold water washing, abrasion, and dehulling are used to remove these compounds.

Saponins are still considered a waste byproduct by most quinoa processors. However, in a review by Ruiz et al. (2014), the authors critically evaluated the literature and found a substantial body of evidence in support of dietary saponins with positive biological activity such as (1) anti-inflammatory properties; (2) anti-carcinogenic traits; (3) antimicrobial activities; and (4) abilities to lower plasma cholesterol. Therefore, future commercial applications for saponins are possible.

In the pharmaceutical industry, SAPs (saponins) are used to synthesize hormones, contraceptives, anti-inflammatories, expectorants, and diuretics (Vincken et al. 2007; Augustin et al. 2011). In addition, SAPs have found wide applications in food and cosmetic production (Park, Maeda, and Morita 2005). SAPs have potential as biopesticides and growth promoters (Andresen and Cedergreen 2010; Andresen et al. 2015). For example, molluscicidal and vermicidal activity has been reported in complex mixtures of SAPs derivatives obtained from alkali-treated quinoa seed coats (San Martín, Ndjoko, and Hostettmann 2008) and SAPs-rich byproducts of tea oil production (Potter et al. 2010). Increased consumer demand for natural products with beneficial physicochemical (e.g., surfactant) and biological (e.g., biocidal, antimicrobial) properties makes steroidal and triterpenoid SAPs promising compounds for industrial applications.

The level of SAPs in quinoa is higher than most other species, and by considering quinoa SAPs as a waste product with commercial value, we can overcome the problems associated with forest depletion caused by the use of quillay (*Quillaja saponaria* Molina) SAPs and with the suboptimal rheological and biocidal properties of triterpene SAPs from other sources. The broad genotypic diversity of quinoa (Ruiz et al. 2014) and, consequently, the diversity in its SAPs composition suggests that optimizing the pharmacological, industrial, and agricultural uses of quinoa seed coat SAPs requires a broader and appropriate selection of genotypes (Ruiz et al. 2017).

Derivated Compounds from the Coat (Husk)

Quinoa grains contain high fiber amounts that enrich their nutritional properties, and their agro-wastes are used in material science and engineering. Cellulose, hemicellulose, and lignin are contained in the coats removed by dehulling process (Angeli et al. 2020). Synthesis of bio-nanocomposites of carbon nanotube/ZnO was studied to reutilize residues of quinoa plants. Extracts of natural cellulose fibers were obtained from quinoa wastes to reinforce multi-walled nanotubes, which have catalytic activities and high stabilities (Naeimi, Khoshkam, and Eslaminejad 2021).

Nano-cellulose obtained from quinoa husk has an innovative potential application for removing dyes from wastewaters. Quinoa husk was treated by enzymatic and acid hydrolysis to remove nanocellulose particles as carrier material to immobilized laccase capable of acting as biocatalyst (PersiLac1/NCs) and degrade malachite-green and congo-red dyes from water. This composite showed high efficiency, with superior reuse up to 18 consecutive runs during its application in real dye-polluted effluents (Ariaeenejad, Motamedi, and Salekdeh 2022).

Outer layers from six quinoa seed varieties were obtained by mechanical decortication and were analyzed by Ruiz et al. (2017). Methanol extracts (80% v/v) from coats were saponified by acid and basic hydrolysis and were analyzed by GC-MS and LC-MS/MS. Three essential compounds in aglycone forms were detected: oleanolic acid, hederagenin, and phytolaccagenic acid. Those compounds' relative abundance and concentrations strictly depended on the genotype, environmental and agronomic factors. Nevertheless, the phytochemicals that constituted the seed coat of quinoa could be used as bio-pesticides and in producing foods, feeds, and pharmaceuticals.

Antioxidants

Quinoa by-side products are also considered an excellent source of biomolecules with nutraceutical properties. Quinoa leaves contain higher amounts of micronutrients such as copper, manganese, and potassium than amaranth (Pathan et al. 2019). Besides, their total polyphenol (131.8 mg•100 g^{-1}) content and flavonoids concentration (62 mg⓾100 g^{-1}) are considerably higher to be considered an exciting source of antioxidants. Primary polyphenols identified in quinoa leaves are gallic acid, morin, kaempferol, and catechin (Vazquez-Luna et al. 2019).

Recent reviews on other bioactive (Fuentes and Paredes-Gonzalez 2015) reported naturally occurring phytochemicals such as phytosterols, betalains, squalene, phytoecdysteroids, and phenolics, which may have future nutraceutical applications. In addition, future studies on quinoa's role in treating and preventing complex human diseases are merited based on the mounting evidence on quinoa's health benefits.

Polyphenols

Ethanolic extracts of leaves were studied *in vitro* to evaluate their antioxidant and anti-carcinogenic potential by Gawlik-Dziki et al. (2013). A complete profile of quinoa leaf extracts was described, including qualitative and quantitative data. Results showed that quinoa leaf extracts contained high phenolic compounds with nutraceutical activity against rat prostate cancer proliferation in cellular lines (AT2 and MAT-LyLu) by a synergistic effect at intracellular levels.

Water effluents from quinoa transformation processes are a rich source of flavonoids, mainly quercetin and kaempferol glycosides. Phytoecdysteroids and flavonoid glycosides were reported in those effluents at 10.03 mg⓾g^{-1} and 25.89 mg⓾g^{-1}, respectively. Those compounds can reduce cellular inflammation due to their anti-diabetic properties (Jeong et al. 2012, Graf et al. 2014).

Applications and Uses

There is some available information that some studies have included agro-wastes from quinoa processing to produce byproducts. Applications and some of the compounds and biomolecules with interesting approaches are depicted in Figure 1. The number of investigations has increased in the last ten years (Skendi et al. 2020), the production of biofuels being the more extensive research field. More research will undoubtedly be conducted shortly to improve and optimize bioprocessing and recovery of compounds with beneficial impacts to increase sustainability issues.

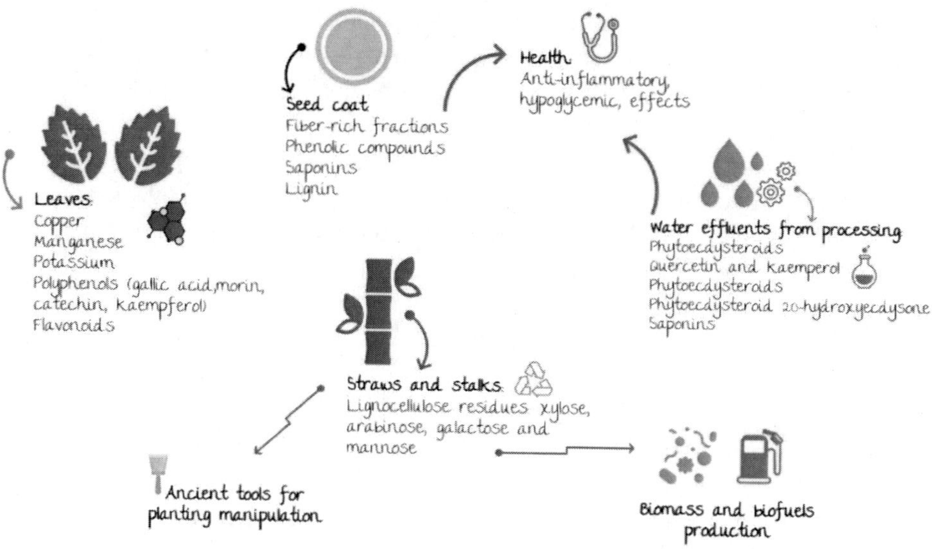

Figure 1. Alternative applications of agro-wastes and byproducts from quinoa.

Food and Feeds

During the transformation of quinoa, some co-products are discarded and have been considered food waste. In a study conducted by Fernández-López et al. (2020), the fiber-rich fraction obtained from black quinoa produced by the wet-milling process was analyzed. This fraction was incorporated in the formulation (2.5%) of Bologna-type sausages, and its effects during the shelf life at low temperatures were monitored. The main objective of the investigation was to reduce the incorporation of synthetic additives (avoiding high concentration of nitrates) and exploit the remaining fraction of quinoa wet milling. The results were promising, even when the microbiological evaluation did not increase the proliferation of microorganisms in the product. Moreover, the antioxidant compounds contained in the fiber residue could have been considered as influence factors to reduce microbial contamination, and above all, some molecules of the quinoa coat could be acting as natural nitrate sources. The latter will benefit consumers' health, reducing the number of additives used by the food industry.

In the same context, Ortiz-Gómez et al. (2022) characterized the main co-products obtained from quinoa wet milling. The processing factors such as alkali concentration (NaOH) and maceration temperature significantly affect the final starch and protein contents in the co-products. Therefore, the authors suggested that those starch and protein fractions could be used to develop functional formulations such as bakery products. First, however, further studies of the characterization of those fractions are needed (morphological, structural, thermal, and rheological properties, among others).

Cosmetics

Water effluents from quinoa processing are a rich source of biomolecules that could be used as ingredients for the production of personal use, such as shampoo, soaps, toothpaste, and organic insecticides, among others. To reutilize those compounds, Biver et al. (2021) evaluated a biological treatment to meet an environmental standard of the wastewater of the quinoa processing plant. Through different stages of sedimentation, aeration in a biological reactor, second sedimentation, and filtration, it was found that the water residues from this processing industry contained high concentrations of oils and fats, saponins, nitrogen, surfactants, and fecal coliforms. Besides, the chemical oxygen demand (COD) and the biochemical oxygen demand (BOD) in five years were extremely high, but its biodegradability index was 0.6, resulting in an easily degradable effluent.

Health

Some health benefits provided by wastes from side-stream processing of quinoa were found in compounds contained in the residual water during the primary stages of grain malting (germination). An important biomolecule with nutraceutical application, the phytoecdysteroid 20-hydroxyecdysone (20HE), was lixiviated into quinoa wastewater effluents during processing (Graf et al. 2014). Those authors optimized the quinoa leachate extraction and obtained a concentration of 491 μg⑩g^{-1} seed of this molecule. The extract of 20HE was evaluated to observe its hypoglycemic effect *in vivo*. A group of diet-induced obese and hyperglycemic male mice (C57BL/6J) were fed with 250 and 500 mg⑩kg^{-1} doses of quinoa leached. Results showed that biomolecules lixiviated on quinoa wastewater could have anti-diabetic properties to be considered during the treatment of non-communicable diseases and the formulation of natural pharmaceutical products.

Extracts from quinoa waste (coat) presented acute anti-inflammatory effects, according to a study conducted by Lozano et al. (2013). HPLC-DAD was used to identify the compounds of the sapogenin obtained by acid hydrolysis of saponin extracts of quinoa waste. Oleanolic acid, methyl olenate, hederagenin, and phytolaccagenic acid were the major constituents of the sapogenin aliquot. The authors concluded that synergistic effects among biomolecules of the sapogenin extract significantly decrease the inflammation in the croton-induced ear edema model. On the contrary, in the carrageenan-induced paw model, the isolated sapogenins showed a significantly better response against inflammation than the whole sapogenins extract.

Biomass Utilization to Produce Biofuels

South American countries, especially those of the Andean areas, have increased their quinoa production to supply worldwide demand (Jacobsen 2011). Peru is one of the countries that produce this crop in high amounts. However, several land-use changes have emerged, causing socio-economical issues such as displacement, rebound, and cascade effects, compromising the region's sustainability (Bedoya-Perales et al. 2018).

Carrasco et al. (2015) identified quinoa stalks as potential sources of lignocellulose residues. In addition, those authors identified four principal sugars in the hemicellulose fraction: xylose, arabinose, galactose, and mannose. The study's main objective was to evaluate the SO_2 steam pretreatment and enzymatic hydrolysis to obtain monosaccharides for fermentation purposes. Acceptable sugar yields were obtained; however, compared with other agro-wastes such as sugarcane residues and wheat straw, quinoa stalks were not considered the best option to obtain these compounds.

The utilization of quinoa residues has been evaluated to increase the biodegradability of slaughter wastes and produce biogas. Meneses-Quelal et al. (2021a) studied the synergy between mixtures of slaughterhouse residues and agriculture wastes from quinoa and wheat used as inoculum sewage sludge to improve efficient management in the meat industry. The use of quinoa waste as a substrate for co-digestion significantly increased the biodegradation of organic material contained in the agro-industrial byproducts. Biogas production was higher than that produced with amaranth and wheat residues, which increases the high value of quinoa processing to be considered an alternative raw material to generate biofuels.

A similar application was studied by Alarcon et al. (2017). They produced pellets from agricultural biomass obtained from quinoa and other seeds and evaluated their energetic properties. An optimum mixture of 30% starch with 70% quinoa ashes allowed the production of pellets with high mechanical durability (abrasion index) and increased calorific value. As a result, farmers in the Puno region from Peru could easily use these pellets to produce energy.

Quinoa wastes such as stems, leaves, bushes, and husks were used to produce biomass at low costs and were evaluated in a preliminary study conducted by Suárez-Rivero et al. (2019). Pellets with four different compositions of those byproducts were obtained, and their physicochemical and mechanical characteristics were evaluated. The raw materials analyzed maintained some of their characteristics that could increase the quinoa production chain's value and promote the circular economy in regions where this pseudo-cereal is produced in high amounts. Some authors' perspectives included using agricultural wastes to elaborate a high nutritional animal supplement, source of cellulose and hemicellulose in the paper industry, and biomass to produce biofuels.

The use of quinoa stalks and their potential to produce energy was reported recently by Filik (2020). There were two exciting approaches to exploiting this material; the first consisted of producing roughage for animal consumption. Unfortunately, the results showed that these wastes could not be recommended for animal feed due to their poor nutritional content, evaluated as *in vitro* digestibility and metabolizable energy value. The second approach, on the contrary, was successfully evaluated since quinoa stalks showed an electrical energy production of 1,000 kW_{el} approximately. Therefore, the author concluded that according to the classification of the FNR (Fachagentur Nachwachsende Rohstoffe), this waste could be used for a model VIII energy plan facility.

Meneses-Quelal et al. (2021b) evaluated the production of methane and biogas obtained by anaerobic digestion of quinoa waste, manure, and slaughterhouse residues. This problem has increased in the Andean region, and sustainable processes have been explored and assessed to resolve those environmental issues. Two kinetic models were used to predict the performance of the process, showing high regression coefficients ($r^2 > 0.0998$) in the Gompertz model. In addition, the bio-digestion of quinoa straw produced the highest methane production rate at 1:2 substrate-inoculum ratios, showing excellent biodegradability performance.

An integrated and optimized extraction scheme to extract saponins, xylan, and cellulose from quinoa stalks to produce feedstock was evaluated by Gil-Ramirez et al. (2018). Temperature and extraction times were the principal factors to assess the saponin extraction, an alkaline method was utilized to obtain xylan, and finally, cellulose was extracted by purification with acetic and nitric acid. The recovery yields of the compounds were 15.4 mg·g^{-1} raw material, 120 mg·g^{-1} raw material, and 296 mg·g^{-1} raw material for saponin, xylan, and cellulose, respectively. The authors concluded that quinoa stalks are a low-value, highly available agricultural agro-waste to be considered feedstock for bio-refinery.

Quinoa stalks are considered an alternative source of lignin, cellulose, and hemicellulose. Xylans are the major hemicellulose component and can be obtained from quinoa stalks by several extraction methods. These compounds have different applications in chemical, food, and industrial areas. Salas-Veizaga et al. (2017) evaluated the efficiency of obtaining glucuronoarabinoxylan from quinoa stalks using alkaline extraction and recovery of two-thirds of the xylan. They concluded that using enzymes with different specificities and ratios (glycoside hydrolases) could contribute to comprehending the substitution pattern of the polymer.

Ancient Tools

This particular use of quinoa wastes (Babot and Hocsman 2014) was reported in the western location of Argentina. Ancient inhabitants of this region used quinoa wastes of seeds and chunks of both panicles and stalks to construct tools that probably assisted the ancestors in planting manipulation.

Conclusion

It has been documented that the quinoa byproducts obtained from the seed coat, water waste, and whole grain contain various health-enhancing substances (saponins, fiber, peptides, and phenolic compounds). Quinoa byproducts have attracted scientific interest because of the different industrial applications that those compounds have. Byproducts from quinoa have potential applications in the pharmaceutical industry (hormones, contraceptives, anti-inflammatories, expectorants, and diuretics), food industry (bakery, sausage, reduction of additives, addition of fiber), cosmetic industry, biomaterials industry (nanotubes), and they also have demonstrated applications to improve health (anti-diabetic, anti-hypertensive, anti-tumoral compounds), in water waste treatment and biopesticide production. The recovery and evaluation of its efficiency are important challenges for scientists and the evaluation of sensorial and nutritional aspects. Those byproducts could contribute to more sustainable production in the industries mentioned above, providing economic benefits and reducing environmental problems.

References

Alarcon, M., C. Santos, M. Cevallos, R. Eyzaguirre, and S. Ponce. 2017. "Study of the mechanical and energetic properties of pellets produce from agricultural biomass of quinoa, beans, oat, cattail and wheat." *Waste and Biomass Valorization* 8:2881–2888. doi: 10.1007/s12649-017-9983-0.

Aluko, R. E., and E. Monu. 2003. "Functional and bioactive properties of quinoa seed protein hydrolysates." *Journal of Food Science* 68(4):1254-1258. doi: 10.1111/j.1365-2621.2003.tb09635.x.

Andresen, M., and N. Cadergreen. 2010. "Plant growth is stimulated by tea-seed extract: A new natural growth regulator?" *HortScience* 45(12):1848-1853. doi: 10.21273/HORTSCI.45.12.1848.

Andresen, M., E. G. Wulff, E. R. Mbega, M. S. Stokholm, S. E. Glazowska, P. E. Zida, R. B. Mabagala, and O. S. Lund. 2015. "Seed treatment with an aqueous extract of agave sisalana improves seed health and seedling growth of sorghum." *European Journal of Plant Pathology* 141:119-132. doi: 10.1007/s10658-014-0530-6.

Angeli, V., P. M. Silva, D. C. Massuela, M. W. Khan, A. Hamar, F. Khajehei, S. Graeff-Hönninger, and C. Piatti. 2020. "Quinoa (*Chenopodium quinoa* Willd.): An overview of the potentials of the 'Golden Grain' and socio-economic and environmental aspects of its cultivation and marketization." *Foods* 9(2):216. doi: 10.3390/foods9020216.

Ariaeenejad, S., E. Motamedi, and G. H. Salekdeh. 2022. "Highly efficient removal of dyes from wastewater using nanocellulose from quinoa husk as a carrier for immobilization of laccase." *Bioresource Technology* 349:126833. doi: 10.1016/j.biortech.2022.126833.

Augustin, J. M., V. Kuzina, S. B. Andersen, and S. Bak. 2011. "Molecular activities, biosynthesis and evolution of triterpenoid saponins." *Phytochemistry* 72(6):435-457. doi: 10.1016/j.phytochem.2011.01.015.

Babot, M. P., and S. Hocsman. 2014. "La quinua en las sociedades prehispánicas del Noroeste Argentino [Quinoa in the pre-Hispanic societies of the Argentine Northwest]." *Ciencia Hoy* 23(138):24-29.

Bedoya-Perales, N. S., G. Pumi, A. Mujica, E. Talamini, and A. D. Padula. 2018. "Quinoa expansion in Peru and its implications for land use management." *Sustainability* 10(2): 532. doi: 10.3390/su10020532.

Biver, L. C., M. P. Mabel, and H. C. Segundo. 2021. "Biological water treatment residuals from plants quinoa processors." *The Ecuadorian Journal of S.T.E.A.M.* 1(3):1089–1097. doi: 10.18502/ESPOCH.V1I3.9535.

Carrasco, C., D. Cuno, K. Carlqvist, M. Galbe, and G. Lidén. 2015. "SO_2-Catalysed steam pretreatment of quinoa stalks." *Journal of Chemical Technology and Biotechnology* 90(1):64–71. doi: 10.1002/JCTB.4286.

Dakhili, S., L. Abdolalizadeh, S. M. Hosseini, S. Shojaee-Aliabadi, and L. Mirmoghtadaie. 2019. "Quinoa protein: Composition, structure and functional properties." *Food Chemistry* 299:125161. doi: 10.1016/j.foodchem.2019.125161.

Fernández-López, J., R. Lucas-González, A. Roldán-Verdú, M. Viuda-Martos, E. Sayas-Barberá, J. Ballester-Sánchez, C. M. Haros, and J. A. Pérez-Álvarez. 2020. "Effects of black quinoa wet-milling coproducts on the quality properties of bologna-type sausages during cold storage." *Foods* 9(3):274. doi: 10.3390/foods9030274.

Filik, G. 2020. "Biodegradability of quinoa stalks: The potential of quinoa stalks as a forage source or as biomass for energy production." *Fuel* 266:117064. doi: 10.1016/j.fuel.2020.117064.

Fuentes, F. F., and X. Paredes-Gonzalez. 2015. "Nutraceutical perspectives of quinoa: Biological properties and functional applications." In *The State of the World's Quinoa,* eds. Didier Bazile, Daniel Bertero, and Carlos Nieto, 286-299. Regional Office for Latin America and Caribbean at Food and Agriculture Organization (FAO). doi: 10.13140/RG.2.1.4294.2565.

Gawlik-Dziki, U., M. Świeca, M. Sułkowski, D. Dziki, B. Baraniak, and J. Czyż. 2013. "Antioxidant and anticancer activities of *Chenopodium quinoa* leaves extracts - *In vitro* study. " *Food Chemical Toxicology* 57:154–160. doi: 10.1016/j.fct.2013.03.023.

Gil-Ramirez, A., D. M. Salas-Veizaga, C. Grey, E. Nordberg Karlsson, I. Rodriguez-Meizoso, and J. A. Linares-Pastén. 2018. "Integrated process for sequential extraction of saponins, xylan and cellulose from quinoa stalks (*Chenopodium quinoa* Willd.)." *Industrial Crops and Products* 121:54–65. doi: 10.1016/J.INDCROP.2018.04.074.

Graf, B.L., A. Poulev, P. Kuhn, M. H. Grace, M. A. Lila, and I. Raskin. 2014. "Quinoa seeds leach phytoecdysteroids and other compounds with anti-diabetic properties." *Food Chemistry* 163:178–185. doi: 10.1016/j.foodchem.2014.04.088.

Guo, H., Y. Hao, X. Yang, G. Ren, and A. Richel. 2021. "Exploration on bioactive properties of quinoa protein hydrolysate and peptides: A review." *Critical Reviews in Food Science and Nutrition* 28: 1–14. doi: 10.1080/10408398.2021.1982860.

Httiarachy, N. S., and U. Kalapathy. 1998. "Functional properties of soy proteins." In *Functional Properties of Proteins and Lipids*, eds. J. R. Whitaker, F. Shahidi, A. Lopez-Munguia, R. Y. Yada, and G. Fuller, 80-95. Washington: American Chemical Society.

Jacobsen, S. E. 2011. "The situation for quinoa and its production in southern Bolivia: From economic success to environmental disaster." *Journal of Agronomy and Crop Science* 197(5):390–399. doi: 10.1111/j.1439-037X.2011.00475.x.

Jeong, S. M., M.J . Kang, H. N. Choi, J. H. Kim, and J. I. Kim. 2012. "Quercetin ameliorates hyperglycemia and dyslipidemia and improves antioxidant status in type 2 diabetic db/db mice." *Nutrition Research and Practice* 6(3):201–207. doi: 10.4162/nrp.2012.6.3.201.

Kuktaite, R., R. Repo-Carrasco-Valencia, C. C. H. de Mendoza, T. S. Plivelic, S. Hall, and E. Johansson. 2021. "Innovatively processed quinoa (*Chenopodium quinoa* Willd.) food: Chemistry, structure and end-use characteristics." *Journal of the Science of Food and Agriculture* 102:5065-5076. doi:10.1002/jsfa.11214.

Lozano, M., E. Gonzales R., Y. Flores, and G. R. Almanza. 2013. "Effect in acute inflammation of sapogenin extract and isolated sapogenins from quinoa waste (*Chenopodium quinoa* Willd)." *Revista Boliviana de Química* 30(2):115–21.

Meneses-Quelal, W. O., B. Velázquez-Martí, J. Gaibor-Chávez, Z. Niño-Ruiz, and A. Ferrer-Gisbert. 2021a. "Anaerobic co-digestion of slaughter residues with agricultural waste of amaranth quinoa and wheat." *BioEnergy Research* 15::1649-1663. doi:10.1007/s12155-021-10350-9.

Meneses-Quelal, W. O., B. Velázquez-Martí, J. Gaibor-Chávez, and Z. Niño-Ruiz. 2021b. "Biochemical potential of methane (BMP) of camelid waste and the andean region agricultural crops." *Renewable Energy* 168:406–415. doi: 10.1016/j.renene.2020.12.071.

Montesano, D., M. Gallo, F. Blasi, and L. Cossignani. 2020. "Biopeptides from vegetable proteins: New scientific evidences." *Current Opinion in Food Science* 31::31–37. doi: 10.1016/j.cofs.2019.10.008.

Naeimi, A., S. Khoshkam, and T. Eslaminejad. 2021. "Natural cellulose fibers from quinoa wastes reinforced carbon nanotube/ZnO bio-nanocomposite as a novel recyclable catalyst for oxidation reaction." *Polymer Bulletin* 79: 7795-7808. doi:10.1007/s00289-021-03876-7.

Oleszek, W. A. 2002. "Chromatographic determination of plant saponins." *Journal of Chromatography A* 967(1):147–162. doi: 10.1016/S0021-9673(01)01556-4.

Ortiz-Gómez, V., A. Fernández-Quintero, D. F. Roa-Acosta, J. E. Bravo-Gómez, and J. F. Solanilla-Duque. 2022. "Physicochemical characterization of quinoa (*Chenopodium quinoa* Cv. Nariño) co-products obtained by wet milling." *Frontiers in Sustainable Food Systems* 6:1-12. doi:10.3389/fsufs.2022.851433.

Park, S. H., T. Maeda, and N. Morita. 2005. "Effect of whole quinoa flours and lipase on the chemical, rheological and breadmaking characteristics of wheat flour." *Journal of Applied Glycoscience* 52(4):337-343. doi:10.5458/jag.52.337.

Pathan, S., F. Eivazi, B. Valliyodan, K. Paul, G. Ndunguru, and K. Clark. 2019. "Nutritional composition of the green leaves of quinoa (*Chenopodium quinoa* Willd.)." *Journal of Food Research* 8(6): 55–65. doi: 10.5539/jfr.v8n6p55.

Pathan, Safiullah, and Rafat A. Siddiqui. 2022. "Nutritional composition and bioactive components in quinoa (*Chenopodium quinoa* Willd.) greens: A review." Nutrients 14(3):558. doi:10.3390/nu14030558.

Potter, D. A., C. T. Redmond, K. M. Meepagala, D. W. Williams. 2010. "Managing earthworm casts (Oligochaeta: Lumbricidae) in turfgrass using a natural byproduct of tea oil (*Camellia* sp.) manufacture." *Pest Management Science* 66(4):439-446. doi: 10.1002/ps.1896.

Quiroga, C., R. Escalera, G. Aroni, A. Bonifacio, J. A. González, M. Villca, R. Saravia, and A. Ruiz. 2017. "Traditional processes and technological innovations in quinoa harvesting, processing and industrialization" In *State of the Art Report on Quinoa around the World in 2013*, 218–249, eds. D. Bazile, D. Bertero, C. Nieto. FAO:Santiago.

Repo-Carrasco-Valencia, R. A. M., and L. Astuhuaman Serna. 2011. "Quinoa (*Chenopodium quinoa*, Willd.) as a source of dietary fiber and other functional components." *Ciência e Technologia de Alimentos Campinas* 31(1):225–230.

Ruiz, K. B., B. Khakimov, S. B. Engelsen, S. Bak, S. Biondi, and S. E. Jacobsen. 2017. "Quinoa seed coats as an expanding and sustainable source of bioactive compounds: An investigation of genotypic diversity in saponin profiles." *Industrial Crops and Products* 104: 156–163. doi:10.1016/j.indcrop.2017.04.007.

Ruiz, K. B., S. Biondi, R. Oses, I. S. Acuña-Rodríguez, F. Antognoni, E. A. Martinez-Mosqueira, A. Coulibaly, A. Canahua-Murillo, M. Pinto, A. Zurita-Silva, D. Bazile, S. E. Jacobsen, and M. A. Molina-Montenegro.

2014. "Quinoa biodiversity and sustainability for food security under climate change. A review." *Agronomy for Sustainable Development* 34:349–359. doi:10.1007/s13593-013-0195-0.

Salas-Veizaga, D. M., Rodrigo Villagomez, J. A. Linares-Pastén, C. Carrasco, M. T. Álvarez, P. Adlercreutz, and E. Nordberg Karlsson. 2017. "Extraction of glucuronoarabinoxylan from quinoa stalks (*Chenopodium quinoa* Willd.) and evaluation of xylooligosaccharides produced by GH10 and GH11 xylanases." *Journal of Agricultural and Food Chemistry* 65(39):8663–8673. doi:10.1021/acs.jafc.7b01737.

San Martín, R., K. Ndjoko, and K. Hostettmann. 2008. "Novel molluscicide against *Pomacea canaliculata* based on quinoa (*Chenopodium quinoa*) saponins." *Crop Protection* 27(3-5:310-319. doi:10.1016/j.cropro.2007.03.015.

Scanlin, L., and K. A. Lewis. 2017. "Quinoa as a Sustainable Protein Source." In *Sustainable Protein Sources,* 223-38. Elsevier. https://doi.org/10.1016/B978-0-12-802778-3.00014-7.

Sharma, S., R. Singh, and S. Rana. 2011. "Bioactive peptides: A review." *International Journal Bioautomation* 15(4):223–250. doi:10.1093/fqs/fyx006.

Shen, Y., X. Tang, and Y. Li. 2021. "Drying methods affect physicochemical and functional properties of quinoa protein isolate." *Food Chemistry* 339:127823. doi:10.1016/j.foodchem.2020.127823.

Shi, Z., Y. Hao, C. Teng, Y. Yao, and G. Ren. 2019. "Functional properties and adipogenesis inhibitory activity of protein hydrolysates from quinoa (*Chenopodium quinoa* Willd.)." *Food Science and Nutrition* 7(6): 2103–2112. doi:10.1002/fsn3.1052.

Skendi, A., K. G. Zinoviadou, M. Papageorgiou, and J. M. Rocha. 2020. "Advances on the valorisation and functionalization of byproducts and wastes from cereal-based processing industry." *Foods* 9(9):1243. doi:10.3390/foods9091243.

Suárez-Rivero, D., P. A. Patiño-Puentes, J. Ortiz-Aguilar, M. Suárez-Rivero, O. Marín-Mahecha, and C. M. Zuluaga-Domínguez. 2019. "Preliminary evaluation of the by-products of quinoa as alternative source of usable biomass." *Chemical Engineering Transactions* 74: 187–192. doi:10.3303/CET1974032.

Vazquez-Luna, A., V. Pimentel Cortés, F. Fuentes Carmona, and R. Díaz-Sobac. 2019. "Quinoa leaf as a nutritional alternative." *Ciencia e Investigación Agraria* 46(2):137–143. doi:10.7764/rcia.v46i2.2098.

Vega-Gálvez, A., M. Miranda, J. Vergara, E. Uribe, L. Puente, and E. A. Martínez. 2010. "Nutrition facts and functional potential of quinoa (*Chenopodium quinoa* Willd.), an ancient andean grain: A review." *Journal of the Science of Food and Agriculture* 90(15):2541–2547. doi:10.1002/jsfa.4158.

Vega-Gálvez, A., Ricardo San Martín, Mariela Sanders, Margarita Miranda, and Elena Lara. 2010b. "Characteristics and mathematical modeling of convective drying of quinoa (*Chenopodium quinoa* Willd.): Influence of temperature on the kinetic parameters." *Journal of Food Processing and Preservation* 34(6):945–963. doi:10.1111/j.1745-4549.2009.00410.x.

Vincken, J. P., L. Heng, A. de Groot, and H. Gruppen. 2007. "Saponins, classification and occurrence in the plant kingdom." *Phytochemistry* 68(3):275-297. doi: 10.1016/j.phytochem.2006.10.008.

Wu, G.. 2015. "Nutritional properties of quinoa." In *Quinoa: Improvement and Sustainable Production,* 193–210. https://doi.org/10.1002/9781118628041.ch11.

Zheng, Y., X. Wang, Y. Zhuang, Y. Li, H. Tian, P. Shi, and G. Li. 2019. "Isolation of novel ACE-Inhibitory and antioxidant peptides from quinoa bran albumin assisted with an in silico approach: Characterization, *in vivo* antihypertension, and molecular docking." *Molecules* 24(24):4562. doi:10.3390/molecules24244562.

Chapter 13

Products of Amaranth Processing

J. E. Borges-Martínez
D. del R. Moguel-Concha
M. S. Cid-Gallegos
D. I. Tellez-Medina
and Cristian Jiménez-Martínez*
National School of Biological Sciences, National Polytechnic Institute,
Mexico City, Mexico

Abstract

Amaranth is an approximately 4,000-year-old pseudocereal. It is one of the most important crops for the Aztec and Inca civilizations and has highly nutritious cultivation. Only the seed is utilized during the amaranth harvest, while the plant remains in the ground until it withers and is merely used as compost. Several studies have analyzed these wastes and have isolated various bioactive compounds, such as bioactive peptides and phenolic compounds. Proteins, peptides, squalene, phenolic compounds, and fiber can be extracted from the amaranth seed during different treatments, such as enzymatic hydrolysis, germination, and fermentation. These compounds can be treated to obtain bioactive compounds with biological activities such as antioxidant, anti-inflammatory, and anticancer, among others. In this chapter, we review some of the compounds that have been obtained from amaranth, seeking to elucidate the nutraceutical effect of amaranth.

Keywords: amaranth, bioactive compounds, biological activities

Origin and Distribution

Recently, consumer awareness has changed substantially regarding the role of food in lifestyle. The food contribution goes beyond satisfying hunger and providing essential nutrients; it is now focused on disease prevention, reducing health risks, and improving human well-being

* Corresponding Author's Email: crisjm_99@yahoo.com.

In: Food Byproducts
Editors: S. Teresita Martín-del-Campo, A. Cardador-Martínez and J. del Pilar Ramírez-Anaya
ISBN: 979-8-88697-612-0

(Coelho et al. 2018). Amaranth (*Amaranthus* spp.) is a pseudocereal whose food crop was the main crop of the Aztec and Inca civilizations. Its presence dates to 4,000 BC, making it one of the oldest crops. Most of the species of the genus *Amaranthus* spp. (50-70 species in total) are native to the Americas (Corke, Cai, and Wu 2015a).

In prehistoric times, the oldest evidence was found in an archaeological excavation in a cave in Tehuacan, Mexico, where seeds dated 6000 years old. This fact suggests that people used amaranth as a staple food in Central America during the Mayan and Aztec periods (Joshi et al. 2018).

Amaranth is a non-graminaceous dicotyledonous plant that produces grain-like seeds, which is why it has been called a pseudocereal. Amaranth is classified into grain amaranth and foliaceous amaranth because of its type of use. The former is recognized as originating in the Americas and comprise three species: *A. cruentus*, *A. hypocondriacus* (Figure 1), and *A. caudatus*. At the same time, leaf-producing amaranths include *A. dubius*, *A. hybridus,* and *A. tricolor*, all cultivated mainly in Asia and Africa (Shukla et al. 2006, D'Amico and Schoenlechner 2017).

Amaranth grain also has wide adaptability and can grow in temperate and tropical conditions. In addition, it can adapt to environmental stresses such as high temperatures, drought, and low-input crops (Huerta-Ocampo and Barba de la Rosa 2011).

In recent decades, demand for amaranth cultivation has increased due to its high nutritional value and agronomic advantages, as it has become a nutritionally balanced cosmopolitan food. Which could be a nutritional substitute in food formulations, and its bioactive compounds present in both the grain and the leaves could have some pharmaceutical use against chronic degenerative diseases (Park, Sharma, and Lee 2020).

On the other hand, amaranth is an important pseudocereal that can prevent malnutrition, especially in low-income food-deficient countries (Martinez-Lopez et al. 2020). Today, amaranth is considered one of the few polyvalent crops that can be used in different ways, as both the grain and its leaves are suitable for human consumption and farm animals (Cheng 2018, Corke et al. 2015, Wu et al. 2019).

Figure 1. Inflorescence of two species of amaranth (grain and foliaceous): a) *A. hypochondriacus*; b) *A. dubius* (Joshi et al. 2018)

Nutritional Composition of *Amaranthus* spp.

Amaranth is one of the few multi-purpose plants that can provide grains and leaves of high nutritional quality for use as food. Amaranth grain is characterized by a higher content of protein, lipid, and lower starch content than cereals (Corke et al. 2015, Perales-Sanchez et al. 2015). This chemical and nutritional composition varies according to varieties, species, geographical origin, and climatic conditions (Joshi et al. 2018).

Proteins

Amaranth has a higher protein composition than wheat (13.3%), corn (9.2%), or rice (7.0%). The proteins present in more significant quantities in amaranth are albumins and globulins (Martinez-Lopez et al. 2020). Depending on the amaranth species, there are differences in protein content, *A. caudatus* shows 13.5 g of protein per 100 g, which is lower compared to other species such as *A. cruentus* (15.7/100 g) and *A. hypochondriacus* (15.5/100 g) (Montoya-Rodríguez et al. 2015, Mota et al. 2016) Also, amaranth has an essential composition concerning essential amino acids (Table 1); most pseudocereals have a low lysine content, while amaranth has a significant amount of lysine in its protein composition (5.2-6.1 g/100 g of protein) (Joshi et al. 2018).

Importantly, amaranth contains a significant amount of sulfur amino acids (2.6-5.5 g/ 100 g protein) compared to most cereals, making amaranth a good source of high-quality protein and a nutritious substitute for some grains such as maize or wheat in functional foods (Pinto 2018, Mota et al. 2016).

Table 1. Nutritional composition of three grain amaranth species in comparison to major cereals

Components	*A. cruentus*	*A hypochondriacus*	*A. caudatus*	Rice	Wheat	Maize
Proximate composition (% dry weight basis)						
Crude Protein	13.80-21.50	15.00-16.60	13.10-21.00	6.8	13.2	10
Amino acids (% of total protein)						
Tryptophan	0.9	1.8	1.4	1	1.2	0.6
Methionine	4.6	1.6	4.2	3	3.5	3.2
Threonine	3.9	3.3	3.4	3.7	2.7	4
Isoleucine	4	2.7	3.6	4.5	4.2	4.6
Valine	4.5	3.9	4.2	6.7	4.3	5.1
Lysine	6.1	5.9	5.2	3.8	2.6	1.9
Phenylalanine	8.5	8.52	6	9.1	8.1	10.6
Leucine	6.1	4.2	5.1	8.2	6.3	13

(Das 2016, Joshi et al. 2018).

Carbohydrates

Starch is the main macromolecule in plants and occurs in granular form in different sizes and shapes. It is part of the physiological energy of humans and is found in the perisperm, as a single entity or in its composite structure (Olawoye and Gbadamosi 2020). Kong et al. (2009)

revealed that amaranth grain starch is embedded in a protein matrix, thus forming a complex with the protein, and hindering starch hydrolysis by enzymes, decreasing starch digestibility.

The carbohydrate content of amaranth starch grain varies depending on the grain species (Table 2), a value lower than that of some common cereals (Repo-Carrasco-Valencia et al. 2010, Corke et al. 2015). Within the components of starch, amylose is found in lesser proportion than amylopectin. A study by Roa et al. (2015) showed that amaranth starch granule diameter is smaller than wheat and maize. This characteristic causes a low gelatinization temperature, as its amaranth grain gelatinization temperature ranges between 54-62°C.

The gelatinization properties of amaranth grain starch depend not only on the size of the granules but also on the degree of crystallinity (Das et al. 2021). Another essential characteristic of amaranth starch is its higher water and oil absorption capacity and higher swelling capacity. This property makes amaranth starch suitable for food applications, especially where high binding capacity is required (Coelho et al. 2018b).

Table 2. Starch and amylose content in different species of amaranth (%)

Species	Starch (%)	Amylose (%)
A. cruentus	48-63	22.1
A. hybridus	55	21.1
A. hypochondriacus	52-62	7.8
A. tricolor	51	29

(Corke et al. 2015).

Lipids

Amaranth seed has a much higher lipid content compared to other cereals. The fatty acid content varies depending on the species (Table 3). It is characterized by a higher concentration of unsaturated fatty acids (75%), of which the most abundant in amaranth are linoleic acid (33-55. 9%), oleic acid (18.7-38.9%), palmitic acid (14.04-26%) and stearic acid (3.11-4.47%). Linolenic acid is also present in the seed but in a minimal proportion (0.20-1.97%) compared to the above fatty acids (Gimplinger et al. 2007, Singh and Singh 2011, Das 2016).

Table 3. Oil and squalene content of amaranth grains and other seeds (%)

Species	Oil (%)	Squalene (%)
A. cruentus	5.57-7.72	5.0-8.0
A. hybridus	6.4	5.23
A. hypochondriacus	5.35-7.05	3.62-5.01
A. tricolor	5.08	6.14
Maize	4.0	0.03
Rice	1.0-3.0	0.3
Peanut	47.0	0.03

(Corke et al. 2015, Das 2016, Coelho et al. 2018b).

Similarly, a significant amount of squalene was reported in amaranth (2.26-5.94%). Squalene is a non-saponifiable lipid that acts as a biosynthetic precursor of all plant and animal steroids. Several studies suggest that amaranth has hypocholesterolemic, antioxidant and anticarcinogenic effects due to its unsaponifiable profile and unsaturated fatty acid content (Tang et al. 2016, Palombini et al. 2013, Yelisyeyeva et al. 2014).

On the other hand, the higher concentration of polyunsaturated fatty acids in amaranth grain affects managing cardiovascular diseases, and unsaturated fatty acids and other lipids such as sterols and phospholipids, tocopherols, triglycerides, and glycolipids (Dus-zuchowska et al. 2019).

Vitamins and Minerals

Although vitamins and minerals are in small quantities in the amaranth grain (Pilco-Quesada et al. 2020), according to a study published by Morales et al. (2021), amaranth is particularly rich in vitamin E and folic acid. In addition, it is an excellent source of riboflavin, and vitamin C. Vitamin E is present in amaranth in the form of tocopherol and tocotrienols. This vitamin has an antioxidant effect and has been reported to increase the stability of amaranth oil. Amaranth is reported to be an excellent source of minerals such as potassium, magnesium, and calcium. In addition, the high calcium content found in amaranth makes it suitable for formulating food products and treating celiac disease. People with celiac disease also present osteoporosis and osteopenia (Gliszczyńska-Świgło, Klimczak, and Rybicka 2018).

Amaranthus spp. Leaves

Amaranth leaves contain between 27.8 and 48.6% total protein, making them slightly higher in protein content than spinach. The fat content of the leaves ranges between 2.9 and 7.1%, with linoleic acid being the polyunsaturated fatty acid with the highest presence in the leaves. They are also a good source of fiber (11.-22.9%). They also have high concentrations of minerals such as P, Fe, Mg, and Ca. Vitamins A and C are found in significant concentrations (0.420-0.250 mg/kg). Lysine is an amino acid found in a higher proportion in amaranth leaves (6.2-6.8%). In addition, the red-colored vegetative tissues of amaranth leaves have been found to contain a high level of betalain pigments as natural dyes (Corke et al. 2015, Kaur et al. 2010, Cornejo et al. 2019).

Food Product

Amaranth is an important nutritional crop. It is a rare plant whose leaves are eaten as a vegetable while the seeds are eaten as a cereal (Oke 1983, Saunders and Becker 1984, Rastogi and Shukla 2013). Every day more food products made with amaranth seed, both artisanal and industrial, are found on the market (Paredes-Lopez 2018). In Mexico, amaranth is commonly known as "alegría," a name given to the sweet made with honey and broken amaranth seeds. Alegrias have a good market in different areas of Mexico, especially in regions that include states such as Tlaxcala, Puebla, Mexico, Morelos, and Mexico City (Paredes-López 1994).

The merchants of amaranth products are, in general, the same farmers who grow the grain; this is particularly the case in the rural area of Mexico City. Amaranth can be used in a wide variety of products, such as soups, pancakes, breakfast cereal, granola, buns, crepes, toast, tortillas, fried foods, cookies, empanadas, pasta, snacks, drinks (such as pulque with amaranth or fresh waters with fruits), ice creams made by hand, bases to prepare atoles, breaded-beef steak, tamales, pizzas, pancakes, and confectionery (Paredes-Lopez 2018).

Because amaranth lacks functional gluten, it is mostly used in the production of bakery products that do not need expansion, in addition to being added to mixtures with flours from other cereals (Gebreil, Ali, and Mousa 2020). This product can be used in the form of refined white flour or wholemeal flour (Coţovanu and Mironeasa 2021). This is obtained by grinding the whole or popped grain and is usually used for confectionery and biscuit products, in which amaranth replaces up to 20% of wheat flour. Short pasta is also made, which contains 40% amaranth flour; in the production of long pasta, such as noodles, up to 20% can be used (Schoenlechner et al. 2010). Tovar et al. (2018) observed an improvement in flavor, consistency, and acceptance of the product when the flour was derived from the roasted seed. Aderibigbe et al. (2022), indicate that there are many forms of consumption of amaranth, but that most of them are in mixtures or used as a complement, although the best option is to use it in products that are made entirely with amaranth with the purpose of taking advantage of the maximum its nutritional properties.

Morales et al. (2009) mention that the San Miguel de Proyectos Agropecuarios company is one of the few worldwide firms dedicated to the comprehensive exploitation of amaranth cultivation. This company developed technology to obtain amaranth flours and by-products that concentrate the protein and functional oils (nutraceuticals) of amaranth and the amyloprotein components in differentiated fractions. The Copram product (protein concentrate of amaranth) has 32.5% of amaranth protein (wholemeal flour has 15 to 16%) and 14 to 15.5% of the lipid fraction of amaranth. Copramyl (amyloprotein concentrate of amaranth) is a raw or precooked fraction with 22% protein and an oil profile like that of regular whole meal flour. Both products can be used as texturizers, fortifiers, extenders, flavor enhancers, and basic ingredients for nutraceutical or functional products with a growing market. On the other hand, the company San Miguel has developed powder preparations aimed at specific consumers (high-performance athletes, people who want to lose weight, older adults), with notable added value and the necessary profile to compete against leading brands in the world markets.

It is known that the consumption of amaranth grain and leaf facilitates the functions of the digestive system because it is a product rich in fiber, even surpassing the content of other common cereals such as wheat, barley, and rice (Zhu 2020); Therefore, a dietary fiber and laxative drink has been developed from the leaf from Amaranth (Paredes-López 2018). In addition, the vegetable amaranth has been used as a traditional vegetable for several centuries in China. In sub-Saharan Africa, Central and South America, it is also an important leafy vegetable, eaten as a cooked green vegetable or used in soup (Corke, Cai, and Wu 2015a).

Amaranth Grain Processing Techniques

Pseudocereals are not usually consumed raw but are processed before consumption. Cooking is a typical processing method; however, soluble nutrients such as minerals, phenolic compounds, or amino acids are sometimes lost. Repo-Carrasco-Valencia et al. (2010) showed that minerals such as Fe and phenolic compounds decrease during the cooking process. Another technique used for the consumption of amaranth grain is popping to improve its flavor and make it edible; it has been demonstrated that during this process, the content of minerals such as Ca decreases as well as the phenolic compounds present in the grain (Muyonga, Andabati, and Sepuuya 2014). Another process used to consume amaranth seed is germination, which has been reported to increase the content of phenolic compounds 72 hours after germination and amino acids such as tyrosine, lysine, leucine, and serine (Perales-Sánchez et al. 2014). The bioactive compounds present in amaranth can also be modified during the fermentation process. Amare et al. (2015) showed that some amino acids such as proline, glutamic acid, and tyrosine increased during fermentation. They attributed it to the fact that protein hydrolysis occurs during the fermentation process and significantly affects the amino acid composition of the seed.

Bioactive Compounds of *Amarathus* spp.

Different extracts from all parts of the Amaranth plant appear to have medicinal benefits. Recent studies have focused on identifying therapeutic compounds obtained from different parts of the Amaranthus plant. Among the biological activities that amaranth extracts may exhibit, antioxidant and anti-inflammatory activities stand out, which has led to increased interest to investigate the nutraceutical and clinical potential of amaranth as a functional food. Phytochemical analysis of the different parts of the *Amaranthus* spp. has established the presence of different bioactive compounds, including alkaloids, flavonoids, glycosides, phenolic acids, steroids, saponins, amino acids, bioactive peptides, vitamins, minerals, terpenoids, lipids, carotenoids, and betalains (Peter and Gandhi 2017, Nana et al. 2012, Perales-Sánchez et al. 2014, Klimczak, Małecka, and Pachołek 2002, Paśko et al. 2008).

Antioxidants are molecules that diminish the effect of free radicals and are essential for protection against diseases such as cancer and other degenerative disorders. The antioxidant potential of amaranth has been attributed to the presence of phenolic compounds and flavonoids. Amaranth leaves, flowers, and seeds have been shown to possess significant antioxidant activity (Table 4); rutin is the main phenolic compound found in amaranth extracts, to which free radical scavenging activities are attributed (Kraujalis et al. 2013, Peter and Gandhi 2017).

Table 4. Biochemical validation studies of the antioxidant properties of *Amaranthus* spp.

Species	Part of the plant used	Extract	Antioxidant activity
A. spinosus	Leaves	Chloroform, n-hexane, ethyl- acetate	DPPH
	Leaves	Methanolic	Alpha-amylase inhibition assay by CNPG3
A. caudatus	Whole plant	Methanolic	Alpha–amylase, ABTS, DPPH, SOD, OH–, NO
A. hybridus *A. cruentus*	Aerial parts	Hydroacetonic, methanolic, aqueous	Xanthine oxidase inhibitory
	Seeds	3 novel purified peptides	HMG-CoA reductase assay in terms of NADPH
A. tricolor	Leaves	Dry leaf powder	Ascorbic acid

(Peter and Gandhi 2017, Perales-Sánchez et al. 2014, Kraujalis et al. 2013).

Inflammation generated by oxidative stress and macromolecule damage is a prominent physiological feature of all chronic degenerative diseases. Studies with different amaranth extracts using in vivo models have shown the activity of the extract (Table 5) to stabilize the lipid bilayer membrane of mast cells, thus preventing membrane disruption, histamine release, and mast cell degranulation, suggesting a role in the prophylaxis and management of anaphylactic reactions (Patil et al. 2012).

Table 5. *In vivo* evaluation of Pharmacological activity of *Amaranthus* spp.

Species	Plant part used	Extract type/ active component	Model	Pharmacological activity	Dosage/toxicity
A. spinosus	Whole plant	Methanolic	Mouse	Anti-nociceptive, anti-inflammatory	500 mg/kg
	Whole plant	50% ethanolic	Rats	Hepatoprotective	400 mg/kg
A. caudatus	Leaves	Aqueous	Albino Wistar rats	Anticarcinogenic	1 ml of 0.2 g/ml for 14 d
	Leaves	Methanolic	Rats	anti-diabetic, hypocholestro-lemic	400 mg/kg for 21 d
A. hypochondriacus	Seeds	Whole	Rats	Hepatoprotective	–
	Seed and sprouts	Whole	Mouse	Hypocholesterolemic, hypoglycemic	1 g/kg body weight
A. cruentus	Seeds	Flour & protein concentrate	Wistar rats	Hepatoprotective	–

(Patil et al. 2012, Peter and Gandhi 2017; Corzo-Ríos et al. 2021).

Bioactive Peptides

Other compounds present in *Amaranthus spp.* that have been studied and present biological activities are bioactive peptides. These are defined as amino acid sequences that are not active in the original protein structure but can exert biological functions in the organism after their release through hydrolysis *in vitro* or *in vivo* (Miner-Williams, Stevens, and Moughan 2014). The bioactive peptides present in amaranth (Table 6) include antioxidant, antimicrobial, anticancer, and anti-inflammatory activity. The bioactivity of these peptides depends on their length, the constitution of the amino acids, the degree of hydrolysis, the protein used as a substrate, and the enzyme specificity (Morales, Miguel, and Garcés-Rimón 2021; Park, Sharma, and Lee 2020).

Research and Potential Uses in the Industry

Amaranth proteins concentrates, or isolates are characterized by excellent solubility, as well as foaming, emulsifying, and stabilizing properties (Constantino and García Rojas 2022) and therefore have a high potential for functional foods, particularly where gel formation is desired (Avanza, Puppo, and Añón 2005).

Amaranth starch is characterized by small granule size, low amylose content, high water-holding capacity, good freeze stability, resistance to mechanical shear, and excellent gelling and thickening properties (Zhu 2017). Both the properties obtained from amaranth proteins and starch mainly offer a wide range of uses in food applications (Santra and Schoenlechner 2017).

Table 6. Antioxidant activity of hydrolysates and peptides derived from various *Amaranthus* spp.

Species	Country	Enzyme /Bacteria	Peptides /Hydrolysates	Peptide Sequence	Findings
A. hypochondriacus	Argentina	Pepsin and pancreatin	Hydrolysate	-	ABTS = Seed sprout isolate (SI) undigested (0.32) SI digested (0.72) µmol/mg
A. mantegazzianus	Argentina	Pepsin and pancreatin	Hydrolysate	-	IC50 = ESR-OH (25%), ORAC (20%), peroxynitrites (20%), and ORAC (50%)
	Argentina	Pepsin and pancreatin	Hydrolysate and peptides	AWEEREQGSR (1) YLAGKPQQEH (2) IYIEQGNGITGM (3) TEVWDSNEQ (4)	IC50 ORAC = (1): 6.7 µg/mL, (2): 16 µg/mL, (3): 17µg/mL, (4): 20 µg/mL
A. caudatus	Ecuador	Pepsin and pancreatin	Hydrolysate and peptides	F1 (YESGSQ, GGEDE, NRPET), F2 (FLISCLL, TALEPT, HVIKPPS, SVFDEELS, ASANEPDEN, DFIILE)	ORAC values = F2 fraction 4.47 µmol TE/mg peptide; F1 fraction 1.56 µmol TE/mg peptide
A. hypochondriacus	USA	Alcalase	Hydrolysate	-	DPPH and ABTS (40%)
	Mexico	Alcalase and flavourzyme	Hydrolysate H1, H2, and H3	LVRW, DPKLTL	DPPH = 76.6 ± 1.6–388.9 ± 2.7 µmol TE/100 g; ABTS = 115.6 ± 10.3–425.8 ± 0.6 mg TE/100 g; FRAP = 63.3 ± 5.7–592.5 ± 29.2 µmol Fe2+/100 g

(Morales et al.2021, Park, Sharma, and Lee, 2020).

Products Derived from Amaranth Starch

The characteristics described above make amaranth starch unique for some food and non-food applications, such as fat substitutes, manufacturing biomaterials with novel properties, as well as carrier and encapsulation materials (Coelho et al. 2018).

Food Applications

Native starch: This has been used as a flour compound ingredient for fried snacks to reduce oil absorption (Bhosale and Singhal 2006) and as a fat substitute for up to 50% in frozen desserts (Malinski et al. 2003).

Modified starch: Phosphorylated starch has been applied in mayonnaise as a fat substitute (Zhu 2017), while hydroxypropylated and oxidized starch was used to encapsulate flavors (Kshirsagar and Singhal 2008). In a study conducted by Tapia-Blácido, Sobral, and Menegalli (2005), films were made from *A. caudatus* using glycerol as a plasticizer. The biofilms obtained presented a good appearance, color, and great flexibility, in addition to lower permeability to oxygen and water when compared to films obtained from other sources of proteins and polysaccharides.

These are used as edible coatings (a thin layer of edible material formed on the food), whose purpose in food preservation is to inhibit the exchange of moisture, oxygen, and carbon dioxide, aroma loss, and lipid migration. Native or modified starch has great potential to form edible coatings since it constitutes an efficient barrier against oxygen, which would make it interesting in extending the shelf life of various fruits.

Non-Food Applications

Native starches have been used as functional and lower-cost ingredients in the formation of low-density films, thermoplastics, superabsorbents, and dye thickeners in textiles (Zhu 2017). In addition, amaranth starch may be a suitable moisture management agent where freeze-thaw cycles exist, be used as a possible substrate to efficiently produce smaller particles (such as nanoparticles) for new biomaterials, be applied in new emulsification systems, and as vehicles in the production of cosmetics. Using this component can reduce the cost of production and impart new and desired properties to products. Despite this, the applications of amaranth starch in the food and non-food industries need to be further explored in depth (Zhu 2017).

Gluten-Free Products

The absence of gluten in amaranth makes it suitable for products without this component. The production of gluten-free bread and bakery products or pasta with amaranth flour is of great interest now, as this need is increasing due to a higher prevalence of celiac disease and its sensitivity (Pellegrini and Agostoni 2015). If we concentrate all the people affected, it can be estimated that the market is around 1% of the world population. In addition, the challenge of these foods is to offer an adequate nutritional balance (Rustagi, Khan, and Jain 2022). Amaranth, as well as other pseudocereals, is nutritionally superior to many gluten-free raw materials. In fact, the growing market can be seen as one of the main driving forces behind the increased production and use of amaranth. However, the production of gluten-free foods, in particular amaranth, is very challenging and cannot be carried out without the addition of other ingredients or specific adaptation of the relevant processing steps (V. Singh et al. 2021).

Agrobiotechnology

One of the applications of amaranth also takes place in the field of genetically modified products. The following investigations are some of the most important within the studies where amaranth gene transfer techniques are applied to other crops of high economic value.

Chakraborty, Chakraborty, and Datta (2000), did an investigation about the cloning of the albumin gene extracted from the seed of *A. hypochondriacus* (AmA1), rich in essential amino acids for the improvement of the nutritional value of potatoes, in which lysine, tyrosine, and the sulfur amino acids, methionine and cysteine are limited, reducing their nutritional value. The results reported that the expression of AmA1 in transgenic tubers results in a significant increase in some essential amino acids (lysine, methionine, cysteine, and tyrosine). The transgenic plants presented a higher protein content (14.57 to 16.12 mg/g) compared to the control plants (11.15 mg/g).

Rascón-Cruz et al. (2004) used amaranthine (11s globulin) from *A. hypochondriacus*, one of the most important storage proteins of the seed and with a high content of essential amino acids, to transform the CIMMYT tropical maize genotype. Analyzes of transgenic maize indicated that recombinant amaranthine precursors were processed to their mature form and stably accumulated in the maize endosperm. The total protein content in the transgenic maize seed had a significant change (from 6 to 32% increase), which correlates with the concentrations of amino acids, which increased from 8 to 44% compared to samples that were not transformed. Likewise, there were specific increases of 18 and 22% in lysine and tryptophan, respectively, in which common corn is deficient; the amino acid isoleucine increased about 36%, suggesting that the increase in the concentration of these amino acids is due to the expression of amaranthine driven by the osGtl promoter in the seeds. Amaranthine digestibility in transgenic tropical maize was confirmed by immunological Western blot technique, indicating high molecular weight of several proteins, including amaranthine, which was completely digested within 20 minutes in simulated gastric fluids and degraded by complete in 15 minutes by simulated intestinal fluid. These findings demonstrate the feasibility of applying genetic engineering to improve the amino acid composition in grains.

Future Prospects of *Amaranthus* spp.

Shortly, due to the high demand and cost of animal protein, many countries are urged to look for other plant-based sources from which the food industry could choose to extract protein from amaranth seed, which is a cheap and high protein source. Amaranth grain can also be processed into compounds with biological activities, such as phenolic compounds or bioactive peptides, which could serve as a substitute for synthetic drugs due to the side effects. Therefore, the development of food and nutraceuticals from amaranth grains could serve in the future to combat chronic degenerative diseases without causing side effects. Another use of amaranth grain would be in genetic engineering, as it has been reported that amaranthine, a protein gene, contains a high amount of lysine and tryptophan, which could help reduce cases of protein malnutrition. Amaranth starch could also be used to produce biodegradable films or wall materials to protect bioactive compounds. The perspective of the use of the amaranth plant and seed is wide. The different secondary metabolites, including squalene, show it as a plant with excellent prospects for exploration in the cosmetic and pharmaceutical industry, and its proteins and carbohydrates in the food industry.

References

Aderibigbe, O. R., O. O. Ezekiel, S. O. Owolade, J. K. Korese, B. Sturm, and O. Hensel. 2022. "Exploring the Potentials of Underutilized Grain Amaranth (*Amaranthus* spp.) along the Value Chain for Food and Nutrition Security: A Review." *Critical Reviews in Food Science and Nutrition*. Taylor and Francis Ltd. doi:10.1080/10408398.2020.1825323.

Amare, Endale, Claire Mouquet-Rivier, Adrien Servent, Gilles Morel, Abdulaziz Adish, and Gulelat Desse Haki. 2015. "Protein Quality of Amaranth Grains Cultivated in Ethiopia as Affected by Popping and Fermentation." *Food and Nutrition Sciences* 06 (01). Scientific Research Publishing, Inc,: 38–48. doi:10.4236/fns.2015.61005.

Avanza, M. v., M. C. Puppo, and M. C. Añón. 2005. "Rheological Characterization of Amaranth Protein Gels." *Food Hydrocolloids* 19 (5). Elsevier: 889–98. doi:10.1016/j.foodhyd.2004.12.002.

Chakraborty, Subhra, Niranjan Chakraborty, and Asis Datta. n.d. "*Increased Nutritive Value of Transgenic Potato by Expressing a Nonallergenic Seed Albumin Gene from Amaranthus Hypochondriacus*." www.pnas.orgcgidoi10.1073pnas.050012697.

Cheng, Acga. 2018. "Review: Shaping a Sustainable Food Future by Rediscovering Long-Forgotten Ancient Grains." *Plant Science*. Elsevier Ireland Ltd. doi:10.1016/j.plantsci.2018.01.018.

Coelho, Laylla Marques, Pedro Miguel Silva, Joana T. Martins, Ana C. Pinheiro, and António A. Vicente. 2018. "Emerging Opportunities in Exploring the Nutritional/Functional Value of Amaranth." *Food and Function*. Royal Society of Chemistry. doi:10.1039/c8fo01422a.

Corke, Harold, Y. Z. Cai, and H. X. Wu. 2015a. "Amaranth: Overview." *Encyclopedia of Food Grains: Second Edition* 1–4: 287–96. doi:10.1016/B978-0-12-394437-5.00032-2.

———. 2015b. "Amaranth: Overview." *Encyclopedia of Food Grains: Second Edition* 1–4: 287–96. doi:10.1016/B978-0-12-394437-5.00032-2.

Cornejo, Fabiola, Geovanna Novillo, Elena Villacrés, and Cristina M. Rosell. 2019. "Evaluation of the Physicochemical and Nutritional Changes in Two Amaranth Species (*Amaranthus quitensis* and *Amaranthus caudatus*) after Germination." *Food Research International* 121 (Elsevier: 933–39. doi:10.1016/j.foodres.2019.01.022.

Corzo-Ríos, Luis, Leticia Garduño, Xariss Sánchez-Chino, Jorge Martínez-Herrera, Anaberta Cardador-Martínez and Cristian Jiménez-Martínez. 2021. "Effect of the consumption of amaranth seeds and their sprouts on alteration of lipids and glucose metabolism in mice." *International Journal of food science and technology* 3269 (53). doi:10.1111/ijfs.15014.

Coţovanu, Ionica, and Silvia Mironeasa. 2021. "Impact of Different Amaranth Particle Sizes Addition Level on Wheat Flour Dough Rheology and Bread Features." *Foods* 10 (7). MDPI AG. doi:10.3390/foods10071539.

D'Amico, Stefano, and Regine Schoenlechner. 2017. *Amaranth: Its Unique Nutritional and Health-Promoting Attributes. Gluten-Free Ancient Grains: Cereals, Pseudocereals, and Legumes: Sustainable, Nutritious, and Health-Promoting Foods for the 21st Century*. Elsevier Ltd. doi:10.1016/B978-0-08-100866-9.00006-6.

Das, Dipak, Nisar A. Mir, Narender Kumar Chandla, and Sukhcharn Singh. 2021. "Combined Effect of PH Treatment and the Extraction PH on the Physicochemical, Functional and Rheological Characteristics of Amaranth (*Amaranthus hypochondriacus*) Seed Protein Isolates." *Food Chemistry* 353 (August). Elsevier Ltd. doi:10.1016/j.foodchem.2021.129466.

Das, Saubhik. 2016. *Amaranthus: A Promising Crop of Future*. Singapore: Springer.

Dus-zuchowska, Monika, Jaroslaw Walkowiak, Anna Morawska, Patrycja Krzyzanowska- Jankowska, Anna Miskiewicz-chotnicka, Juliusz Przyslawski, and Aleksandra Lisowska. 2019. "Amaranth Oil Increases Total and LDL Cholesterol Levels without Influencing Early Markers of Atherosclerosis in an Overweight and Obese Population: A Randomized Double-Blind Cross-over Study in Comparison with Rapeseed Oil Supplementation." *Nutrients* 11 (12). MDPI AG. doi:10.3390/nu11123069.

Gebreil, Seham Yehia, Maha Ibrahim Kamal Ali, and Esraa Ahmed Mohamed Mousa. 2020. "Utilization of Amaranth Flour in Preparation of High Nutritional Value Bakery Products." *Food and Nutrition Sciences* 11 (05). Scientific Research Publishing, Inc.: 336–54. doi:10.4236/fns.2020.115025.

Gimplinger, D. M., G. Dobos, R. Schönlechner, and H. P. Kaul. 2007. "Yield and Quality of Grain Amaranth (*Amaranthus* sp.) in Eastern Austria." *Plant, Soil and Environment* 53 (3): 105–12. doi:10.17221/2224-pse.

Gliszczyńska-Świgło, Anna, Inga Klimczak, and Iga Rybicka. 2018. "Chemometric Analysis of Minerals in Gluten-Free Products." *Journal of the Science of Food and Agriculture* 98 (8). John Wiley and Sons Ltd: 3041–48. doi:10.1002/jsfa.8803.

Huerta-Ocampo, Jose Angel, and Ana Paulina Barba de la Rosa. 2011. "Amaranth: A Pseudo-Cereal with Nutraceutical Properties." *Current Nutrition & Food Science* 7 (1): 1–9. https://www.ingentaconnect.com/content/ben/cnf/2011/00000007/00000001/art00001.

Joshi, Dinesh C., Salej Sood, Rajashekara Hosahatti, Lakshmi Kant, A. Pattanayak, Anil Kumar, Dinesh Yadav, and Markus G. Stetter. 2018. "From Zero to Hero: The Past, Present and Future of Grain Amaranth Breeding." *Theoretical and Applied Genetics*. Springer Verlag. doi:10.1007/s00122-018-3138-y.

Kaur, Shubhpreet, Narpinder Singh, and Jai Chand Rana. 2010. "*Amaranthus hypochondriacus* and *Amaranthus caudatus* Germplasm: Characteristics of Plants, Grain and Flours." *Food Chemistry* 123 (4). Elsevier Ltd: 1227–34. doi:10.1016/j.foodchem.2010.05.091.

Klimczak, I., M. Małecka, and B. Pachołek. 2002. "Antioxidant Activity of Ethanolic Extracts of Amaranth Seeds." *Nahrung - Food* 46 (3): 184–86. doi:10.1002/1521-3803(20020501)46:3<184::AID-FOOD184>3.0.CO;2-H.

Kong, Xiangli, Harold Corke, and Eric Bertoft. 2009. "Fine Structure Characterization of Amylopectins from Grain Amaranth Starch." *Carbohydrate Research* 344 (13): 1701–8. doi:10.1016/j.carres.2009.05.032.

Kraujalis, Paulius, Petras R. Venskutonis, Vaida Kraujalienė, and Audrius Pukalskas. 2013. "Antioxidant Properties and Preliminary Evaluation of Phytochemical Composition of Different Anatomical Parts of Amaranth." *Plant Foods for Human Nutrition* 68 (3). Kluwer Academic Publishers: 322–28. doi:10.1007/s11130-013-0375-8.

Kshirsagar, Amol C., and Rekha S. Singhal. 2008. "Preparation of Hydroxypropyl Corn and Amaranth Starch Hydrolyzate and Its Evaluation as Wall Material in Microencapsulation." *Food Chemistry* 108 (3): 958–64. doi:10.1016/j.foodchem.2007.11.074.

Malinski, Edmond, James R. Daniel, Xia X. Zhang, and Roy L. Whistler. 2003. "*Isolation of Small Starch Granules and Determination of Their Fat Mimic Characteristics*." Vol. 80.

Martinez-Lopez, Alicia, Maria C. Millan-Linares, Noelia M. Rodriguez-Martin, Francisco Millan, and Sergio Montserrat-de la Paz. 2020. "Nutraceutical Value of Kiwicha (*Amaranthus caudatus* L.)." *Journal of Functional Foods*. Elsevier Ltd. doi:10.1016/j.jff.2019.103735.

Montoya-Rodríguez, Alvaro, Mario A. Gómez-Favela, Cuauhtémoc Reyes-Moreno, Jorge Milán-Carrillo, and Elvira González de Mejía. 2015. "Identification of Bioactive Peptide Sequences from Amaranth (*Amaranthus hypochondriacus*) Seed Proteins and Their Potential Role in the Prevention of Chronic Diseases." *Comprehensive Reviews in Food Science and Food Safety* 14 (2): 139–58. doi:10.1111/1541-4337.12125.

Morales, Diego, Marta Miguel, and Marta Garcés-Rimón. 2021. "Pseudocereals: A Novel Source of Biologically Active Peptides." *Critical Reviews in Food Science and Nutrition*. Bellwether Publishing, Ltd. doi:10.1080/10408398.2020.1761774.

Mota, Carla, Mariana Santos, Raul Mauro, Norma Samman, Ana Sofia Matos, Duarte Torres, and Isabel Castanheira. 2016. "Protein Content and Amino Acids Profile of Pseudocereals." *Food Chemistry* 193 (February). Elsevier Ltd: 55–61. doi:10.1016/j.foodchem.2014.11.043.

Muyonga, John H., Brian Andabati, and Geoffrey Ssepuuya. 2014. "Effect of Heat Processing on Selected Grain Amaranth Physicochemical Properties." *Food Science & Nutrition* 2 (1). Wiley: 9–16. doi:10.1002/fsn3.75.

Nana, Fernand W., Adama Hilou, Jeanne F. Millogo, and Odile G. Nacoulma. 2012. "Phytochemical Composition, Antioxidant and Xanthine Oxidase Inhibitory Activities of *Amaranthus cruentus* L. and *Amaranthus hybridus* L. Extracts." *Pharmaceuticals* 5 (6). MDPI AG: 613–28. doi:10.3390/ph5060613.

Olawoye, Babatunde, and Saka O. Gbadamosi. 2020. "Influence of Processing on the Physiochemical, Functional and Pasting Properties of Nigerian *Amaranthus viridis* Seed Flour: A Multivariate Analysis Approach." *SN Applied Sciences* 2 (4). Springer Nature. doi:10.1007/s42452-020-2418-8.

Palombini, Sylvio Vicentin, Thiago Claus, Swami Arêa Maruyama, Aline Kirie Gohara, Aloisio Henrique Pereira Souza, Nilson Evelázio de Souza, Jesuí Vergílio Visentainer, Sandra Terezinha Marques Gomes, and Makoto Matsushita. 2013. "Evaluation of Nutritional Compounds in New Amaranth and Quinoa Cultivars." *Food Science and Technology* 33 (2): 339–44. doi:10.1590/S0101-20612013005000051.

Paredes-López, Octavio. 1994. *Amaranth Biology, Chemistry and Technology*. Edited by Octavio Paredes-López. Chemical Ruber Company. 1st ed. Irapuato, México: Taylor & Francis Group.

Park, Seon Joo, Anshul Sharma, and Hae Jeung Lee. 2020. "A Review of Recent Studies on the Antioxidant Activities of a Third-Millennium Food: *Amaranthus* spp." *Antioxidants* 9 (12). MDPI: 1–22. doi:10.3390/antiox9121236.

Paśko, P., M. Sajewicz, S. Gorinstein, and Z. Zachwieja. 2008. "Analysis of Selected Phenolic Acids and Flavonoids in *Amaranthus cruentus* and *Chenopodium quinoa* Seeds and Sprouts by HPLC." *Acta Chromatographica* 20 (4): 661–72. doi:10.1556/AChrom.20.2008.4.11.

Patil, Savita D., Milap R. Patel, Sachin R. Patel, and Sanjay J. Surana. 2012. "*Amaranthus spinosus* Linn. Inhibits Mast Cell-Mediated Anaphylactic Reactions." *Journal of Immunotoxicology* 9 (1): 77–84. doi:10.3109/1547691X.2011.631609.

Pellegrini, Nicoletta, and Carlo Agostoni. 2015. "Nutritional Aspects of Gluten-Free Products." *Journal of the Science of Food and Agriculture*. John Wiley and Sons Ltd. doi:10.1002/jsfa.7101.

Perales-Sánchez, Janitzio X.K., Cuauhtémoc Reyes-Moreno, Mario A. Gómez-Favela, Jorge Milán-Carrillo, Edith O. Cuevas-Rodríguez, Angel Valdez-Ortiz, and Roberto Gutiérrez-Dorado. 2014. "Increasing the Antioxidant Activity, Total Phenolic and Flavonoid Contents by Optimizing the Germination Conditions of Amaranth Seeds." *Plant Foods for Human Nutrition (Dordrecht, Netherlands)* 69 (3): 196–202. doi:10.1007/s11130-014-0430-0.

Perales-Sanchez, X, C Reyes, and Roberto Gutierrez. 2015. *Efecto de La Germinación Sobre El Potencial Antioxidante de Amaranto* [*Effect of Sprouting on the Antioxidant Potential of Amaranth*]. 1st ed. Culiacan, Sinaloa: Publicia.

Peter, Kavita, and Puneet Gandhi. 2017. " Rediscovering the Therapeutic Potential of *Amaranthus* Species : A Review ." *Egyptian Journal of Basic and Applied Sciences* 4 (3). Informa UK Limited: 196–205. doi:10.1016/j.ejbas.2017.05.001.

Pilco-Quesada, Silvia, Ye Tian, Baoru Yang, Ritva Repo-Carrasco-Valencia, and Jukka Pekka Suomela. 2020. "Effects of Germination and Kilning on the Phenolic Compounds and Nutritional Properties of Quinoa (*Chenopodium quinoa*) and Kiwicha (*Amaranthus caudatus*)." *Journal of Cereal Science* 94 (April). Elsevier Ltd: 102996. doi:10.1016/j.jcs.2020.102996.

Pinto, Suneeta. 2018. "Amaranth - A Functional Food." *Concepts of Dairy & Veterinary Sciences* 1 (3). Lupine Publishers LLC. doi:10.32474/cdvs.2018.01.000112.

Rascón-Cruz, Q., S. Sinagawa-García, J. A. Osuna-Castro, N. Bohorova, and O. Paredes-López. 2004. "Accumulation, Assembly, and Digestibility of Amarantin Expressed in Transgenic Tropical Maize." *Theoretical and Applied Genetics* 108 (2): 335–42. doi:10.1007/s00122-003-1430-x.

Rastogi, Anu, and Sudhir Shukla. 2013. "Amaranth: A New Millennium Crop of Nutraceutical Values." *Critical Reviews in Food Science and Nutrition* 53 (2): 109–25. doi:10.1080/10408398.2010.517876.

Repo-Carrasco-Valencia, Ritva, Jarkko K. Hellström, Juha Matti Pihlava, and Pirjo H. Mattila. 2010. "Flavonoids and Other Phenolic Compounds in Andean Indigenous Grains: Quinoa (*Chenopodium quinoa*), Kañiwa (*Chenopodium pallidicaule*) and Kiwicha (*Amaranthus caudatus*)." *Food Chemistry* 120 (1). Elsevier Ltd: 128–33. doi:10.1016/j.foodchem.2009.09.087.

Roa, Diego F., Rosa I. Baeza, and Marcela P. Tolaba. 2015. "Effect of Ball Milling Energy on Rheological and Thermal Properties of Amaranth Flour." *Journal of Food Science and Technology* 52 (12). Springer India: 8389–94. doi:10.1007/s13197-015-1976-z.

Rustagi, S., S. Khan, and T. Jain. 2022. "Comparative Study of Nutritional and Functional Characteristics of Pearl Millet, Buckwheat, Amaranth and Unripe Banana Flours for Gluten-Free Bakery Products." *Food Research* 6 (3): 411–19. doi:10.26656/fr.2017.6(3).624.

Schoenlechner, Regine, Julian Drausinger, Veronika Ottenschlaeger, Katerina Jurackova, and Emmerich Berghofer. 2010. "Functional Properties of Gluten-Free Pasta Produced from Amaranth, Quinoa and Buckwheat." *Plant Foods for Human Nutrition* 65 (4): 339–49. doi:10.1007/s11130-010-0194-0.

Shukla, Sudhir, Atul Bhargava, A. Chatterjee, J. Srivastava, N. Singh, and S. P. Singh. 2006. "Mineral Profile and Variability in Vegetable Amaranth (*Amaranthus tricolor*)." *Plant Foods for Human Nutrition* 61 (1): 21–26. doi:10.1007/s11130-006-0004-x.

Singh, Narpinder, and Prabhjeet Singh. 2011. "Amaranth: Potential Source for Flour Enrichment." *Flour and Breads and Their Fortification in Health and Disease Prevention*, 101–11. doi:10.1016/B978-0-12-380886-8.10010-8.

Singh, Vijaya, Yashvir Chauhan, Ram Dalal, and Susanne Schmidt. 2021. "Chickpea." In *The Beans and the Peas*, 173–215. Elsevier. doi:10.1016/B978-0-12-821450-3.00003-2.

Tang, Yao, Xihong Li, Peter X. Chen, Bing Zhang, Ronghua Liu, Marta Hernandez, Jamie Draves, Massimo F. Marcone, and Rong Tsao. 2016. "Assessing the Fatty Acid, Carotenoid, and Tocopherol Compositions of Amaranth and Quinoa Seeds Grown in Ontario and Their Overall Contribution to Nutritional Quality." *Journal of Agricultural and Food Chemistry* 64 (5): 1103–10. doi:10.1021/acs.jafc.5b05414.

Tapia-Blácido, Delia, Paulo J. Sobral, and Florencia C. Menegalli. 2005. "Development and Characterization of Biofilms Based on Amaranth Flour (*Amaranthus caudatus*)." *Journal of Food Engineering* 67 (1–2): 215–23. doi:10.1016/j.jfoodeng.2004.05.054.

Wu, Tao, Yufang Gao, Junyu Hao, Jinjin Yin, Wen Li, Jieting Geng, Rui Liu, Wenjie Sui, and Min Zhang. 2019. "Lycopene, Amaranth, and Sorghum Red Pigments Counteract Obesity and Modulate the Gut Microbiota in High-Fat Diet Fed C57BL/6 Mice." *Journal of Functional Foods* 60 (February). Elsevier: 103437. doi:10.1016/j.jff.2019.103437.

Yelisyeyeva, O. P., K. O. Semen, G. V. Ostrovska, D. V. Kaminskyy, T. V. Sirota, N. Zarkovic, D. Mazur, O. D. Lutsyk, K. Rybalchenko, and A. Bast. 2014. "The Effect of Amaranth Oil on Monolayers of Artificial Lipids and Hepatocyte Plasma Membranes with Adrenalin-Induced Stress." *Food Chemistry* 147. Elsevier Ltd: 152–59. doi:10.1016/j.foodchem.2013.09.119.

Zhu, Fan. 2017. "Structures, Physicochemical Properties, and Applications of Amaranth Starch." *Critical Reviews in Food Science and Nutrition* 57 (2): 313–25. doi:10.1080/10408398.2013.862784.

———. 2020. "Dietary Fiber Polysaccharides of Amaranth, Buckwheat and Quinoa Grains: A Review of Chemical Structure, Biological Functions and Food Uses." *Carbohydrate Polymers*. Elsevier Ltd. doi:10.1016/j.carbpol.2020.116819.

Chapter 14

Chickpea (*Cicer arietinum* L.): Processing and Release of Bioactive Compounds

D. del R. Moguel-Concha
J. E. Borges-Martínez
M. S. Cid-Gallegos
F. M. Juárez-Chairez
D. I. Téllez-Medina
and Cristian Jiménez-Martínez*
Escuela Nacional de Ciencias Biológicas, Instituto Politécnico Nacional, Ciudad de México, México

Abstract

Chickpea is a legume belonging to the *Fabaceae* family and is among the most widely consumed beans and peas. It is also a seed rich in carbohydrates, protein, fiber, and other micronutrients. Different biological activities have been reported *in vitro*, such as antihypertensive, antioxidant, and anticarcinogenic. This chapter focuses on the nutritional composition and the bioactive components released during the different processes that can be carried out on this seed. These bioactive components have the potential to be included in products for the food and pharmaceutical industry, and this confers an essential value to the chickpea seed.

Keywords: chickpea, biological activities, bioactive compounds

Introduction

Legumes belong to the Fabaceae family and are classified into two groups: grain and oilseeds. Grain legumes are usually consumed after different heat treatments such as boiling, steaming, roasting, frying, and pressure cooking, while oilseeds are mainly used to extract oil (Juárez Chairez et al., 2020). Legumes are of high interest due to their high nutrient density and health

* Corresponding Author's Email: crisjm_99@yahoo.com.

In: Food Byproducts
Editors: S. Teresita Martín-del-Campo, A. Cardador-Martínez and J. del Pilar Ramírez-Anaya
ISBN: 979-8-88697-612-0

benefits. They are high in dietary fibre, resistant starch, protein, vitamins, and minerals. The high lysine content of legume protein and high folate content make legume meals ideal for blending with other products (Hall, Hillen, and Robinson 2017). Specifically, chickpea (*Cicer arietinum* L.) is the second most cultivated legume in the world. India is the world's largest chickpea producer, followed by Australia and Turkey (Kaur and Prasad 2021).

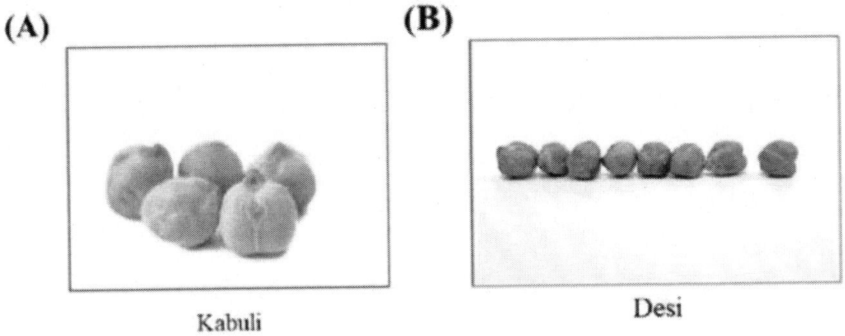

Figure 1. Chickpea seed varieties (Wang et al., 2021).

There are two types or varieties of this seed (Figure1): Kabuli (macrosperma) and Desi (microsperma). The Kabuli type is consumed in the Mediterranean region. In the Americas, the seeds are large (100-750 mg), round, with a smooth surface and beige, and have an energy value of approximately 365 kcal/100 g. Seeds of type Desi are mainly cultivated in semi-arid soils, are small, angular, with a rough, striated surface, and thick, dark-coloured testa, with an energy value of 327 kcal/100 g (Juárez Chairez et al., 2020, Jukanti et al., 2012). Desi type seed accounts for about 80-85% of the total area of chickpea cultures in Asia and Africa. In contrast, the Kabuli type is mainly produced in West Asia, North Africa, North America, and Europe (Jukanti et al., 2012).

Chemical Composition

Chickpea is a good source of carbohydrates and proteins, which together constitute about 80% of the total seed mass on a dry basis. Chickpea contains no cholesterol and is a good source of dietary fiber, vitamins, and minerals (Jukanti et al., 2012). Carbohydrates are the primary component in chickpea seed, ranging 62-70%. Chickpea seed contains monosaccharides such as ribose, glucose, galactose and fructose, disaccharides such as sucrose and maltose, and oligosaccharides such as stachyose, cycleritol, raffinose and verbascose. Cyceritol and stachyose are the most abundant in chickpea seed, constituting 36-43% and 25%, respectively, of the total sugars in chickpea seed (Juárez Chairez et al., 2020, Jukanti et al., 2012).

The other compounds correspond to polysaccharides such as starch (35% resistant and 65% available) and dietary fiber (18-22%), of which 4-8% is soluble, and 10-18% is insoluble (Juárez Chairez et al., 2020). The carbohydrates profile is a special characteristic that makes legumes a particular food group. However, when present in significant proportions, legume carbohydrates are digested slowly, which is why legumes have been considered low glycaemic index foods (Sánchez-Chino et al., 2015). When α-galactosides reach the colon, they are fermented by bacteria that produce hydrogen, methane, and $CO2$, which later cause flatulence because they are not absorbed by the human digestive system, which lacks the enzyme α-galactosidase, responsible for the degradation of these polysaccharides. The concentration of α-galactosides in legumes is 2-10 g/100 g of dry sample (Sánchez-Chino et al., 2015).

On the other hand, chickpea proteins have a high bioavailability and good digestibility (48-89.01%) (Juárez Chairez et al., 2020). Specifically, chickpea seeds of the type Kabuli have higher protein digestibility compared to the Desi type and in vitro protein digestibility ranging 34-76% (Kaur and Prasad 2021). In general, chickpea proteins include globulins (~56 g/100 g), albumins (~12 g/100 g), glutelins (~18 g/100 g), prolamins (~3 g/100 g) and residual proteins. Among legumes, chickpea proteins are reported to have a higher content of glutelins. Globulins are the main storage proteins of chickpea and are composed of two main groups: 11S proteins referred to as legumins (320-400 kDa) and 7S proteins referred to as vicilin (145-190 kDa) (Boukid 2021). Legumins are oligomeric proteins consisting of six αβ-subunits (56-60 kDa), where disulfide bonds link the α and β chains. This is the main storage protein and accounts for 97% of all globulins (Boukid 2021, Juárez Chairez et al., 2020).

Vicilin are trimeric proteins that lack cysteines and thus disulfide bonds. The albumin fraction plays an essential role in seeds because they include the most enzymatic and metabolic proteins. Albumins are a rich source of essential amino acids, as are other legume proteins, particularly sulfur-containing amino acids (tryptophan, threonine, and lysine) and therefore have a higher nutritional value compared to globulins. Glutelins belong to the 11-12S globulin family, structurally like globulin (Boukid 2021). Chickpea protein is rich in essential amino acids such as isoleucine, lysine, total aromatic amino acids, and tryptophan. Leucine (8.7% of protein) has been reported as the highest concentration, followed by arginine (8.3% of protein) and lysine (7.2% of protein) (Boukid 2021). However, leucine, total sulphur amino acids, methionine, cystine, threonine and valine have been reported as deficient amino acids (Boukid 2021).

Lipids are found in low concentrations in the chickpea seed (4-10%). The fatty acid composition of chickpea is 66% polyunsaturated fatty acids (PUFA), 19% monounsaturated fatty acids (MUFA) and saturated fatty acids (SFA) are about 15%. The most dominant fatty acid in chickpea is linoleic (51.2%), followed by oleic (32.6%) (Kaur and Prasad 2021), linolenic (0.5-0.9%), palmitic (18.9-20.4%) and stearic (1.3-1.7%) (Juarez-Chairez et al., 2020). The linoleic acid metabolism at the tissue level is responsible for producing prostaglandins, a hormone responsible for smooth muscle constriction and lowering blood pressure (Kaur and Prasad 2021). Finally, this seed is an outstanding source of B vitamins (riboflavin, thiamin, niacin, and pyridoxine), vitamin B9 (folic acid) and vitamin E (tocopherol) (Juárez Chairez et al., 2020). Chickpea seed carries Fe, Zn, Ca, Mg, K, Cu and P. About 100 g of chickpea seed can meet the dietary requirements of Fe (1.05 mg/day in men and 1.46 mg/day in women) and Zn (4. 2 mg/day in men and 3.0 mg/day in women) and 200 g can satisfy those of Mg (260 mg/day in men and 220 mg/day in women) (Juarez Chairez et al., 2020, Jukanti et al., 2012).

Bioactive Compounds from Chickpea through Different Processing

In plants, phenolic compounds (Figure 2) are responsible for several functions, including structural components, UV protection and antioxidant, signaling and defense molecules.

Lignins and lignans are the phenolic polymers that make up the secondary cell wall of plants, providing the physical structure and being closely related to the growth of the various plant parts, mainly the stem, roots, and seeds. These polymers are responsible for providing a protective barrier against herbivores, fungi, bacteria, and viruses, as well as forming an intricate network of biochemical compounds that perform scavenging of reactive oxygen species (ROS) to antimicrobial and wound-signaling functions. In addition, hydrolysable tannins are known as anti-herbivore molecules, causing proteins in the digestive tract of insects to precipitate, impairing enzymatic functions and reducing the digestibility of ingested proteins. Phenolic acids are important signaling molecules in interactions with plant microbes, e.g., salicylic acid, and can enhance nutrient uptake and protect against infections (De Camargo et al., 2019).

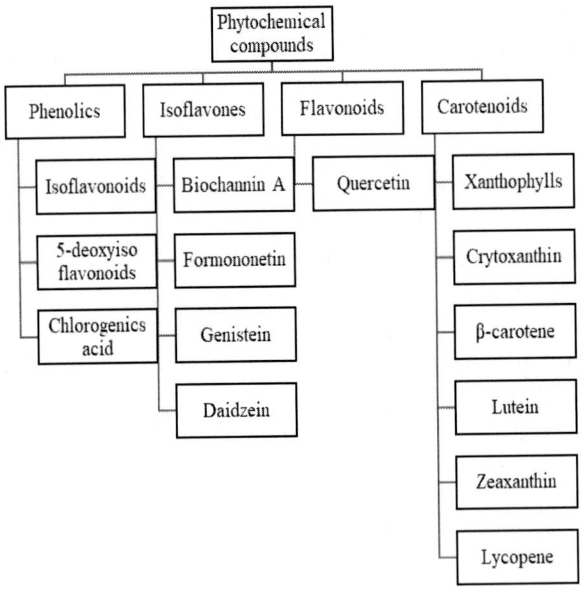

Figure 2. Phytochemical compounds identified in chickpea seed.

The synthesis of phenolic compounds takes place from phenylalanine and tyrosine, and phenols regulate lignin metabolism and synthesis (Sánchez-Chino et al., 2015). Phe and Tyr give rise to cinnamic and ρ-hydroxycinnamic acids, which generate the cinnamol structure of flavonoids upon condensation with acetate. Subsequently, glycosylated, or sulphated derivatives are formalized. In this group, proanthocyanidins (condensed tannins) can be identified, which are the major fraction in seeds such as chickpea. Phenolic compounds have various beneficial effects on the health of consumers; the main biological activities of these molecules are antioxidation, anti-inflammation and inhibition of α-amylase and aminoglycosidase enzymes involved in glucose regulation; they have also been reported to have an inhibitory effect on angiotensin-converting enzyme-1 (ACE-1) (Sánchez-Chino et al., 2015). Finally, they have been reported to account for 0.72-1.81 mg/g of seed (Juárez Chairez et al., 2020).

Regarding isoflavones, these are secondary metabolites that can be isolated from plant tissues, including leaves, stems, roots, flowers, seeds, and shoots; in the latter two, these polyphenolic compounds are found in high amounts and appear to regulate physiological processes important for plant growth. Isoflavones are widely present as glycosides and are therefore highly polar molecules. Chemically, they are derived from heterocyclic phenols that have a closed structure like that of estrogens. These compounds are potent antioxidants and show anticancer activity in vivo. Moreover, when used in in vitro experiments with human cell lines, proliferative (estrogenic) and anti-proliferative (anti-estrogenic) effects have been reported.

Genistein, one of the most important isoflavones, and perhaps other phytoestrogens, have been proposed by several authors to inhibit tumor cell growth through tyrosine kinase at activated growth factor receptors, and inhibition of DNA topoisomerases, which are essential for mitogenic signal transduction. Soy isoflavones have been shown to have effects against breast and prostate cancer cell lines (Sanchez-Chino et al., 2015). During oxidative stress, interleukins and inflammatory mediators are known to be produced. Natural product isoflavones have been shown to have anti-inflammatory properties in vitro and in vivo. Neuroinflammation, a phenomenon mediated primarily by microglial cells, can culminate in Parkinson's and Alzheimer's diseases, multiple sclerosis, and amyotrophic lateral sclerosis, among other chronic diseases.

The potential of genistein in the treatment and/or prevention of neuroinflammation has already been reported. The modulatory effects induced by flavonoids are mediated by their impact mainly on mitogen-activated protein kinases (MAPKs) and nuclear factor kappa B (NK-κB). Furthermore, biochanin A, a derivative of genistein, has also been reported to reduce neuronal inflammatory damage to joint cartilage. Its mechanism of action has been attributed to its ability to reduce phosphorylation of MAPKs, c-Jun N-terminal kinases (JNK), extracellular signal-regulated kinase (EPK), and p38. However, biochanin A acts simultaneously on other targets. It suppresses the expression of laminin and adhesion molecules, namely vascular cell adhesion molecule-1 (VCAM-1), intercellular adhesion molecule-1 (ICAM-1), and E-selectin molecules in human umbilical vein endothelial cells. The effects of biochanin A are related to the activation of peroxisome proliferator-activated receptor (PPAR)-γ. In addition, its ability to reduce NF-κB activation has been reported. E-selectin depletion is a marker of improved endothelial function, control of metabolic status, and reduction of atherosclerosis and inflammation in diabetic patients (De Camargo et al., 2019).

Daidzein, another isoflavone aglycone, has demonstrated anti-inflammatory potential; inhibition of the TRL4-MyD88-NF-κB signaling pathway is possibly involved in the protective action of daidzein. By reducing the NF-κB pathway, daidzein and other flavonoids such as quercetin have been shown to simultaneously affect the STAT-1 pathway. Finally, formononetin, a derivative of daidzein, has antinociceptive effects in vivo in mouse models and reduces neutrophil migration to the peritoneal cavity, increases PPAR-γ gene expression and enhances superoxide dismutase activity, and inhibits myeloperoxidase activity (De Camargo et al., 2019).

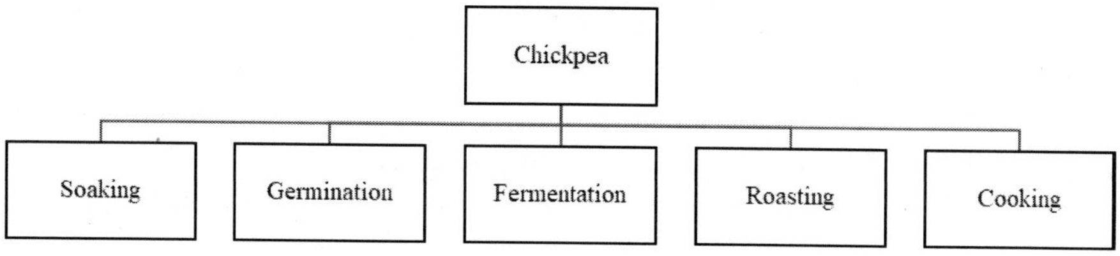

Figure 3. Processes used for the consumption of chickpea seed.

On the other hand, chickpea contains flavonoids, the main source of the non-enzymatic antioxidant apparatus in plants; they provide a UV-screening effect and play a central role in the elimination of ROS derived from photorespiration under high light conditions, as well as under low/high temperature stress or other oxidative stress (De Camargo et al., 2019). There are also plant carotenoids which are lipid-soluble antioxidants/pigments responsible for the bright colours (usually red, yellow, and orange) of different plant tissues (Jukanti et al., 2012). They are of great importance as they increase the bioavailability of iron in the human body by acting as promoters of iron absorption (Kaur and Prasad 2021). They are classified into two types: oxygenated, called xanthophylls, which include lutein, violaxanthin and neoxanthin; and non-oxygenated, called carotenoids, which include β-carotene and lycopene. The carotenoids reported in chickpea are β-carotene, lutein, zeaxanthin, β-cryptoxanthin, lycopene and a-carotene. Carotene is the most important and widely distributed carotenoid in plants and is converted to vitamin A more efficiently than the other carotenoids (Jukanti et al., 2012).

With respect to their health-promoting potential, phenolic compounds are known for their capability to scavenge ROS (such as superoxide, hydrogen peroxide, and hydroxyl radicals). Also, the antioxidant activity of phenolic compounds can be explained by their reducing power and/or their metal chelating activity towards ferrous and ferric ions, respectively (De Camargo et al., 2019, Kaur and Prasad 2021). Related to structure/activity, phytochemicals with one or more methoxy substitutions are considered strong chelating agents. Also, isoflavones have been reported to bind metals at the 4-keto and 5-OH sites (De Camargo et al., 2019). Biochanin A, found in chickpeas, reduces oxidative stress by decreasing malondialdehyde (MDA) levels and increasing catalase (CAT), superoxide dismutase (SOD) and total antioxidant status in diabetic rats; and has reported a protective effect against oxidative stress in a rat model of Parkinson's disease by inhibiting nicotinamide adenine dinucleotide adenine phosphate (NADH oxidase) activation (De Camargo et al., 2019).

Legumes are traditionally consumed after several processes, including soaking, cooking, and germination; these treatments improve the palatability of legumes and increase the bioavailability of nutrients, due to the inactivation of protease inhibitors and hemagglutinins, among other compounds (Sánchez-Chino et al., 2015). In the specific case of chickpea seed, there are various treatments (Figure 3) to which it is subjected for consumption, such as soaking, germination, fermentation, roasting, and cooking.

Soaking is the process of hydration of the grain, allowing water to be absorbed by the cells and increasing the moisture content (Kaur and Prasad 2021). This treatment, using hot or cold water, softens the outer layers of the seeds, thus facilitating wet shelling (Boukid 2021). Both soaking and dehulling improve protein extraction and reduce anti-nutritional factors (Boukid 2021). Soaking the kernels is the prerequisite for the germination process. Seed dormancy is broken during soaking due to the absorption of water by the seeds. Seed germination occurs in three stages: The first is the imbibition stage, during which rapid absorption of water by the seeds occurs, in the second stage activation of biochemical processes occurs and the rate of water absorption is reduced, and during the third stage of germination emergence of the seed radicle occurs (Kaur and Prasad 2021). Germination has been found to be associated with improved nutritional quality by increasing protein and starch digestibility, vitamin content and mineral bioavailability, and reduction of anti-nutritional components such as phytic acid, tannins and oligosaccharides.

During germination, the formation of enzymes that convert complex carbohydrates into simpler forms that are easily metabolized occurs and the number of oligosaccharides mainly raffinose and stachyose are reduced to a greater extent influenced by germination time compared to the presence or absence of light (Kaur and Prasad 2021). In a study by Wu et al., (2012), they reported that germination increased the content of phenolic compounds (91.7 mg GAE/100 g of fresh sample). Antioxidant capacity was assessed by oxygen radical absorbance capacity (ORAC) assay, reporting that germination leads to almost a threefold increase in chickpea seed (874 μmol Trolox equivalents/100 g of fresh sample in ungerminated seed vs. 2457 μmol Trolox equivalents/100 g of fresh sample at 3 days of germination).

With respect to isoflavones, the content of biochanin A and formononetin changed during germination from 19 and 1.0 μg/g of a fresh sample of ungerminated seeds to 702 and 1484 μg/g of fresh sample in seeds germinated for 4 days, respectively. And, with respect to flavonoids, an increase of approximately 90-fold was found (83 μg/g fresh sample in non-germinated seeds to 7568 μg/g of fresh sample in germinated seeds). On the other hand, Haileslassie, Henry, and Tyler (2019) determined the effects of soaking, germination, cooking and their combinations on the content of nutrients and anti-nutrients in chickpea. Soaking and sprouting significantly reduced the phytate and tannin contents of chickpea samples and, overall, polyphenol contents were lower after soaking-cooking than after sprouting-cooking.

For most chickpea samples, longer germination times (72 h) gave superior results in terms of decreased cooking time, tannin content and phytate:zinc and phytate:iron molar ratios. On the other hand, Sofi et al., (2020) studied the effect of sprouted chickpea flour and chickpea protein isolate on the qualities of rice-based noodles showing that after increasing level of sprouted chickpea flour and protein isolate, the protein content (7.22-14.35%), antioxidant activity (22.75-33.79%) and total phenolic content (117.65-223.35 mg GAE/100 g) of rice-based noodles increased significantly. The in vitro protein digestibility (78.68-87.59%) of the noodles increased significantly with the added chickpea ingredients (30% sprouted flour and 1% protein isolate), this increase could be due to a greater number of active sites for protease action by chickpea proteins and protein denaturation induced during noodle processing, which enhances proteolytic enzymatic hydrolysis digestion.

The glycemic index (from 70.83 to 57.49) of rice-based noodles showed a significant decrease with increasing level of chickpea ingredients (sprouted flour and protein isolate). Overall, they reported that germination improved the proteins' nutritional, functional and antioxidant properties. Also, Bruno et al., (2019) reported that the phenolic content and antioxidant potential were significantly higher in germinated chickpea flour (8.4 ± 0.22 mg GAE/g flour and 2.36 ± 0.03 mmol Trolox equivalent/100 g flour, respectively) than those in ungerminated chickpea flour (7.3 ± 0.08 mg GAE/g flour and 2.06 ± 0.04 mmol Trolox equivalent/100 g flour, respectively). In addition, Ferreira et al., (2019) also reported that germination succeeded in increasing protein content and solubility without altering its digestibility (above 85.4%). It also reduced carotenoid and tocopherol content; however, the unsaturated fatty acid profile remained unchanged. The content of phenolic compound levels for the ungerminated seed was 369.47 mg/100 g free phenols, and 109.01 mg/100 g bound phenols, while for the germinated seed, an increase of 20.10% at 36 h and 19. 05% at 48 h with respect to free phenols; however, with respect to bound phenolic compounds, a reduction of 17.90% at 12 h and 45.95% at 24 h was found, followed by an increase of 6% at 36 h and 31.9% at 48 h.

Regarding the antioxidant activity, in ungerminated seed, it was 77.26 mg/100 g (ABTS scavenging), whereas, after germination, it reached 71.23% (36 h) and 67.66% (48 h) of ABTS antioxidant activity. Finally, the γ-aminobutyric acid content (6.42 mg/100 g, seed without germination at 245.76 mg/100 g during 48 h of germination), the increase in γ-aminobutyric acid content is probably associated with increased activity of the enzyme glutamate decarboxylase, which catalyzes the decarboxylation of L-glutamic acid, thus releasing carbon dioxide. During germination, amylases, proteases, and other enzymes responsible for cell wall hydrolysis are synthesized, thus promoting the solubilization of bound phenolic compounds. It is also possible that during germination, the enzyme phenylalanine ammonylase is very active, thus promoting the synthesis of phenolic compounds, justifying the increase in the levels of free and bound fractions of phenolic compounds at the end of 48 h of germination. Domínguez-Arispuro et al., (2018) performed optimization of chickpea germination with different temperatures and times, concluding that the optimal germination conditions for chickpea were 171 h and 33.7 °C, with an increase of 286.21% and 213.14% in the levels of free and bound phenolic compounds, respectively, leading to an increase in antioxidant activity by ABTS. During germination, the plant reactivates secondary metabolism, promoting the synthesis of bioactive compounds such as phenolics, which are enhanced when compared to the ungerminated seed. Likewise, Domínguez-Arispuro et al., (2021) evaluated isoflavone synthesis and the impact of soluble isoflavone on cellular antioxidant activity and antiproliferative activity of breast cancer cells during germination of type-Desi chickpea seed. They identified six isoflavones (formononetin, biochanin-A and its glycosides) and the total isoflavone content increased (from 0.31 to 35.72 μgBA/mg extract). Germination activates the central phenylpropanoid pathway and isoflavonoid branching pathways in legumes by increasing phenylalanine ammonia-lyase activity.

This enzyme is known as the initiator of isoflavone synthesis. Cell antioxidant activity (HepG2) increased fivefold, from 137.2 to 788.2 μmol of quercetin equivalents/100 g of sample. These compounds could act at the cell membrane level and break the free radical chain reaction at the cell surface, or they might cross the membrane and react with intracellular ROS. In the sample of germinated chickpea flour, the malonyl glucoside of biochanin-A was the main isoflavone; hence, the antioxidant mechanism may be related to this compound. In addition, phenolic compounds that act as antioxidants, such as glycosylated isoflavones, bind strongly to the HepG2 cell membrane and act at the cell surface level. This behavior has been explained by their molecular size, polarity and solubility and they cannot easily enter the cell. The bioactive compounds from the germinated chickpea decreased the proliferation of the MDA-MB-231 cell line. They also caused morphological changes such as cell shrinkage, rounding and nuclear fragmentation.

Related to fermentation, digestibility, nutritional value, and palatability are improved with the decrease of anti-nutritional factors. Fermentation significantly increases the concentration of vitamins such as riboflavin, thiamine, niacin, and ascorbic acid, and improves protein digestibility. It has been reported that trypsin and chymotrypsin activity decreases, as well as the level of phytates, because of fermentation. During the fermentation process, the enzyme α-glucosidase is activated, which is responsible for the hydrolysis of the oligosaccharides, verbascose, stachyose and raffinose and converts them into digestible sugars. For example, Xing et al., (2020) applied spontaneous solid-state fermentation (SSF) with daily setback at 37 °C. The dominant indigenous lactic acid bacteria strains in the chickpea flour and enriched fractions were *Pediococcus pentosaceus* and *Pediococcus acidilactici*. The strains were selected based on their ability to metabolise α-galactosides, causing flatulence. At 72 h, raffinose and stachyose concentrations were diminished by 88.3-99.1%, while verbascose became undetectable. In addition, phytic acid decreased by 17%, while total phenolic content increased by 119%, likely since plant phenols are known to vary in their molecular structure as free or conjugated structures.

Furthermore, fermented chickpea flour has demonstrated an improved composition of essential amino acids such as methionine, cysteine, phenylalanine, tyrosine, and threonine (Rachwa-Rosiak, Nebesny, and Budryn 2015). Concerning cooking, Cid-Gallegos et al., (2020) induced colon cancer in male mice by administering azoxymethane/dextran sodium sulfate (AOM/DSS) and fed a standard diet with 10 and 20% cooked chickpea; for evaluation, they were sacrificed 1, 7 and 14 weeks after induction. The results show that cooking does not significantly ($p < 0.05$) modify nutritional compounds; however, it decreases the concentration of non-nutritional compounds and, consequently, the in vitro antioxidant activity, which was evaluated with ORAC, decreasing by 48%, with hydroxyl radical by 32%, and with superoxide radical by 39%. Total phenolic compounds were reduced by 25%. This was due to their thermolability, water solubility, and hydrolysability. In vivo evaluation showed that animals administered with AOM/DSS had higher concentrations of nitric oxide (NO), oxidized carbonyl groups (CO), malondialdehyde (MDA), and 4-hydroxy-2 nonenal (4-HNE) expression than animals without AOM/DSS administration. However, at all three times evaluated, these markers were significantly reduced ($p < 0.05$) by the consumption of cooked garbanzo. The best effect on oxidation markers occurred with the 20% cooked chickpea diet, demonstrating the antioxidant potential of cooked chickpea.

On the other hand, Dida Bulbula and Urga (2018) evaluated the effect of various traditional processing methods (boiling, wet roasting, dry roasting, germination, and fermentation) on nutritional composition and anti-nutritional factors (phytate, condensed tannin) in chickpea. The ranges of moisture content, protein, fat, fibre, total ash, utilized carbohydrate and energy were 5.9-9.4, 13.8-16.7, 4.3-6.1, 3.4-5.9, 2.3-2.7, 61.22-68.67 and 359.26-382.95% respectively. They reported that as the fermentation period increased, from 0 h (pH = 6.28) to 72 h (pH = 4.25), the pH decreased with a concomitant increase in acidity. And the ranges of condensed phytate and tannin contents were 72.07-97.46 and 99.26-175.23 mg/100 g, respectively. They concluded that fermentation and germination were the most effective in reducing phytate and condensed tannin content. Also, Milán-Noris et al., (2018) evaluated the anti-inflammatory activity of sprouted and cooked protein concentrates (GC and CC, respectively) and their relationship with protein and isoflavone composition before and after digestion and gastrointestinal absorption in vitro.

The anti-inflammatory activity of GC digests was almost 2 times higher than that of CC, which was associated with higher peptide, formononetin, and biochanin A content. The anti-inflammatory activity of the phenolic fraction in the digests was 7 times higher than that of the protein fraction. The most active peptide fraction of the GC digest ($IC50=93$ µg/mL) contained a total of 24 peptides derived from legumin and vicilin. They reported the potential of sprouted chickpea protein concentrates on exerting anti-inflammatory effects in the lower intestine, which could contribute to the prevention of inflammatory bowel diseases.

Finally, Fares and Menga (2012) studied the effects of the roasting process on the carbohydrate profile and antioxidant properties of chickpea flour, as well as the cooking behavior and antioxidant and nutritional properties of pasta enriched with chickpea flour. The roasting process increased the resistant starch (3-fold), insoluble dietary fiber (17%), and antioxidant properties of the flour. The addition of chickpea flour (raw and roasted) to durum wheat semolina changed the carbohydrate profile in the uncooked and cooked enriched pasta, especially with the roasted chickpea, and worsened the overall quality of the pasta. The increase in total phenolic content and total free phenolic acid content in the uncooked pasta was due to the positive effects of the addition of chickpea flours, while the increase in the fraction of bound phenols in the cooked pasta came from durum wheat, which was crucial for its high concentrations of ferulic acid. The increased free fraction of Trolox-equivalent antioxidant capacity in the cooked pasta was consistent with adding chickpea.

Conclusion

Chickpea seed is one of the most consumed legumes worldwide. In this chapter, a brief review of thermal and non-thermal processes such as soaking, germination, fermentation, roasting and cooking that help to improve the biological value and enhance the effect of its components, such as phenolic compounds, was carried out. However, it is important to mention that, with respect to chickpea seed, not only in vivo studies are needed to demonstrate its biological effects, but also more in vitro studies that help to continue knowing the advantages that are caused by consuming chickpea seed on its own, added to other foods or isolating components of it, allowing to obtain the best benefit from this seed.

References

Boukid, Fatma. 2021. "Chickpea (*Cicer arietinum* L.) Protein as a Prospective Plant-Based Ingredient: A Review." *International Journal of Food Science and Technology* 56 (11): 5435–44. https://doi.org/10.1111/ijfs.15046.

Bruno, Joseph A, Konas David W., Matthews Evan L., Feldman Charles H., Pinsley Kate M., and Kerrihard Adrian L.. 2019. "Sprouted and Non-Sprouted Chickpea Flours: Effects on Sensory Traits in Pasta and Antioxidant Capacity." *Polish Journal of Food and Nutrition Sciences* 69 (2): 203–9. https://doi.org/10.31883/pjfns/109280.

Camargo, Adriano Costa De, Trevenzoli Favero Bruno, Caldeira Morzelle Maressa, Franchin Marcelo, Alvarez-Parrilla Emilio, De La Rosa Laura A., Vilar Geraldi Marina, Maróstica Mário Roberto, Shahidi Fereidoon, and Schwember Andrés R.. 2019. *Is Chickpea a Potential Substitute for Soybean? Phenolic Bioactives and Potential Health Benefits. International Journal of Molecular Sciences*. Vol. 20. https://doi.org/10.3390/ijms20112644.

Cid-Gallegos, María S., Sánchez-Chino Xariss M., Álvarez-González Isela, Madrigal-Bujaidar Eduardo, Vásquez-Garzón Verónica R., Baltiérrez-Hoyos Rafael, Villa-Treviño Saúl, Dávila-Ortíz Gloria, and Jiménez-Martínez Cristian. 2020. "Modification of in Vitro and in Vivo Antioxidant Activity by Consumption of Cooked Chickpea in a Colon Cancer Model." *Nutrients* 12 (9): 1–22. https://doi.org/10.3390/nu12092572.

Dida Bulbula Dejene, and Urga Kelbessa. 2018. "Study on the Effect of Traditional Processing Methods on Nutritional Composition and Anti Nutritional Factors in Chickpea (*Cicer arietinum*)." *Cogent Food & Agriculture* 4 (1). https://doi.org/10.1080/23311932.2017.1422370.

Domínguez-Arispuro, D. M., Canizalez-Román A., Reyes-Moreno C., Milán-Noris A. K., Milán-Carrillo J., Acosta-Smith E., and Cuevas-Rodríguez E. O. 2021. "Isoflavones from Black Chickpea (*Cicer arietinum* L) Sprouts with Antioxidant and Antiproliferative Activity." *Saudi Journal of Biological Sciences* 28 (1): 1141–46. https://doi.org/10.1016/j.sjbs.2020.11.048.

Domínguez-Arispuro, D. M., Cuevas-Rodríguez E. O., Milán-Carrillo J., León-López L., Gutiérrez-Dorado R., and Reyes-Moreno C. 2018. "Optimal Germination Condition Impacts on the Antioxidant Activity and Phenolic Acids Profile in Pigmented Desi Chickpea (*Cicer arietinum* L.) Seeds." *Journal of Food Science and Technology* 55 (2): 638–47. https://doi.org/10.1007/s13197-017-2973-1.

Fares, Clara, and Menga Valeria. 2012. "Effects of Toasting on the Carbohydrate Profile and Antioxidant Properties of Chickpea (*Cicer arietinum* L.) Flour Added to Durum Wheat Pasta." *Food Chemistry* 131 (4): 1140–48. https://doi.org/10.1016/j.foodchem.2011.09.080.

Ferreira, Cristiano Dietrich, Bubolz Kern, da Silva Jennifer, Dittgen Vanessa Lambrecht, Ziegler Caroline Valmor, de Oliveira Raphaelli Chirle, and de Oliveira Maurício. 2019. "Changes in the Chemical Composition and Bioactive Compounds of Chickpea (*Cicer arietinum* L.) Fortified by Germination." *Lwt* 111 (January): 363–69. https://doi.org/10.1016/j.lwt.2019.05.049.

Haileslassie, Hiwot Abebe, Henry Carol J., and Tyler Robert T. 2019. "Impact of Pre-Treatment (Soaking or Germination) on Nutrient and Anti-Nutrient Contents, Cooking Time and Acceptability of Cooked Red Dry Bean (*Phaseolus vulgaris* L.) and Chickpea (*Cicer arietinum* L.) Grown in Ethiopia." *International Journal of Food Science and Technology* 54 (8): 2540–52. https://doi.org/10.1111/ijfs.14165.

Hall, Clifford Hillen Cassandra and Garden Robinson Julie. 2017. "Composition, Nutritional Value, and Health Benefits of Pulses." *Cereal Chemistry* 94 (1): 11–31. https://doi.org/10.1094/CCHEM-03-16-0069-FI.

Juárez, Chairez Faridy, Cid Gallegos María Stephanie, Meza-Márquez Ofelia Gabriela, and Jiménez-Martínez Cristian. 2020. "Biological Activities of Chickpea in Human Health (*Cicer arietinum* L.). A Review." *Plant Foods for Human Nutrition* 75 (2): 142–53. https://doi.org/10.1007/s11130-020-00814-2.

Jukanti, Aravind Kumar, Gaur P. M., Gowda C. L.L., and Chibbar R. N.. 2012. "Nutritional Quality and Health Benefits of Chickpea (*Cicer arietinum* L.): A Review." *British Journal of Nutrition* 108 (SUPPL. 1). https://doi.org/10.1017/S0007114512000797.

Kaur, Ravneet, and Kamlesh Prasad. 2021. "Technological, Processing and Nutritional Aspects of Chickpea (*Cicer arietinum*) - A Review." *Trends in Food Science and Technology* 109 (September 2020): 448–63. https://doi.org/10.1016/j.tifs.2021.01.044.

Milán-Noris, Ada Keila, Gutiérrez-Uribe Janet Alejandra, Santacruz Arlette, Serna-Saldívar Sergio O., and Martínez-Villaluenga Cristina. 2018. "Peptides and Isoflavones in Gastrointestinal Digests Contribute to the Anti-Inflammatory Potential of Cooked or Germinated Desi and Kabuli Chickpea (*Cicer arietinum* L.)." *Food Chemistry* 68 (March): 66–76. https://doi.org/10.1016/j.foodchem.2018.06.068.

Rachwa-Rosiak, Danuta, Ewa Nebesny, and Grażyna Budryn. 2015. "Chickpeas—Composition, Nutritional Value, Health Benefits, Application to Bread and Snacks: A Review." *Critical Reviews in Food Science and Nutrition* 55 (8): 1135–43. https://doi.org/10.1080/10408398.2012.687418.

Sánchez-Chino, Xariss, Jiménez-Martínez Cristian, Dávila-Ortiz Gloria, Álvarez-González Isela, and Madrigal-Bujaidar Eduardo. 2015. "Nutrient and Nonnutrient Components of Legumes, and Its Chemopreventive Activity: A Review." *Nutrition and Cancer* 67 (3): 401–10. https://doi.org/10.1080/01635581.2015.1004729.

Sofi, Sajad Ahmad, Singh Jagmohan, Chhikara Navnidhi, and Panghal Anil. 2020. "Effect of Incorporation of Germinated Flour and Protein Isolate from Chickpea on Different Quality Characteristics of Rice-Based Noodle." *Cereal Chemistry* 97 (1): 85–94. https://doi.org/10.1002/cche.10192.

Wang, Junyu, Li Yonghui, Li Ang, Hai Liu Rui, Gao Xin, Li Dan, Kou Xiaohong, and Zhaohui Xue. 2021. "Nutritional Constituent and Health Benefits of Chickpea (*Cicer arietinum* L.): A Review." *Food Research International* 150 (May). https://doi.org/10.1016/j.foodres.2021.110790.

Wu, Ziyun, Song Lixia, Feng Shengbao, Liu Yuancai, He Guangyuan, Yioe Yoecelyn, Shao Quan Liu, and Dejian Huang. 2012. "Germination Dramatically Increases Isoflavonoid Content and Diversity in Chickpea (*Cicer arietinum* L.) Seeds." *Journal of Agricultural and Food Chemistry* 60 (35): 8606–15. https://doi.org/10.1021/jf3021514.

Xing, Qinhui, Dekker Susanne, Kyriakopoulou Konstantina, Boom Remko M., Smid Eddy J., and Schutyser Maarten A. I.. 2020. "Enhanced Nutritional Value of Chickpea Protein Concentrate by Dry Separation and Solid State Fermentation." *Innovative Food Science and Emerging Technologies*. https://doi.org/10.1016/j.ifset.2019.102269.

Section 3. Animal Origin and Diverse

Chapter 15

Bioactive Compounds in Dairy Byproducts

Sandra T. Martín-del-Campo
and R. Vázquez-García*
Tecnologico De Monterrey, Querétaro, Mexico

Abstract

Milk and dairy products, when produced on a large scale, produce a series of byproducts that promote a loss of money, and dairy leads to environmental problems as they are considered waste. One of the most common examples in this sense is whey, which is currently used as an additive in products that require high protein concentrations. Although the macromolecules' presence is well known, another aspect that has been increasing in recent years is the study of compounds that have some bioactivity and that have been analyzed in byproducts such as whey, buttermilk, and colostrum, products that originally have been taken as waste. This chapter seeks to make these products of dairy origin visible by recounting the uses that can be given to them as they are products with high nutritional value and contain proteins and lipids that can be used as additives or derived foods.

Keywords: dairy byproducts, buttermilk, colostrum, phospholipids, MFGM

Introduction

Food waste has become a major challenge in large-scale food production since it represents a significant economic loss and a sustainability problem causing pollution increase. According to FAO, food waste is all those residual food discarded for nutrition or food security and is considered an imbalance in the production chain. Diverse alternatives are currently being sought to avoid food waste, which would improve the economy and society using thousands of tons of wasted food worldwide (Iriondo-DeHond, Miguel, and Del Castillo 2018).

Milk and dairy products are widely consumed worldwide, and their production has increased exponentially over the last forty years. It is estimated that by 2025 about 177 million

---

* Corresponding Author's Email: rosa.vazquez.g@tec.mx.

In: Food Byproducts
Editors: S. Teresita Martín-del-Campo, A. Cardador-Martínez and J. del Pilar Ramírez-Anaya
ISBN: 979-8-88697-612-0

tons of milk will be produced, marketed, and consumed by around 6 billion people. Dairy production represents an important source of food waste; although the industry has tried to reduce economic losses and be sustainable, the pollution produced is considerable thanks to this activity. Using byproducts derived from milk not only represents an improvement in economic returns but also promotes sustainability and social improvement by taking full advantage of the nutrients contained in this food (Oliveira, Fox, and O'Mahony 2019).

Whey

Whey is obtained from milk precipitation after cheese production. About 85% of the total milk used in this process is transformed into whey, considered a waste product generally disposed of in water bodies, causing significant environmental damage due to the number of salts and organic compounds it contains (Osorio-González et al., 2018).

In general, the use of cheese whey is usually the most common way to waste dairy products use. It is known that the presence of a high amount of water-soluble proteins can lead to the development of derived products to obtain higher profits with the use of waste, which mostly represents a significant loss of value and environmental destruction. Table 1 shows the diverse studies in whey that have shown evidence of bioactive peptides after hydrolysis with different enzyme complexes.

Table 1. Bioactivity in cheese whey

Bioactivity	Source	Proteinase	References
Antihypertensive activity	β-lactoglobulin goat cheese whey	Pepsin, proteinase K, actinase E, thermolysin, and papain	(Abubakar et al., 1998)
ACE inhibitory activity	α-lactoalbumin and β-lactoglobulin goat Comté ripened cheese whey	Pepsin, trypsin, and chymotrypsin	(Didelot et al., 2006)
	β-lactoglobulin sheep cheese whey	Trypsin	(Chobert et al., 2005)
	α-lactoalbumin and β-lactoglobulin cheese whey	Native cheese microbiota	(Mazorra-Manzano et al., 2020)
	α-lactoalbumin cow queizuar cheese whey	Cardiosins A and B from *C. cardunculus*	(Estévez et al., 2020)
	Goat whey cheese	Pepsin	(Ibrahim, Ahmed, and Miyata 2017)
ACE inhibitory, dipeptidyl peptidase IV (DPP-IV) inhibitory, and cytotoxic activities	Bovine raw sweet whey	Protease preparation from *Enterococcus faecalis*	(Worsztynowicz, Białas, and Grajek 2020)
Antioxidant activity	β-lactoglobulin goat cheese whey	Pepsin	(Ahmed et al., 2015)
	Not specified	Alcalase, flavourzyme, neutrase, and protamex	(Dryáková et al., 2010)
	α-lactoalbumin and β-lactoglobulin bovine cheese whey	Pepsin, trypsin, chymotrypsin, thermolysin, and Corolase PP	(Hernández-Ledesma et al., 2005)
	Not specified	Pepsin	(Jiang et al., 2017)

Bioactivity	Source	Proteinase	References
Antioxidant activity	Camel cheese whey	Pepsin	(Ibrahim, Isono, and Miyata 2018)
Antioxidant and calcium-binding activities	Colostrum cow whey	Pepsin	(Fajardo-Espinoza, Romero-Rojas, and Hernández-Sánchez 2020)
Antioxidant and modulatory activities	Mozzarella di Bufala cheese whey	Not specified	(De Simone et al., 2009)
Antioxidant and antibacterial activity	Bovine cheese whey	Alcalase	(da Cruz et al., 2020)
Antihypertensive and antioxidant activity	Sheep cheese whey	Protease preparation from *Bacillus* sp. P7	(Corrêa et al., 2014)
	Bovine cheese whey	Alcalase	(López et al., 2022)
	Sheep cheese whey	Bacterial and fungal food-grade protease preparation	(Welsh et al., 2017)
Antibacterial activity	Goat cheese whey	Human gastrointestinal enzymes	(Almaas et al., 2011)
Antibacterial and immunomodulatory activity	Colostrum buffalo whey	Fermented using *L. rhamonosus* C25	(Kashyap, Narayan, and Vij 2022)
Antiproliferative activity	Goat cheese whey	Pepsin and trypsin enzymes	(Sahna, Cakir, and Tunali-Akbay 2022)
Antihypercholesterolemic activity	Camel cheese whey	Pepsin	(Baba et al., 2021)
Anticarcinogenic and antiproliferative activity	Goat cheese whey	Pepsin, trypsin, and papain	(Cakir and Tunali-Akbay 2021)

Other compounds of nutritional importance found in whey and its proteins are sphingomyelins, which are important components in the neuronal development of any mammal. Moloney et al., (2018) identified these compounds in bovine whey samples, using α-lactoglobulin concentrate after adding it to infant formula to supplement concentrations required for human infant nutrition.

Recent studies found many bioactivities in whey proteins, and the presence of lactoglobulin and lactalbumin make whey a highly profitable product after cheese production, and it should not be discarded after production. Current extraction and purification techniques for these compounds ensure the permanence of bioactivity without damaging the component and that it can be used later as an additive for food enrichment (Hernández-Ledesma, Ramos, and Gómez-Ruiz 2011; Hernández-Ledesma, Recio, and Amigo 2008).

Buttermilk

Buttermilk is a product obtained from butter production, specifically from the churning step. It is generally considered waste due to the high lipids concentrations contained, although, in some regions of the world, its fermentation leads to the production of a beverage. In Europe, the consumption of fresh buttermilk remains marginal. In some cases, buttermilk is dried to be added to dairy or bakery products as an emulsifier or destinated to animal feed (Vanderghem et al., 2010). Some authors have determined the presence of bioactive compounds after fermentation, as is the case of Bourlieu et al., (2018), who reported the presence of bioactive compounds related to ceramides, lecithins, and sphingomyelins. The studies promote a particular interest in using these compounds in food matrices related to neonatal nutrition, which requires special attention to brain and liver protective compounds without being cholesterol-high, as mentioned by Claumarchirant et al., (2016). They used milk fat globules to recreate the polar lipids of human milk to add to infant milk formulas.

Milk fat globule membrane (MFGM) represents 2-6% of the total mass of fat globules and is composed of proteins (25-60% of MFGM), glycoproteins, polar lipids, neutral lipids (56-80% of total lipids), and enzymes, as well as other minor components. The major component of neutral lipids are triglycerides (37-68% of total neutral lipids), while phospholipids and sphingolipids are the major components of polar lipids (15-43%). MFGM isolates could be used as emulsifiers in the dairy industry (Vanderghem et al., 2010).

In the studies of Fuller et al., (2013), the membranes obtained from glycoproteins and glycolipids of bovine buttermilk were used as an antiviral agent as a capsule against rotavirus neuraminidase-sensitive, showing a high percentage of inhibition. It was shown that the highest antiviral activity fraction was protein-free and presented non-polar characteristics, which, when fractionated with non-polar components such as chloroform and methanol, enhanced its antiviral activity.

Fermented Milk

It has been shown that the use of lactic acid bacteria in fermented milk production has led to the appearance of compounds obtained from proteins. On many occasions, using fermented waste milk to produce bioactive peptides has led to the development of products using bacteria. Such is the case of the work done by Aguilar and Klotz (2009). They used *Lactobacillus plantarum* WS4174, *Lactobacillus plantarum* LB279, and *Pediococcus acidilacti* P120 strains to produce bacteriocins derived from fermented milk that managed to control pathogenic bacteria proliferation such as *Escherichia coli*, *Staphylococcus aureus* and *Listeria monocytogenes* from food handlers.

Kadyan et al., (2021) observed the presence of antimicrobial effects against *Escherichia coli* and antioxidants effects using the DPPH activity test in a bovine whey probiotic drink after a fermentation process using *Lactobacillus plantarum* MTCC 5690 and *Kluyveromyces lactis* NCDC 257, after a gas chromatography analysis coupled mass spectroscopy, the presence of 58 metabolites related to bioactivities was determined derived from medium and short chain peptides.

Chaves-López et al., (2014) determined the presence of ACEI activity peptides in milk after 36 hours of fermentation with yeast strains of *Torulaspora delbruekii* KL66A, *Galactomyces geotrichum* KL20B, *Pichia kudriavzevii* KL84A, and *Kluyveromyces marxianus* KL26A, and with lactic acid bacteria strains of *Lactobacillus plantarum* LAT03, *Lactobacillus plantarum* KLAT01 and the not virulent *Enterococcus faecalis* KE06, either growing alone or in different combinations.

Another product widely produced after lactic fermentation is kefir, which is taken as a waste product in several countries, but that has shown the presence of bioactive peptides, as mentioned by Azi et al., (2021) and Azi et al., (2020). They determined the presence of peptide ACE inhibitory and DPPH radical scavenging activity as antioxidant activity in high percentages in the beverage produced with different soy concentrations. Another example is the work elaborated by Sabokbar and Khodaiyan (2016), where kefir was produced by fermenting a mixture of whey and pomegranate juice, also observing the presence of antioxidant activity, which increases with fermentation time at a constant temperature.

Colostrum

Colostrum is defined as the initial secretion of the mammary gland after parturition. It is considered an important source of nutrients and immunological compounds the newborn calf uses. It is known that the first 24 hours after birth are crucial for feeding the newborn since colostrum contains high concentrations of immunoglobulins that provide immune support during the first days of life. This substance is generally not commercially used since it is exclusively for breeding. However, its use has been considered a controversial issue, mainly in infant milk formula. Elsohaby et al., (2016) observed IgG high concentration in bovine colostrum using infrared spectroscopy, which may lead to the development of the molecule for industrial use.

It has been shown that this product contains high lipid concentrations required for newborn growth and nutrition. In the case of bovine milk, Li et al., (2020) identified at least seven more lipidic compounds in colostrum than in mature milk. They highlight cardiolipins, glycerophosphatidyl ethanolamines, glycerophosphatidyl glycerols, and ceramides, important components in brain development that can be used in the infant feeding industry.

It has been determined that colostrum contains compounds with higher concentrations than mature milk; as previously mentioned, high immunoglobulins and peptides contents and elevated β-lactoglobulins, lactoferrin, and albumin serum concentrations have also been reported, and all bioactive compounds precursors (Korhonen and Pihlanto 2007).

Recovery and Use Systems

During products from waste processing, it is important to follow recovery procedures that ensure the adequate obtaining of usable fractions and good yields for good profits in the case of industrialized use (Rathnakumar et al., 2021). Aguirre-Eskauriatza (2008) determined that the protein recovered from goat's milk whey, using an ultrafiltration system, followed by lyophilization with a protein/carbohydrate fraction recovery of 7:1 reached a yield of up to 12 kg/m^3 of whole whey. The same procedure was used by Espejo-Carpio et al., (2014) to obtain ACE-inhibitory peptides concentrate from goat cheese and milk whey. At the same time, Athira et al., (2015) observed that this same peptide purification technique can preserve peptides sensitive to high temperatures, as is the case of peptides with antioxidant activity.

The protein-rich fraction obtained using this process can be significative used in the food additives and supplements market. It has also been observed that certain high-pressure recovery systems promote the preservation of peptide structures, preserving rheological qualities in temperature ranges that prevent their denaturation (Ahmed and Ramaswamy 2003).

On the other hand, to obtain functional fatty compounds from milk lipid fractions such as cream and buttermilk, using methodologies that combine physicochemical and technological treatments promotes industrial development for these functional compounds. The treatment proposed by Gassi et al., (2016) and Konrad, Kleinschmidt, and Lorenz (2013) include skimming, thermal treatment, acid precipitation of caseins, concentration, purification, centrifugation, ultrafiltration, and diafiltration. Following this procedure, it was possible to achieve a yield of up to 62% of dairy polar lipids for food application. Another treatment that has been shown to achieve up to 70% yield during fat globule membranes purifying from buttermilk is through coagulation and six diafiltration steps, thus ensuring the complete elimination of traces of fat-soluble proteins retained in fat (Holzmüller and Kulozik 2016a, b; Olabi et al., 2015).

Whey is also an important phospholipid source used as a food additive. The concentrates can be obtained following extraction procedures such as those described by Price et al., (2018), where ethanol is used for protein fractionation and remaining phospholipid concentrate by selectivity, after five extractions and obtaining a yield of up to 58% of phospholipid extract. For these extractions, conventional extraction, promoting the obtention of phospholipid concentrate from dairy byproducts, has been used, such Folch method. This method uses chloroform and methanol concentrations with high yields of up to 98%, but when used for food products, it is desired to use more innocuous solvents. It was observed by Rathnakumar et al., (2021) that using ternary amines as hydrophilic solvents has given good results in phospholipids extraction from buttermilk, cream, and cheese whey; even though the chemical structures can be modified, there is an advance in these extraction methods.

Ubeyitogullari and Rizvi (2020) used a phospholipid extraction method from buttermilk using 20% concentrated ethanol modified with supercritical carbon dioxide in temperatures between 50 and 60 degrees Celsius at high pressures. This method presented a 76% yield by obtaining phospholipid extract composed of sphingomyelin, phosphatidyl ethanolamine, phosphatidyl serine, and phosphatidylcholine.

Prospects for Industrial Use

Currently, the trend in the use of food additives of natural origin has obtained relevance at the international level, looking to have safer food with minimum use of chemical substances that are harmful to health. In the case of industrial uses of bioactive peptides obtained from dairy byproducts, they are a natural and economical option for developing food additives. This is the case of bacteriocins produced from fermented milk to eliminate pathogenic bacteria in industrialized foods (Aguilar and Klotz 2009).

Another approach for these products is their industrialization into new dairy products, such as buttermilk for quark cheeses production using successive centrifugations, showing high concentrations of phospholipids such as phosphatidylethanolamine, phosphatidylinositol, phosphatidylcholine, phosphatidylserine and sphingomyelin, all necessary in brain development (Ferreiro et al., 2016).

Another use that has been observed is hydrolysis to obtain nano peptides, as reported by Balandrán-Quintana et al., (2013). They observed the presence of nanostructures obtained from whey after hydrolysis by *Staphylococcus aureus*. These structures can be used in the future to produce transport structures for nutraceuticals in food matrices. Another industrial buttermilk utilization is the application of fractionated lipids as an additive in foods, mainly cheeses, since they give the final product texture, aroma, and flavor, as well as add a certain functionality by containing a high proportion of sphingolipids, such as the case reported by Hickey et al., (2018).

They mentioned that the sensorial conditions of Cheddar cheeses were modified with an acceptance of up to 10% higher than those made in an artisanal way after additive buttermilk powder addition.

Conclusion

The applications shown in this text describe the importance of byproducts uses not only for economic recovery but as a derived product that provides nutritional qualities to foods that use it as an additive. The analysis of the biological activities in waste products is very useful for the industry, for the development of functional foods and possible nutraceuticals that lead us to the possible future treatment or elimination of ailments that afflict a large number of people in the present.

References

Abubakar, Amhar, Tadao Saito, Haruki Kitazawa, Yasushi Kawai, and Takatoshi Itoh. 1998. "Structural Analysis of New Antihypertensive Peptides Derived from Cheese Whey Protein by Proteinase K Digestion." *Journal of Dairy Science* 81 (12):3131-3138. doi: 10.3168/jds.S0022-0302(98)75878-3.

Aguilar, Catalina, and Bernadette Klotz. 2009. "Bacteriocin-like inhibitory substances (BLIS) of acid lactic bacteria: a biocontrol alternative for the food industry." *Canilac Industrial*, 13-22.

Aguirre-Eskauriatza, JE. 2008. "*Producción de proteína liofilizada y biomasa de Lactobacillus casei BPG4 liofilizada, a partir de suero de leche de cabra* [*Production of lyophilized protein and biomass of lyophilized Lactobacillus casei BPG4, from goat milk whey*]." Tesis de Maestría. Tecnológico de Monterrey, México.

Ahmed, Ahmed S., Tawfik El-Bassiony, Laila M. Elmalt, and Hisham R. Ibrahim. 2015. "Identification of potent antioxidant bioactive peptides from goat milk proteins." *Food Research International* 74:80-88. doi: 10.1016/j.foodres.2015.04.032.

Ahmed, J., and H. S. Ramaswamy. 2003. "Effect of High-Hydrostatic Pressure and Temperature on Rheological Characteristics of Glycomacropeptide." *Journal of Dairy Science* 86 (5):1535-1540. doi: 10.3168/jds.S0022-0302(03)73738-2.

Almaas, Hilde, Ellen Eriksen, Camilla Sekse, Irene Comi, Ragnar Flengsrud, Halvor Holm, Einar Jensen, Morten Jacobsen, Thor Langsrud, and Gerd E Vegarud. 2011. "Antibacterial peptides derived from caprine whey proteins, by digestion with human gastrointestinal juice." *British Journal of Nutrition* 106 (06):896-905. doi: 10.1017/S0007114511001085.

Athira, Syamala, Bimlesh Mann, Prerna Saini, Rajan Sharma, Rajesh Kumar, and Ashish Kumar Singh. 2015. "Production and characterisation of whey protein hydrolysate having antioxidant activity from cheese whey." *Journal of the Science of Food and Agriculture* 95 (14):2908-2915. doi: 10.1002/jsfa.7032.

Azi, Fidelis, Chuanhai Tu, Ling Meng, Li Zhiyu, Mekonen Tekliye Cherinet, Zahir Ahmadullah, and Mingsheng Dong. 2021. "Metabolite dynamics and phytochemistry of a soy whey-based beverage bio-transformed by water kefir consortium." *Food Chemistry* 342:Article 128225. doi: 10.1016/j.foodchem.2020.128225.

Azi, Fidelis, Chuanhai Tu, Hafiz Abdul Rasheed, and Mingsheng Dong. 2020. "Comparative study of the phenolics, antioxidant and metagenomic composition of novel soy whey-based beverages produced using three different water kefir microbiota." *International Journal of Food Science & Technology* 55 (4):1689-1697. doi: 10.1111/ijfs.14439.

Baba, Waqas N, Priti Mudgil, Bincy Baby, Ranjit Vijayan, Chee-Yuen Gan, and Sajid Maqsood. 2021. "New insights into the cholesterol esterase-and lipase-inhibiting potential of bioactive peptides from camel whey hydrolysates: Identification, characterization, and molecular interaction." *Journal of Dairy Science* 104 (7):7393-7405. doi: 10.3168/jds.2020-19868.

Balandrán-Quintana, René R., Miguel A. Valdéz-Covarrubias, Ana M. Mendoza-Wilson, and Rogerio R. Sotelo-Mundo. 2013. "α-Lactalbumin hydrolysate spontaneously produces disk-shaped nanoparticles." *International Dairy Journal* 32 (2):133-135. doi: 10.1016/j.idairyj.2013.05.011.

Bourlieu, Claire, David Cheillan, Marielle Blot, Patricia Daira, Michèle Trauchessec, Séverine Ruet, Jean-Yves Gassi, Eric Beaucher, Benoit Robert, Nadine Leconte, Saïd Bouhallab, Frédéric Gaucheron, Geneviève Gésan-Guiziou, and Marie-Caroline Michalski. 2018. "Polar lipid composition of bioactive dairy co-products buttermilk and butterserum: Emphasis on sphingolipid and ceramide isoforms." *Food Chemistry* 240:67-74. doi: 10.1016/j.foodchem.2017.07.091.

Cakir, Bilal, and Tugba Tunali-Akbay. 2021. "Potential anticarcinogenic effect of goat milk-derived bioactive peptides on HCT-116 human colorectal carcinoma cell line." *Analytical biochemistry* 622:114166. doi: 10.1016/j.ab.2021.114166.

Chaves-López, Clemencia, Annalisa Serio, Antonello Paparella, Maria Martuscelli, Aldo Corsetti, Rosanna Tofalo, and Giovanna Suzzi. 2014. "Impact of microbial cultures on proteolysis and release of bioactive peptides in fermented milk." *Food microbiology* 42:117-121. doi: 10.1016/j.fm.2014.03.005.

Chobert, Jean-Marc, Khaled El-Zahar, Mahmoud Sitohy, Michèle Dalgalarrondo, François Métro, Yvan Choiset, and Thomas Haertlé. 2005. "Angiotensin I-converting-enzyme (ACE)-inhibitory activity of tryptic peptides of ovine beta-lactoglobulin and of milk yoghurts obtained by using different starters." *Le Lait* 85 (3):141-152. doi: 10.1051/lait:2005005.

Claumarchirant, Lorena, Antonio Cilla, Esther Matencio, Luis Manuel Sanchez-Siles, Pilar Castro-Gomez, Javier Fontecha, Amparo Alegría, and María Jesús Lagarda. 2016. "Addition of milk fat globule membrane as an ingredient of infant formulas for resembling the polar lipids of human milk." *International Dairy Journal* 61:228-238. doi: 10.1016/j.idairyj.2016.06.005.

Corrêa, Ana Paula Folmer, Daniel Joner Daroit, Roberta Fontoura, Stela Maris Meister Meira, Jeferson Segalin, and Adriano Brandelli. 2014. "Hydrolysates of sheep cheese whey as a source of bioactive peptides with antioxidant and angiotensin-converting enzyme inhibitory activities." *Peptides* 61:48-55. doi: 10.1016/j.peptides.2014.09.001.

da Cruz, Clariana Zanutto Paulino, Ricardo José de Mendonça, Luís Henrique Souza Guimaraes, Matheus Aparecido dos Santos Ramos, Saulo Santesso Garrido, Ariela Veloso de Paula, Rubens Monti, and Gabriella Massolini. 2020. "Assessment of the bioactive potential of cheese whey protein hydrolysates using immobilized alcalase." *Food and Bioprocess Technology* 13 (12):2120-2130. doi: 10.1007/s11947-020-02552-4.

De Simone, Carmela, Gianluca Picariello, Gianfranco Mamone, Paola Stiuso, Alessandra Dicitore, Daniela Vanacore, Lina Chianese, Francesco Addeo, and Pasquale Ferranti. 2009. "Characterisation and cytomodulatory properties of peptides from Mozzarella di Bufala Campana cheese whey." *Journal of peptide science: an official publication of the European Peptide Society* 15 (3):251-258. doi: 10.1002/psc.1093.

Didelot, Sandrine, Stéphanie Bordenave-Juchereau, Eric Rosenfeld, Ingrid Fruitier-Arnaudin, Jean-Marie Piot, and Frédéric Sannier. 2006. "Preparation of angiotensin-I-converting enzyme inhibitory hydrolysates from unsupplemented caprine whey fermentation by various cheese microflora." *International Dairy Journal* 16 (9):976-983. doi: 10.1016/j.idairyj.2005.09.009.

Dryáková, Adriena, Anne Pihlanto, Pertti Marnila, Ladislav Curda, and Hannu Korhonen. 2010. "Antioxidant properties of whey protein hydrolysates as measured by three methods." *European Food Research and Technology* 230:865-874. doi: 10.1007/s00217-010-1231-9.

Elsohaby, Ibrahim, J. Trenton McClure, Siyuan Hou, Christopher B. Riley, R. Anthony Shaw, and Gregory P. Keefe. 2016. "A novel method for the quantification of bovine colostral immunoglobulin G using infrared spectroscopy." *International Dairy Journal* 52:35-41. doi: 10.1016/j.idairyj.2015.08.004.

Espejo-Carpio, Francisco Javier, Raúl Pérez-Gálvez, María del Carmen Almécija, Antonio Guadix, and Emilia M Guadix. 2014. "Production of goat milk protein hydrolysate enriched in ACE-inhibitory peptides by ultrafiltration." *Journal of Dairy Research* 81 (4):385-393. doi: 10.1017/S0022029914000284.

Estévez, Natalia, Pablo Fuciños, Clara Fuciños, Paula Jauregi, Clara A Tovar, and M Luisa Rúa. 2020. "Hydrolysis of whey protein as a useful approach to obtain bioactive peptides and a β-Lg fraction with different biotechnological applications." *Food Hydrocolloids* 109:106095. doi: 10.1016/j. foodhyd. 2020.106095.

Fajardo-Espinoza, FS, A Romero-Rojas, and H Hernández-Sánchez. 2020. "Production of bioactive peptides from bovine colostrum whey using enzymatic hydrolysis." *Revista Mexicana de Ingeniería Química* 19 (1):1-9. doi: 10.24275/rmiq/Alim525.

Ferreiro, T., S. Martínez, L. Gayoso, and J. L. Rodríguez-Otero. 2016. "Evolution of phospholipid contents during the production of quark cheese from buttermilk." *Journal of Dairy Science* 99 (6):4154-4159. doi: 10.3168/jds.2016-10861.

Fuller, K. L., T. B. Kuhlenschmidt, M. S. Kuhlenschmidt, R. Jiménez-Flores, and S. M. Donovan. 2013. "Milk fat globule membrane isolated from buttermilk or whey cream and their lipid components inhibit infectivity of rotavirus in vitro." *Journal of Dairy Science* 96 (6):3488-3497. doi: 10.3168/jds.2012-6122.

Gassi, Jean Yves, Marielle Blot, Eric Beaucher, Benoit Robert, Nadine Leconte, Bénédicte Camier, Florence Rousseau, Claire Bourlieu, Julien Jardin, Valérie Briard-Bion, Sophie Lambert, Geneviève Gésan-Guiziou, Christelle Lopez, and Frédéric Gaucheron. 2016. "Preparation and characterisation of a milk polar lipids enriched ingredient from fresh industrial liquid butter serum: Combination of physico-chemical modifications and technological treatments." *International Dairy Journal* 52:26-34. doi: 10.1016/j.idairyj.2015.08.012.

Hernández-Ledesma, B, I Recio, and L Amigo. 2008. "β-Lactoglobulin as source of bioactive peptides." *Amino acids* 35 (2):257-265. doi: 10.1007/s00726-007-0585-1.

Hernández-Ledesma, Blanca, Alberto Dávalos, Begoña Bartolomé, and Lourdes Amigo. 2005. "Preparation of antioxidant enzymatic hydrolysates from α-lactalbumin and β-lactoglobulin. Identification of active peptides by HPLC-MS/MS." *Journal of Agricultural and Food Chemistry* 53 (3):588-593. doi: 10.1021/ jf048626m

Hernández-Ledesma, Blanca, Mercedes Ramos, and José Ángel Gómez-Ruiz. 2011. "Bioactive components of ovine and caprine cheese whey." *Small Ruminant Research* 101 (1):196-204. doi: 10.1016/j.smallrumres. 2011.09.040.

Hickey, C. D., M. G. O'Sullivan, J. Davis, D. Scholz, K. N. Kilcawley, M. G. Wilkinson, and J. J. Sheehan. 2018. "The effect of buttermilk or buttermilk powder addition on functionality, textural, sensory and volatile characteristics of Cheddar-style cheese." *Food Research International* 103:468-477. doi: 10.1016/j.foodres.2017.09.081.

Holzmüller, Wolfgang, and Ulrich Kulozik. 2016a. "Isolation of milk fat globule membrane (MFGM) material by coagulation and diafiltration of buttermilk." *International Dairy Journal* 63:88-91. doi: 10.1016/ j.idairyj.2016.08.002.

Holzmüller, Wolfgang, and Ulrich Kulozik. 2016b. "Technical difficulties and future challenges in isolating membrane material from milk fat globules in industrial settings – A critical review." *International Dairy Journal* 61:51-66. doi: 10.1016/j.idairyj.2016.03.013.

Ibrahim, Hisham R, Ahmed S Ahmed, and Takeshi Miyata. 2017. "Novel angiotensin-converting enzyme inhibitory peptides from caseins and whey proteins of goat milk." *Journal of advanced research* 8 (1):63-71. doi: 10.1016/j.jare.2016.12.002.

Ibrahim, Hisham R, Hiroki Isono, and Takeshi Miyata. 2018. "Potential antioxidant bioactive peptides from camel milk proteins." *Animal Nutrition* 4 (3):273-280. doi: 10.1016/j.aninu.2018.05.004.

Iriondo-DeHond, Maite, Eugenio Miguel, and María Dolores Del Castillo. 2018. "Food Byproducts as Sustainable Ingredients for Innovative and Healthy Dairy Foods." *Nutrients* 10 (10):1358. doi: 10.3390/ nu10101358.

Jiang, Bin, Xiaoqing Zhang, Yongqiang Yuan, Yuxiao Qu, and Zhibiao Feng. 2017. "Separation of Antioxidant Peptides from Pepsin Hydrolysate of Whey Protein Isolate by ATPS of EOPO Co-polymer (UCON)/Phosphate." *Scientific Reports* 7 (1):Article 13320. doi: 10.1038/s41598-017-13507-9.

Kadyan, Saurabh, H. M. Rashmi, Diwas Pradhan, Anisha Kumari, Arghya Chaudhari, and Gaurav Kr Deshwal. 2021. "Effect of lactic acid bacteria and yeast fermentation on antimicrobial, antioxidative and metabolomic profile of naturally carbonated probiotic whey drink." *LWT - Food Science and Technology* 142:Article 111059. doi: 10.1016/j.lwt.2021.111059.

Kashyap, Reenu, Kapil Singh Narayan, and Shilpa Vij. 2022. "Identification of antibacterial and immunomodulatory bioactive peptides generated from buffalo (Bubalus bubalis) colostrum whey fermented by Lactobacillus rhamnosus C25: LC-MS/MS-based analysis." *Journal of Functional Foods* 95:105158. doi: 10.1016/j.jff.2022.105158.

Konrad, Gerd, Thomas Kleinschmidt, and Claudia Lorenz. 2013. "Ultrafiltration of whey buttermilk to obtain a phospholipid concentrate." *International Dairy Journal* 30 (1):39-44. doi: 10.1016/j.idairyj.2012.11.007.

Korhonen, H, and Anne Pihlanto. 2007. "Technological options for the production of health-promoting proteins and peptides derived from milk and colostrum." *Current Pharmaceutical Design* 13 (8):829-843. doi: 10.2174/138161207780363112.

Li, Mohan, Qilong Li, Shimo Kang, Xueyan Cao, Yan Zheng, Junrui Wu, Rina Wu, Junhua Shao, Mei Yang, and Xiqing Yue. 2020. "Characterization and comparison of lipids in bovine colostrum and mature milk based on UHPLC-QTOF-MS lipidomics." *Food Research International* 136:109490. doi: 10.1016/j.foodres.2020.109490.

López, Emilse C, Agustina Eberhardt, Fernanda Marino, Enrique J Mammarella, Guillermo A Sihufe, and Ricardo M Manzo. 2022. "Physicochemical characterisation of ACE-inhibitory and antioxidant peptides from Alcalase® whey protein hydrolysates using fractionation strategies." *International Journal of Dairy Technology*. doi: 10.1111/1471-0307.12872.

Mazorra-Manzano, Miguel A, Glen R Robles-Porchas, Daniel A González-Velázquez, María J Torres-Llanez, Marcel Martínez-Porchas, Celia O García-Sifuentes, Aarón F González-Córdova, and Belinda Vallejo-Córdoba. 2020. "Cheese whey fermentation by its native microbiota: Proteolysis and bioactive peptides release with ACE-inhibitory activity." *Fermentation* 6 (1):19. doi: 10.3390/fermentation6010019.

Moloney, Cian, Emma Walshe, Martha Phelan, Francesca Giuffrida, Flavia Badoud, Emmanuelle Bertschy, and Jonathan O'Regan. 2018. "Sphingomyelin content of dairy protein ingredients and infant formula powders, and identification of bovine sphingomyelin species." *International Dairy Journal* 78:138-144. doi: 10.1016/j.idairyj.2017.11.006.

Olabi, A., S. Jinjarak, R. Jiménez-Flores, J. H. Walker, and Hamza Daroub. 2015. "Compositional and sensory differences of products of sweet-cream and whey buttermilk produced by microfiltration, diafiltration, and supercritical CO21." *Journal of Dairy Science* 98 (6):3590-3598. doi: 10.3168/jds.2014-9141.

Oliveira, Diana, Patrick Fox, and James A. O'Mahony. 2019. "Byproducts from Dairy Processing." In *Byproducts from Agriculture and Fisheries*, 57-106.

Osorio-González, C.S, F. Sandoval-Salas, F. Hernández-Rosas, J.V. Hidalgo-Contreras, F.C. Gómez-Merino, and D.A. Ávalos de la Cruz. 2018. "Potencial de aprovechamiento del suero de queso en México [Potential use of cheese whey in Mexico]." *Agroproductividad*, 101-106.

Price, Nathan, Tao Fei, Stephanie Clark, and Tong Wang. 2018. "Extraction of phospholipids from a dairy by-product (whey protein phospholipid concentrate) using ethanol." *Journal of Dairy Science* 101 (10):8778-8787. doi: 10.3168/jds.2018-14950.

Rathnakumar, Kaavya, Joana Ortega-Anaya, Rafael Jimenez-Flores, and Sergio I. Martínez-Monteagudo. 2021. "Understanding the switchable solvent extraction of phospholipids from dairy byproducts." *Food and Bioproducts Processing* 126:175-183. doi: 10.1016/j.fbp.2021.01.003.

Sabokbar, N., and F. Khodaiyan. 2016. "Total phenolic content and antioxidant activities of pomegranate juice and whey based novel beverage fermented by kefir grains." *Journal of food Science and Technology* 53 (1):739-47. doi: 10.1007/s13197-015-2029-3.

Sahna, Kıymet Ozlem, Bilal Cakir, and Tugba Tunali-Akbay. 2022. "Antiproliferative Activity of Whey and Casein Bioactive Peptides on Breast Cancer: An In Vitro and In Silico Study." *International Journal of Peptide Research and Therapeutics* 28 (4):1-12. doi: 10.1007/s10989-022-10436-8.

Ubeyitogullari, Ali, and Syed S. H. Rizvi. 2020. "Production of high-purity phospholipid concentrate from buttermilk powder using ethanol-modified supercritical carbon dioxide." *Journal of Dairy Science* 103 (10):8796-8807. doi: 10.3168/jds.2020-18697.

Vanderghem, Caroline, Pascal Bodson, Sabine Danthine, Michel Paquot, Claude Deroanne, and Christophe Blecker. 2010. "Milk fat globule membrane and buttermilks: from composition to valorization." *Biotechnologie, Agronomie, Société et Environnement* 14 (3):485-500.

Welsh, Grace, Kate Ryder, Jodi Brewster, Christina Walker, Sonya Mros, Alaa El-Din A Bekhit, Michelle McConnell, and Alan Carne. 2017. "Comparison of bioactive peptides prepared from sheep cheese whey using a food-grade bacterial and a fungal protease preparation." *International Journal of Food Science & Technology* 52 (5):1252-1259. doi: 10.1111/ijfs.13392.

Worsztynowicz, Paulina, Wojciech Białas, and Włodzimierz Grajek. 2020. "Integrated approach for obtaining bioactive peptides from whey proteins hydrolysed using a new proteolytic lactic acid bacteria." *Food chemistry* 312:126035. doi: 10.1016/j.foodchem.2019.126035.

Index

B

C

D

E

F

G

H

I

J

K

L

M

N

O

P

Q

R

About the Editors

Sandra Teresita Martín-del-Campo

Sandra Teresita Martín-del-Campo obtained a PhD in food science and a PhD in bioprocess areas. She was a research-professor at the Tecnologico de Monterrey, now she is an independent consultant for different food enterprises. Her research focuses on the dairy products metabolomic studies, preservation of food properties during transformation, recuperation of biocompounds from food waste. She is a member of the National System of Researchers Level 1 in the Biotechnology area.

Anaberta Cardador-Martínez

Anaberta Cardador-Martínez is a PhD in the food science area. Her research work is focused on the investigation of nutraceutical properties of nutraceuticals. She also works with essential oils extraction and evaluation of biological characteristics.

Jessica del Pilar Ramírez-Anaya

Jessica del Pilar Ramírez-Anaya is a Biologist with Master's and Doctor degrees in Biotechnology, Nutrition, and Food Sciences and Technology. She works at the University of Guadalajara studying the effects of processing on the functional compounds of avocado and pomegranate products and by-products and, recently, on the antioxidant activity of the population's diet.